From Cotton Belt to Sunbelt

From Cotton Belt to Sunbelt

Federal Policy, Economic Development, and the Transformation of the South, 1938–1980

Bruce J. Schulman

Duke University Press Durham and London 1994

For Angie Arvidson, and for Hoppa

Originally published in 1991 by Oxford University Press

Printed in the United States of America on acid-free paper ∞

Library of Congress Cataloging-in-Publication Data

Schulman, Bruce J.

From Cotton Belt to Sunbelt: federal policy, economic development, and the transformation of the South, 1938–1980 / Bruce J. Schulman. p. cm.

Originally published: New York: Oxford University Press, 1991. Includes bibliographical references and index. ISBN 0-8223-1537-8 (paper: acid-free paper) 1. Southern States—Economic policy. 2. Southern States—Economic conditions—1918– 3. Economic assistance, Domestic—United States—Southern States. 4. Southern States—Politics and government—1865– 5. Southern States—Race relations. I. Title.

HC107.A13S38 1994 338.975—dc20 94-28526 CIP

Contents

❂ Preface to Duke Edition

In 1992, two southern Democrats—Bill Clinton of Arkansas and Albert Gore, Jr. of Tennessee—captured the nation's highest offices. During the campaign, the "Bubba ticket" sparked a national rediscovery of yet another New South, a dynamic region with a youthful, forward-looking leadership, untainted by the racial hostility and economic backwardness of the past. Yet at the same time, the candidates carefully nurtured their southern roots—ballyhooing their southern Baptist faith, their fondness for barbecue, their country lingo. The spectacle prompted cartoonist Doug Marlette, creator of the comic strip *Kudzu,* to identify a new arrival on the national scene, a class of counterfeit crackers and fake good ol' boys. "Due to the homogenization, gentrification and urbanization of their homeland (into something called the Sunbelt)," Marlette jibed, "the ecological imbalance created by the disappearance of genuine Bubbas has prompted the emergence of a compensatory form of Good Ol' Boy: the Faux Bubba." Weekend Billybobs like Clinton and Gore claim all the political capital and accoutrements of a "colorful ethnic identity with none of the unpleasant cultural downside—like getting laid off your job or cut in a knife fight." The economic and political forces unleashed by Franklin D. Roosevelt, all of the "big words ending in 'ization'," seemed to have transformed the South of boll weevils, magnolias, juleps, and sharecroppers into a sunbelt of skyscrapers, dealmakers, military bases and air-conditioned comfort.[1]

Marlette's caricatures evoked the political influence, economic power, and even the cultural cachet of the contemporary South. For decades a launching pad of blacks fleeing racism and whites escaping poverty, the South exerted magnetic power after 1970. Its population swelled by twice the national growth rate as people poured into the region: white suburbanites from the Northeast, industrial workers from the Midwest, retirees, black outmigrants returning to their childhood homes, and, for the first time in southern history, a substantial number of immigrants from foreign shores.[2] In urban Georgia and Alabama, black and white public officials campaign today for the votes of Asians and Latinos; in Tennessee and Mississippi, native-born hoteliers compete with new arrivals from India.

Once the nation's primary economic problem—the laggard stepchild that held back an otherwise prosperous nation—the South has become the nation's most economically vital region. Throughout the severe recession that battered the nation during the late 1980s and early 1990s, the South continued to prosper and grow. The region claims eight of the top ten states in new manufacturing and plant expansions since 1991. In a series of highly publicized decisions, Nissan, Saturn, Toyota, BMW, and Mercedes Benz all opened new facilities in the Southeast, so that the land of Dixie has replaced Michigan as America's carmaking capital.[3]

These indicators, however, only hint at the scale of southern change. The South has emerged, in the words of the *New York Times*, as a "dominant economic, political and cultural force in American life. Suppose just 25 years ago that someone suggested that Atlanta would be home to the world's most influential television news network and would be host of the 1996 Olympics," the nation's paper of record mused in July 1994. "That Charlotte would be the third largest banking center. . . . That country music would be the nation's leading radio format and Nashville flooded with refugees from Los Angeles. That *Fortune* magazine would stick Raleigh-Durham, N.C. on its cover as the nation's best city for business. . . . All of it would have seemed preposterous," the *Times* recalled. "All of it came true."[4]

The South's path from laggard to leader encapsulates the political history of the United States in the years since World War II. The political economy of the Sunbelt South—the region's fondness for economic growth and military procurement and its dislike of bureaucrats, pointy-headed intellectuals, and "welfare Cadillacs"—created a potent national political force and dominated national policy in 1970s and 1980s. Over the past two decades, the South exported political ideology and economic policy along with country music, pickup trucks, and cowboy boots.

It also undergirded a generation of Republican control of the White House. In Franklin Roosevelt's day, few could have imagined the Solid Democratic South as the bedrock of Republican presidential majorities. But since 1964 only one Democratic standardbearer, Georgian Jimmy Carter in 1976, has mustered 50 percent of the region's ballots. (In the three-way 1992 election pitting two Texans against an Arkansan, no candidate won a majority of southern votes.)

President Lyndon B. Johnson had feared such a turn of events. On the day he signed the Civil Rights Act of 1964, LBJ told an aide "I think we just delivered the South to the Republican Party for my lifetime and yours."[5] But while white backlash against the civil rights revolution certainly benefited the GOP, Sunbelt growth also fueled this political realignment. Since 1970, immigration to the South has been heavily white and middle class. Many new arrivals, according to one southern political scientist, "brought their Republican attitudes and loyalties with them, along with the furniture, as they came south from the North and Midwest."[6]

These sojourners sought the "Sunbelt," a new designation encapsulating a novel (if much abused and exaggerated) conception of the region. The idea of the Sunbelt offered the benighted land of Dixie a fresh identity by linking it with the West. "The South," one sagacious analyst of the Sunbelt concept mused, "was known to be hot and muggy; the Sunbelt was portrayed as sunny, mild, and mechanically cooled. The South was unsophisticated and backward; the Sunbelt was seen as cosmopolitan and forward looking. . . . The South was sharecroppers and lintheads; the Sunbelt spoke of high tech. Yankee executives might balk at moving to the South, but they might seek to locate in the Sunbelt."[7]

The idea of the Sunbelt, then, confirmed and reinforced the real economic performance of the South and Southwest after World War II. It also drew attention to the role of government, particularly highway and defense spending, in shifting the regional balance of power. In fact, the military first designated a "Sunshine Belt" across the nation's southern rim a generation before pundits and politicians "discovered" this new region in the 1970s. From the administration of Franklin D. Roosevelt to the presidency of Bill Clinton, federal intervention channeled the South's transformation, catalyzing growth and reform and defining the limits of Sunbelt prosperity.

National policy also shaped southern politics. The relentless competition for federal largesse and new industry domesticated regional politics and tamed the occupants of southern state houses. Political scientist Richard Scher's colorful typology of southern governors records a gradual shift away from the familiar Gothic characters of the not-so-distant past—bourbon-steeped "Good-Time Charleys," inconsequential "Ghosts," racist "Demagogues"—toward sober "Policymakers" and "Moderate Populists." Governors like Jimmy Carter, Lamar Alexander, and Bill Clinton supplanted those like Lester Maddox, Herman Talmadge, Earl Long, and Orval Faubus. This new breed modernized southern governance, amending their state constitutions and allowing for stronger, more responsible chief executives. Southern governors increasingly resembled their counterparts from other sections of the country. They became respectable—and respected—political figures. Two of them captured the White House.

When Franklin D. Roosevelt and his New Dealers dreamed of bringing the South into the nation's mainstream, they never could have envisioned the South as an economic leader, political bellwether, and cultural trendsetter. They could not have imagined the America presided over by Bill Clinton of Arkansas, Oxford, and Yale law school. They expected federal policy to transform the South, but never anticipated that southerners would also turn federal intervention to their own ends and remake the nation in their own image.[8]

Notes

1. Doug Marlette, *Faux Bubba: Bill and Hillary Go to Washington* (N.Y.: Times Books, 1993), 1–2, 14.
2. "A Sweetness Tempers South's Bitter Past," *New York Times,* July 31, 1994, 1.
3. Ibid.
4. Ibid., 20.
5. Lyndon B. Johnson, quoted in Joseph Califano, Jr., *The Triumph and Tragedy of Lyndon Johnson* (N.Y.: Simon and Schuster, 1991), 55.
6. Richard K. Scher, *Politics in the New South* (N.Y.: Paragon House, 1992), 169.
7. Bradley R. Rice, "Searching for the Sunbelt," in Raymond A. Mohl, ed., *Searching for the Sunbelt: Historical Perspectives on a Region* (Knoxville: Univ. of Tennessee Press, 1990), 219.
8. This point is developed by David Goldfield in "The Changing Continuity of the South," *Reviews in American History* 20 (June 1992): 233.

❂ Preface

This study had its genesis in the coincidence of two news items in the winter of 1983. A magazine article applauded the South's new high-technology centers—research parks that received billions of dollars in defense contracts and attracted scientists and engineers from across the nation. Meanwhile, a friend read to me the Labor Department's latest figures on black teenage unemployment. The appalling statistics made me wonder if black youths had always faced such circumstances. Juxtaposed, the two stories suggested some connection between the South's sunbelt boom and the economic distress of the unemployed teenagers, many of whose parents had migrated from the South to the industrial North in the 1940s and 1950s.

Pursuing these questions reignited my longstanding interest in the role of the federal government in American life. It also suggested the need for understanding the neglected history of federal economic policy between 1940 and 1962. The transformation of the southern economy did not proceed from unguided market forces alone. My preliminary investigations revealed that the federal government often sponsored those developments and always channeled them. Government policy not only regulated private economic decision-making, but also shaped the local political environments in which those decisions were taken. State action, as the British historian John Brewer has noted with regard to another time and place, formed "the hidden sinews" that animated the body politic and often restrained the invisible hand of the market.[1]

What follows, then, is an analysis of the South's remarkable odyssey from the economic catastrophe of the 1930s to the Sunbelt of the 1970s, and of the excruciating limits of that emerging prosperity. It is also a study of federal action—of its successes and its failures. By linking the history of the southern people with the history of national public policy, this study seeks to unite two issues which dominate the domestic political economy of postwar America—the emergence of the Sunbelt and the expansion of federal power over the nation's economic life.

This is not a story of "place over time"—of the South's persistent resistance to national norms and the perseverance of America's traditional antistatism in national policy. It is rather a study of "place over people"—of policies designed not so much to uplift poor people as to enrich poor places. That much of the sunbelt South shivers still in the dark cold of poverty was no oversight, but rather the consequence—sometimes intentional, often unintentional—of a set of policies.

Moreover, these policies delineated a pattern of white over black, of what some historians have termed "*Herrenvolk* development." The federal government consistently adopted programs, be they aid to development, agricultural subsidies, or military spending, which benefited white citizens more than black citizens, which favored predominately white areas and institutions over their black counterparts.[2]

Finally, *From Cotton Belt to Sunbelt* focuses attention on the extent and the intensity of federal action in the United States. Historians and social scientists have long underrated the size and power of the American state. It is supposedly underdeveloped, like the ninety-eight-pound weakling of the Charles Atlas ads. But, after 1945 the federal government was neither puny nor weak. The "arm of the federal government," as William E. Leuchtenburg (quoting Alben Barkley) has recently noted, "reached into every home and into every city and into every town in the United States."[3]

The study, after an introduction, begins with the year 1938, a date often associated with the waning of the New Deal. Historians have long agreed on the limits of New Deal reform; they have asserted that the Roosevelt administration lacked both the will and the means to reorder the nation's fundamental socio-economic conditions during the 1930s. And after 1938, any commitment to domestic economic reform was overshadowed by foreign concerns. In President Roosevelt's own words, "Dr. New Deal" gave way to "Dr. Win-the-War." At the same time, conservative advisers like Henry Stimson and James F. Byrnes displaced New Dealers like Rexford Tugwell at FDR's side.[4]

The early New Deal experience of the South confirmed this conventional interpretation. While many New Dealers in Washington desired a federal assault on the southern economy—a desire shared by President Roosevelt—national policymakers were reluctant to challenge the South's existing economic and political arrangements. New Deal programs either made compromises with the region's elite, or, as in the case of the agricultural program, were dominated and administered by southern leaders. But zeal to reform the South, rather than dissipating, emerged in earnest during Roosevelt's second term. It was not so much that the New Deal withered after 1938 as that it headed south. Decrying the South's economic backwardness and political conservatism, the Roosevelt administration launched a series of aggressive programs to reorder the southern economy. A generation of young liberal southerners entered the national government to preside over these policies during the Roosevelt and Truman administrations.

After 1950, however, the federal government's efforts to remold the southern economy appeared to flag. Even during the War on Poverty of the 1960s, the South—still the nation's poorest region—participated only sporadically in national welfare programs. Nonetheless, during this period, the South's dependency on the federal government increased. The region's ability to lure industry and research installations flowed from the largesse of the national defense establishment. Dr. Win-the-War did not displace Dr. New Deal in 1938, or even during World War II. But he assuredly overshadowed his reformist colleague after 1945. Defense-related programs played a large role in shaping southern economic development, while the influence of the federal welfare state waned.

Changes in federal policy wrought a critical transformation in the character of southern political leadership. As Keynesian fiscal policy replaced New Deal reform as the mainstay of national economic policy, and as the national security state supplanted the social welfare state as the South's principal benefactor, young southern liberals fled national service or lost their seats in the elections of the early 1950s. Meanwhile, a group of politicians dedicated to business development came to the fore. These new "Whigs" or "business progressives" eventually dominated the South. Their ability to win military spending, research contracts, and highway and airport funds proved essential both to their political success and to their region's development.

The rise of this new leadership recalled longstanding questions about the nature of the modern South's ruling classes. Southern historians have long disputed the character of the region's post-Civil War elite. But for all the debate on the character of the nineteenth-century southern elite, scholars agree that by the beginning of this century, an alliance of planters and low-wage industrialists maintained a firm grip on the region's economic and political institutions. By 1933, that alliance had consolidated its power. It seemed so impregnable that many scholars assumed that southern opposition to the New Deal in the 1930s, and to desegregation in the 1950s, indicated its continuing reign. My investigation of federal policy in the South questions that view. It suggests instead that federal policy, sometimes deliberately and often unintentionally, rent the planter–mill owner alliance during the 1930s and 1940s and encouraged the emergence of the new Whigs after 1950.[5]

The study's first two chapters investigate the economic, political, and intellectual background of the federal government's assault on the southern economy. They consider the origins of the 1938 *Report on Economic Conditions* and establish the importance of raising wages to the Roosevelt administration's program for a reformed South. The next three chapters analyze federal intervention in the South. Chapter 3 pursues the history of national wage and labor policy in the decade and a half after 1938 and examines its effects on the southern workforce. Chapter 4 considers the role of TVA, wartime, and reconversion policy in the emergence of capital-intensive southern industry. Chapter 5 analyzes the economic and, especially, the political implications of the federal grant system.

The final section carries the study to 1980. Chapters 6 and 7 contrast the activities of the Pentagon and the federal welfare agencies in the contemporary South and assess the contributions of each to sunbelt growth and to the persistence of southern poverty. The conclusion relates that analysis to developments in national politics and race relations.

"Of books about the South," the political scientist V. O. Key wrote in 1949, "there is no end. Nor will there be so long as the South remains the region with the most distinctive character and tradition." Key's prediction holds true today. Books about the South proliferate, although the South which commanded Key's attention, the South of one-party politics, economic backwardness, and segregation has come to an end.[6]

The steady stream of books, however, has never yet yielded a single, universally accepted definition of the region. Some scholars identify the South with the eleven ex-Confederate states. Many others add Kentucky and Oklahoma to form a thirteen-state region. The U.S. Census Bureau has designated a sixteen-state area—the so-called Census South, composed of the ex-Confederate states, Oklahoma, Kentucky, West Virginia, Maryland, Delaware, and the District of Columbia.

This study follows the definition of the South demarcated by the *Report on Economic Conditions of the South* in 1938—the thirteen-state region comprising the eleven ex-Confederate states (Alabama, Arkansas, Florida, Georgia, Louisiana, Mississippi, North Carolina, South Carolina, Tennessee, Texas, and Virginia) plus Kentucky and Oklahoma. The Southwest refers to the four states of Texas, Arkansas, Oklahoma, and Louisiana; the Southeast, to the remaining nine southern states. I define the South in this manner not because this collection of states satisfies any precise criteria regarding economy, polity, or climate, but because New Deal-era policymakers generally did so. In order to study the impact of federal policy in the South, I determined to begin with and to follow the region Franklin Roosevelt called "the Nation's No. 1 economic problem."

The same logic underlies the title of this book. *From Cotton Belt to Sunbelt* is not meant to suggest that the Cotton Belt represented the entire South of 1938—that the region did not contain areas as diverse as the tobacco belt, the pine barrens, the Gulf oil coast, and such distinctive cities as Atlanta, Birmingham, and New Orleans. Nor do I contend that the moniker "Sunbelt," which attached to the South in the 1970s, applies as well to the Deep South or to the Appalachians as to northern Virginia, southern Florida, and Dallas, Texas. The terms instead identify the region's place in the national imagination in 1938 and in 1980. In the 1930s, federal policymakers, southern liberals, and much of the national press came to view the problems and prospects of the South as those of the Cotton Belt writ large. Similarly, while only part of the South basked in the recent sunbelt boom, public officials and political commentators have assigned the southern states to the Sunbelt, hitching both their hopes and their fears about future federal intervention in the region to this conception. History, not geography, made the South, as the great regional sociologist

Rupert Vance noted in 1935, and this book's title, poorly as it captures the region's diversity, was fashioned in the hope of characterizing the history analyzed below.[7]

This study received generous financial support from the Mrs. Giles Whiting Fellowship in the Humanities, the Mabelle McLeod Lewis Memorial Fund, and the UCLA Academic Senate. Mrs. Frances Weaver welcomed me to her home in North Carolina. She and Richard Shrader guided me through the Southern Historical Collection. Jerry N. Hess steered me through the morass of the National Archives in Washington, D.C. Gayle Peters unlocked the mysteries of the Federal Archive and Record Center in East Point, Georgia. Jeremy Greshin, Rich Riehl, Stephen Davis, Jeff Morley, and Rosie Zagarri all provided friendly quarters during my research travels. They offered not only bed and board, but much needed encouragement and counsel.

My principal debts are to the three scholars who supervised the initial version of this book. Carl N. Degler gave this study the benefits of his wisdom and his long familiarity with the topic (Professor Degler lent me his personal copy of the Report on Economic Conditions of the South, which he had purchased in 1938). He warned me against an unsympathetic "northern" perspective on southern history. During my career as a graduate student at Stanford University, I found insight, encouragement, and friendly combat in Professor Degler's office. Gavin Wright introduced me to new ideas and arguments; he first suggested the phrase "place over people" as a way of understanding the developments I was charting. He also cautioned against unqualified generalizations and forced me to consider the full implications of my own statements. I only wish that I could have corrected all of the passages he so justifiably marked "bad sentence." David M. Kennedy served as director of the doctoral dissertation from which this book is adapted. He constantly prodded me to recognize the larger dimensions of my discoveries. Professor Kennedy also taught me respect for the underappreciated art of academic advising. A good adviser must act as critic, coach, and confessor, and David Kennedy handled all those roles masterfully.

I am also indebted to my colleagues, teachers, and students at UCLA and Stanford University. Bob Dallek, Dick Weiss, Tom Hines, Valerie Matsumoto, Gina Morantz-Sanchez, Mary Yeager, and George Sanchez all offered helpful advice. I only hope Tom forgives me for not calling this book "Old Times Forgotten." Dana Comi, Paul Goyne, and Anne Ustach provided expert research assistance. Mary Corey, Celia Applegate, and Stewart Weaver read this manuscript with care and saved me from numerous literary offenses. Conversations with Jim Campbell steered me toward this topic and around many of the obstacles it presented. Morton Sosna read early drafts of this study and offered many insights from his own work on the modern South. All along the way, Julie Reuben rescued this study from disaster and its author from despair. Without Loraine Sinclair, I never could have navigated a safe course through graduate school, and because of her my years at Stanford were a pleasure.

Robert McMath of Georgia Tech gave an early version of this manuscript a particularly penetrating and helpful reading. At Oxford University Press, the suggestions of an anonymous reviewer and the expert editing of Sheldon Meyer, Rachel Toor, and Leona Capeless have helped bring this book to completion.

I never could have completed this project without the support and encouragement of my friends and family. Four people deserve special mention. My mother, Marianne Schulman, struggled much so that her son could pursue this course. Angie Arvidson helped to make writing and living worthwhile. My only sadness is that she cannot witness the completion of this project. My grandfather Jules Schulman taught me to "look out for falling rocks." He made me the scholar and the person that I am. Finally, Alice Killian edited and proofread this manuscript many times. She endured what she called my "obsession with southern agriculture" with good humor and occasionally, with much needed impatience. That was the least of her contributions.

From Cotton Belt to Sunbelt

Chapter 1 ❂ Introduction: Becoming Economic Problem No. 1

In July 1938 President Franklin D. Roosevelt declared the South "the Nation's No. 1 economic problem—the Nation's problem, not merely the South's." The South, the nation's poorest region, could no longer remain isolated, an island of economic deprivation diminishing a once flourishing nation. In 1937, per capita income in the region reached barely half of the standard of the rest of the nation. The South registered the nation's lowest industrial wages, farm income, and tangible assets. Those abstract statistics translated into concrete human suffering. "The low income belt of the South," the official *Report on Economic Conditions of the South* concluded, "is a belt of sickness, misery, and unnecessary death."[1]

Roosevelt convened a conference of prominent southerners to sketch "a picture of the South in relation to the rest of the country, in order that we may do something about it." He intended that portrait to serve as a blueprint for national policy and a manifesto for the upcoming Congressional elections. The ensuing *Report on Economic Conditions of the South* marked the onset of a concerted national effort to restructure the southern economy. Roosevelt pledged his Administration to develop the "despoiled" southern economy, to "address the challenges of the new industrial era," and to raise the living standards of all of the region's inhabitants.[2]

In the minds of national policymakers, the South's woes stemmed from underdevelopment. The Depression South remained an overwhelmingly agrarian region in an urban, industrial nation. In 1930, the majority of Americans dwelled in cities, while only one-third of the southern people resided in urban areas. With 28 percent of the nation's population, the South claimed an imposing 41 percent of its agricultural workers, but just 15 percent of its industrial wage earners.[3]

Agriculture remained the core of the regional economy and southern farmers had struggled since Appomattox. At one stroke, uncompensated emancipation had eradicated many of the region's assets; it eliminated nearly half of the wealth in the plantation belt of the Deep South. When slavery disappeared so did the South's main source of collateral for credit and the principal basis of tax revenues for state and local governments.[4]

The rural South's most vexing problem, however, was how to organize the labor force after the end of slavery. The freedmen rejected landlords' efforts to reconstitute the slave gang system on a wage basis, especially since the cash and credit-poor landlords offered only end-of-season payments, not regular wages. Under those conditions, ex-slaves insisted on managing their own (and their families') labor, and on working their own land. Landlords found their former bondsmen intractable, even rebellious, and used every means within their power, including extra-legal violence, to discipline and intimidate the freedmen.[5]

Still, planters never succeeded in restoring the plantation regimen. Gradually, family-based sharecropping and the neo-plantation form of labor organization came to dominate the postbellum South. By 1900, most large plantations contained family-operated sharecropper plots, and a residual wage labor section, managed by the planter and staffed mainly by single men. These fragmented farms retarded regional economic development. Share plantations simply operated less efficiently than wage-labor farms; tenants, after all, lacked incentives to improve the land they tilled. At the same time, the credit shortage and the small scale of agriculture forced specialization in cotton and other cash crops. Self-sufficiency was lost. In the plantation districts, production of foodstuffs and livestock declined as large holdings divided into small family-operated units. The South began importing large quantities of foodstuffs from the Midwest.[6]

The Cotton Belt bore these burdens most heavily, and the Cotton Belt expanded in the decades after the Civil War. It grew geographically, economically, and culturally, so that, in the American imagination, cotton cultivation became synonymous with southern life. The postbellum South found itself more dependent than ever on the whims of "King Cotton." As southern farmers increasingly specialized in cotton, international cotton demand alternately fueled and strangled the region, leaving the South at the mercy of the world market. The fortunes of the region's farmers fluctuated with world demand, but never rose high; rural southerners never earned more than half the income of their northern counterparts.[7]

Tenants and sharecroppers suffered most from the failures of southern agriculture. They confronted deep and desperate poverty, and they also faced the machinations of powerful landlords. Still, tenants and croppers were never wholly powerless. They extracted shares or rental agreements rather than settling for end-of-season wages. They worked as families, not as members of supervised squads or gangs. Tenants often compelled landlords to offer them year-round contracts, even though the planters would have preferred to hire them only for the planting and harvest seasons. And southern tenants frequently changed landlords despite the vagrancy and anti-enticement laws on the books in many southern states. As one ex-slave explained, freedom "made one difference. You could change places and work for different men."[8]

This mobility vexed planters; they needed a secure supply of labor for the harvest and planting seasons, so they were forced to rely on tenants rather than machines and seasonal wage-labor. The roar of engines, so common in the corn and wheat belts, was almost unknown on southern farms. To some extent, the low level of mechanization reflected the peculiarities of southern crops and the limitations of existing technology. As yet, no suitable mechanical cotton-picker had been developed. But southern agriculture lagged behind the rest of the nation in the introduction of suitable machinery, even where it was available. Tractors, for example, could perform pre-harvest tasks in cotton cultivation. Yet the South possessed barely half the number of tractors per harvested acre as did the rest of the nation. Poverty, of course, prevented many southerners from purchasing tractors. But the South's largest, most prosperous farms, ones that could easily afford machinery, were the least mechanized. Rather than deficient capital, cheap labor and owners' need for a secure peak season supply of hands inhibited the adoption of machinery. The backwardness of southern agriculture was apparent to nearly all observers by the 1930s.[9]

Shifting their gaze from cotton fields to mills and factories, New Dealers saw only more economic underdevelopment in the 1930s. The South's limited industrial capacity was concentrated in low-wage, low-productivity industries. Low-wage manufacturing—textiles, hosiery, and lumber—formed the backbone of southern industry. These industries relied on unskilled labor, often including women and children. They also employed labor-intensive production processes; in 1937, southern manufacturers employed 18 percent more workers per plant than their rivals outside the region, but the value of output per plant tallied 18 percent less. Low labor costs compensated for this inefficiency. Southern industrial workers put in longer hours than the workers of other regions, but their average annual wages amounted to a meager two-thirds of their counterparts' in the rest of the nation.[10]

After the Civil War, industrialization below the Mason-Dixon line had proceeded rapidly. Led by the development of cotton textiles, the South's rate of industrial growth actually outpaced the national average from 1869 to 1929. In North Carolina, manufactures surpassed farm products as the state's most valuable product during the final two decades of the nineteenth century. Towns sprouted up throughout the region's interior, stimulating the vibrant commerce and local economic boosterism that the Old South had lacked.[11]

But regional industrialization, impressive as it may have been, never induced a full-scale industrial revolution along northern lines. The South of the 1930s still lagged behind the rest of the United States; six decades of catching up had not erased the region's debility. For one thing, the South's industrial revolution, unlike the North's, was not accompanied by rapid urbanization. Towns developed, but few major cities emerged. Textile mills, though they often owed their existence to the capital and the promotional efforts of southern towns, usually located outside a community's borders. There they found cheaper land, lower taxes, and less interference with their labor supplies. Lumber and timber products, the region's other principal industry, likewise located on the periphery of southern towns. Scattered across the region's pine and hardwood forests, small, impermanent southern sawmills contributed little to local economic development.[12]

Meanwhile, southern industry remained dependent on labor-intensive production. From the beginning of the postbellum manufacturing boom, cheap labor had been the secret of southern industrial success. The southern states with the severest capital shortages in 1880 actually led the manufacturing boom; those states harbored the most abundant labor supplies. No other region could match the South's supply of cheap rural labor. Relatively few southern farmers owned their own land or possessed the wealth to withstand chronic agricultural crises. And after 1900, impoverished Appalachian mountaineers swelled the migration to mills and factories.[13]

Southern manufacturing organized around these labor reserves. Employers paid low wages and hired mainly unskilled workers, including women and children. Textile mill owners, for instance, only offered jobs to families which provided many workers, usually demanding a minimum of one millhand for each room of housing the family occupied. Choices of industrial equipment and product lines—for instance, the South's concentration in coarse, low-end textile products—were similarly designed to exploit the region's surfeit of unskilled labor.[14]

The 1920s, the pre-New Deal decade, witnessed further tribulations for southern industry. The once-proud southern lumber industry peaked around World War I, thereafter losing profits and market share to the automated sawmills of the Pacific Northwest. Southern textiles also fell on hard times. After World War I, the sudden burst of the wartime boom spread panic through the industry. Mills compensated for their wartime overexpansion by laying off employees, reorganizing production, and most commonly, by demanding greater productivity from their workforce. At the same time, the regional wage differential mounted. The widening gap in wages only reinforced the impression that the South was falling ever farther behind the rest of America, that it needed not merely more industry, but more high-paying, high-skilled manufacturing jobs.[15]

The sources of the region's distress, then, seemed obvious to the Roosevelt administration and to many contemporary southerners. They viewed the South as a colonial economy, a source of raw materials, cheap labor and profits for the industrial North. "The South actually works for the North," Texas Congressman Maury Maverick explained, "mortgage, insurance, industrial and financial corporations pump the money northward like African ivory out of the Congo." The trials of "colonization" extended beyond the economic sphere. They infected the region's social and political relations as well. "The South never had a chance in American life," the New York *Post* maintained. "Its economic relationship to the rest of the nation was always cockeyed and from there it is only a step to cockeyed race relationships." Poverty caused racial friction and they in turn perpetuated the undemocratic, demagogic politics of the South.[16]

Franklin D. Roosevelt shared that assessment of the South as a colonial economy. At the time he took up residence in Warm Springs, Georgia, in 1924, Roosevelt's initial impressions of the region echoed Henry Grady's 1889 account of a Georgia funeral, for which the South provided only "the corpse and the hole in the ground." In similar terms, Roosevelt described his distress at hearing the whistles of the milk trains from Wisconsin in a region with fine dairy pasture, at finding only Washington apples in the local store when the "best apples in the world" grew only 75 miles away. "I went to buy a pair of shoes," the President ruefully concluded, "and the only shoes I could buy had been made in Boston, or Binghamton, New York, or St. Louis."[17]

This vision of the South as a colonial economy carried two corollaries. First, it implied that the lack of industry, especially of highly mechanized durable goods manufacturing, was the source of southern backwardness. "Lacking industries of its own," the *Report on Economic Conditions* explained, "the South has been forced to trade the richness of its soil, its minerals and forests, and the labor of its people for goods manufactured elsewhere." Only advanced manufacturing could provide the high wages, the purchasing power and the tax base to extricate the South from its misery.[18]

For New Dealers, low wages exposed the ills of existing southern industry. The *Report on Economic Conditions* referred to "wicked wage differentials" between the South and the rest of the country. "Wage differentials become in fact differentials in health and life," the *Report* concluded. Low wages explained the region's deficient housing and poor nutrition. They forced families to augment the earnings of adult males by sending children and women to the mills. In 1930, for example, the South claimed three-fourths of the nation's child laborers. Therefore, building more of these sweatshops would not deliver the South from its poverty. The new factories might bring some wealth to the South, but they offered little to distressed southerners. Only high-wage jobs, many southern liberals and Roosevelt administration officials believed, could relieve the people of the South, liberate them from colonial bondage, and truly integrate the region into the American mainstream.[19]

The second assumption of the colonial model located the impetus for economic development in the federal government. As a "colony" of the North, the South did not bear full responsibility for its problems, nor did it possess the resources to eradicate them. The failure of philanthropic foundations further underlined the need for national action. Foundations had attempted with little success to bolster southern education, to ease the torments of sharecroppers, and to promote racial justice. State governments had proved similarly incapable. Without appropriate outside intervention, the region could develop only along lines consonant with its colonial tradition; it would remain a rich land with poor people.[20]

STATE AND SOCIETY IN THE PRE-NEW DEAL SOUTH

The *Report on Economic Conditions* identified the South's travails in purely economic terms. It carefully avoided apportioning blame to the region's own leadership. Yet President Roosevelt and many members of the conference that prepared the *Report* recognized the political component of southern underdevelopment. For Roosevelt and his contemporaries, the South was the nation's recalcitrant region, a bulwark of sectionalism against an emerging national political economy. Roosevelt and his advisers desired federal supervision of regional economic reform not only because the southern states were unable to undertake it themselves, but also because they were unwilling.[21]

This regional political economy had its roots in the "Redemption" of the 1870s. In many southern states, the first post-Reconstruction governments had built *laissez-faire* principles into their state constitutions. Over the ensuing decades, southern governments seldom evinced the desire to evade such constitutional strictures. When the New Dealers of the 1930s looked back at the region's past, they saw an almost unbroken line, a "Solid South" of conservative Democracy and white supremacy unchallenged since the end of Reconstruction.[22]

When the Redeemers assumed power during the 1870s, minimal government largely defined their political philosophy. Decrying the alleged excesses and corruption of Reconstruction, they slashed the salaries of state employees, withdrew funding from educational and social welfare agencies, and shortened the school year. They grasped the reins of power in a poor region, a section which had just emerged from two decades of war and social strife, and established governments largely insensitive to their people's grievous needs. At the same time, the white South returned to the halls of power in Washington. There, southern politicians cooperated with the fiscally conservative Democrats of the Northeast, eschewing, at least for a time, alliance with the agrarian West and its proponents of active government.[23]

The Redemption governments, of course, were not wholly inactive. They promoted the interests of planters by enacting coercive agricultural labor legislation, they wooed business with low taxes and generous subsidies, and they refrained from regulating the work places of industrial laborers. Government spending slowly mounted as well, especially where it could be financed with non-tax revenues.[24]

Initially, the leaders of the post-Reconstruction South claimed the loyalties of the region's small, independent white farmers. In the 1870s and 1880s, this yeomanry responded not only to the Redeemers' invocations of white supremacy and home rule but to their conservative fiscal policies as well. Before the Civil War, southern state governments, especially in the Deep South, had relied heavily on the tax on slaves to finance their operations. Other assessments remained so low that non-slaveholders bore negligible, and in many cases, nonexistent tax burdens. On the other side of the ledger, southern public expenditures never rivaled the systematic programs of internal improvements and free, compulsory education that prevailed in many northern states. But the regimes of the Old South provided rudimentary public services whenever a revenue windfall was at hand.[25]

This antebellum legacy of modest services and low taxes shaped the yeomanry's reaction to Reconstruction and Redemption. After Appomattox, southern states suddenly had to serve an expanded free population, almost doubled by the inclusion of blacks. At the same time, emancipation had eliminated the region's tax base, and war had swelled the ranks of the needy, sick, and homeless. The Reconstruction governments had no choice but to raise property taxes, subjecting many yeomen to taxation for the first time and forcing them to raise cash at a time when finding even small amounts proved difficult. In return, the Reconstruction regimes expanded services and facilities, but these expenditures were diffused over the larger, biracial southern populace. White farmers realized little advantage from government spending. They naturally resented Reconstruction, not only for the perceived humiliations of black rule but also because their taxes soared without a noticeable improvement in public services. White voters happily embraced the Redeemers and their program of retrenchment.[26]

Even so, opposition to the Redeemers and to the planter and business elites they served emerged in the postbellum South. Varying from state to state, coalitions of Republicans, Independent Democrats, Fusionists, and Greenbackers challenged the regional status quo in the two decades after Reconstruction. Blacks, usually within the Republican party, struggled to protect their franchise, their schools, and their economic freedom from electoral fraud, restrictive legislation, and white violence. Upcountry farmers, facing desperate conditions in the 1880s, rose up against fiscal orthodoxy and the domination of the plantation belt. This opposition reached its crescendo with the Populist uprising of the 1890s. In many states the Populists pushed the ruling Democrats to the wall, forcing them to embrace active government and democratic reforms. In North Carolina a "Fusion" ticket of Populists and Republicans actually ousted the Democrats in 1896. The North Carolina Fusionists increased social spending, capped interest rates, and replaced appointive county offices with elective ones.[27]

By the turn of the century, however, the Old Guard had turned back the challenge in North Carolina and throughout the South. Soon thereafter, the regular Democrats consoldiated their rule, fashioning a new regional political order. In a series of white supremacy campaigns, the Democrats harped on racial fears, thereby turning back the Populist tide and discrediting all potential opposition to the "Party of the Fathers." In the process, the southern states, led by the resurgent conservative Democrats, disfranchised black southerners. Disfranchisement secured the power of the regular Democrats throughout the region. It not only eliminated blacks from the voter rolls but reduced the white electorate as well. After 1900, then, only external pressures could threaten the rule of the conservative Democrats.[28]

This closed one-party system also concentrated power in the hands of affluent businessmen and planters. All political competition revolved around the Democratic primary—the de facto election in the one-party South. Since party organizations never financed intra-party contests, candidates had to rely on their own supporters and raise their own funds. Occasionally, this system favored the loudest and flashiest demagogue with the largest personal following. But, more often, it favored the candidate with means or with the support of the wealthy. When the political scientist V. O. Key outlined the effects of disfranchisement and one-party rule in his classic 1949 study of *Southern Politics,* he noted the supremacy of southern society's "upper brackets" and the political weakness of its "lower bracket."[29]

Early twentieth-century public policy served the interests of the region's upper brackets. Planters and industrialists did not always occupy the state houses (although they often did), but the programs of the region's political leadership mirrored the economic interests of its ruling factions. Southern politicians relied on the support of the upper brackets for their election. Both groups desired disorganized, tractable lower classes, which never again would threaten their political power or their economic domination. State officials also shared the economic elites' interest in insulating the region from national political and market forces. Open channels risked both the loss of labor and the inflow of political heresy.[30]

Their policy, however, was not retrenchment but "Progressivism." The first three decades of the twentieth century ushered in a new era in southern fiscal policy, a genuine expansion of government action and social services. Most southern states increased support for education and created public health agencies. Road-building became a regional obsession as southern states appropriated ever-mounting sums to "pull themselves out of the mud." By the 1920s, state debt and state revenues were rising at a faster rate in the eleven ex-Confederate states than in the rest of the nation.[31]

This seeming revolution in state government betokened only a change in tactics for the region's conservative leadership, not a change in strategy. Launched by a cohort which historian George B. Tindall has dubbed "neo-Whigs" or "business progressives," southern progressivism sought efficiency rather than reform. Southern progressives rarely evinced the concern for social uplift and the urge to regulate business which animated other progressives. Business progressives concentrated on the economic problems of the South—building roads to enhance the region's transportation networks, maintaining the South's wage advantages in industry, luring capital for textile mills—not on the problems of the South's impoverished, undereducated people. Low taxes remained a central feature of their program so that, despite the burst of government activity, per capita tax collections remained well below national norms. The South's business progressives erected a minimal infrastructure for industrial growth, without threatening the Redeemers' commitment to low wages, unregulated business, and low taxes.[32]

And they did so while disfranchising nearly half of the region's population. Enhanced services for whites accompanied segregated, inferior public facilities for black southerners. Many states forced blacks to finance their own schools exclusively from their own tax payments, despite their slender economic resources. While southern cities improved municipal services in the 1920s, black neighborhoods remained without parks, paved roads, street lights, and adequate sanitation systems. Evidence from one state indicates that blacks were taxed at higher rates than whites despite the discrimination in services. In the textile states, white mill workers similarly found themselves excluded from public schools and city facilities.[33]

The public health and road campaigns even more sharply delineated the priorities of southern progressivism. The South faced the nation's most acute health problems, but, even after it created its public health commissions, the region trailed far behind the rest of the nation in the quantity and the quality of its medical facilities. Philanthropic foundations, not government agencies, initiated and substantially underwrote the region's battles against hookworm, malaria, and pellagra. On the other hand, southern state governments took the lead from the start in road building. An old Mississippi adage captured this paradox—the southern leadership's emphasis on aid to business over social uplift. Mississippians, the saying went, "ride to the poorhouse on the best roads in the country."[34]

They also rode to some of the poorest schools in the nation. The region had a long heritage of ambivalence about public education. Before the Civil War, some southern states had established school systems, but without support from state tax receipts. Georgia, for instance, chartered its school system in 1859, financing it mainly through nontax revenues. The Reconstruction governments improved southern schools, but they faced the problem of educating nearly twice as many students in a defeated region suddenly shorn of much of its wealth. The Redeemers, not surprisingly, slashed support for the schools and crippled the region's already deficient educational systems. In 1900, all but two states outside the South had enacted compulsory school laws; among the southern states, only Kentucky had done so.[35]

Around the turn of the century, a region-wide educational reform movement began. Like the public health campaign, the educational crusade had its impetus not in the state houses of the South, but in the foundation boardrooms of the North. Philanthropists, principally the General Education Board of the Rockefeller Foundation and the Peabody Foundation, ignited the educational explosion. They financed school construction and trained teachers and administrators. The General Education Board, in one historian's words, played "the role of the central directorate for southern education." Eventually, southern cities levied taxes to match foundation grants, and state governments entered into the educational crusade. Southern progressives boasted about their commitment to education, but it ever remained a partial, piecemeal effort. Nowhere in the South did per pupil expenditures in 1930 exceed two-thirds of the United States average; most southern states continued to wallow at about half the national standard.[36]

Educational reform simultaneously rewarded the most privileged segments of southern society while curtailing education for the least well-off. It provided a handful of good schools for the region, but little good schooling for the southern people. Black schools hardly benefited from the crusade. In fact, by 1930, the South spent more than three times as much per student for whites as for blacks. Poor whites fared better, but still received substandard educations. In the mill villages of the Carolinas, children of operatives attended company schools. These mill schools, staffed and operated by the textile company, provided only primary education. In mill schools, attendance was haphazard and hurry calls to the factory often interrupted classes. And what little education occurred in mill classrooms aimed students toward careers in the cotton factories. Mill owners had little incentive to furnish children with the knowledge and skills that might lead their labor supply out of the mills and even out of the region.[37]

Southern states, then, stepped up their spending and government activity between 1900 and 1930, but they did not fundamentally alter the regional political economy. Educational expenditures mounted as long as they neither required high taxes, nor trained blacks and mill workers, thereby threatening the region's sources of cheap unskilled labor. The South's political leadership embraced "progressivism," removing "corruption" from government by denying citizenship to blacks and eliminating political opposition.[38]

This new order fell hardest on black southerners. For them, political powerlessness and segregation compounded economic distress. Before disfranchisement, black southerners had struggled within the political system, organizing and voting despite the intimidation and violence of whites. They mobilized against segregation and disfranchisement and tasted bitter defeats. Jim Crow diminished blacks' political power, impoverished their public facilities, and handicapped all of their community building and organizing activities.[39]

In the face of these disadvantages, black southerners fought to hold their own in the early years of the twentieth century. Many migrated to the region's cities, and there, under the pressure of Jim Crow, they developed their own business class geared toward the black market. Social clubs, black chapters of the YMCA and YWCA, separate black churches, and other community institutions emerged. But the burdens of segregation were heavy. As one historian of southern black leadership has noted, segregation and disfranchisement not only deprived urban blacks of their fair share of a city's prosperity and political influence, they also robbed black leaders of the power to protect their constituencies. In Charleston, South Carolina, the city's white government discontinued the Labor Day parade that had been a source of pride and achievement in the city's black neighborhoods. "The black workers showed the white workers up so badly that the white people got mad," remembered Mamie Garvin Fields, a black Charlestonian. "They stopped the Labor Day parade after that." Fields and many others labored quietly for black rights, but the pre-Depression South provided only barren ground for black protest.[40]

In the countryside, where most southern blacks resided, the fight for racial justice faced even more desperate odds. Blacks were not without economic opportunities in the rural South. Tenants and croppers frequently changed landlords, and at any one time in the early twentieth century nearly one-fifth of black farm operators owned their own land. Still, not even the few black landowners had a secure hold over their property, and the proportion of blacks operating their own farms fell during the 1910s and 1920s. Furthermore, land ownership and even movement up the lower rung of the tenure ladder from sharecropping to a more secure tenant status required good credit. And, in the rural South's local credit markets, credit ratings were based on personal reputation. Tenant families could move from plantation to plantation, and single men could seek work in distant counties, but to secure a mule, farm implements, or a piece of land, black southerners needed a safe reputation in their home county. Whites controlled access to the means of advancement; they even determined which black farmers would receive the aid of the United States Agricultural Extension Service. For a black tenant, "being acceptable," as Arthur Raper noted in his classic study of Georgia's plantation belt, was "no empty phrase. . . . It means that he is considered safe by the local white people—he knows 'his place' and stays in it." Not all rural blacks deferred to white landlords and merchants, but advancement, and often even economic survival, was impossible without such deference.[41]

In this situation, black southerners nourished ambivalent attitudes about government action. They feared state and local governments, rightfully seeing them as enemies. But, at least in the cities of the South, blacks continued to regard the national government as a friend. The national government certainly did not provoke in southern blacks the same fears and suspicions it invariably aroused among white southerners. Mamie Garvin Fields recalled that she "could frighten some of the ignorant white folks down here by speaking the name of the United States government." She even won concessions from local authorities by playing on those fears.[42]

Ignorant or not, white southerners and their political leaders generally feared the national government. Federal intervention ever raised the specter of "Black Reconstruction." As Mamie Garvin Fields put it, "The last thing they wanted was for some 'foreigners'—*'funnuhs'*—coming from outside to talk to 'their nigras.'" Even when southern whites saw no threat to white supremacy in federal action, they sensed economic discrimination. In the 1880s the Interstate Commerce Commission codified unfavorable regional freight rate differentials, and in 1916 the federal government enacted a short-lived national restriction on child labor. Southerners fought these dispensations from Washington; for instance, they spearheaded the successful Court fight against the national child labor law. That case underlined the antipathy of the region's leadership to federal intervention. Southern manufacturers had submitted gracefully to state regulation of child labor, but they furiously resisted federal regulation. Federal regulation, they believed, meant stricter enforcement and promised more dangerous labor legislation in the future.[43]

The Great Depression temporarily altered the outlook of the region's leaders. The initial blows of the Depression hit hardest in the South, collapsing the already fragile foundation of many southerners' subsistence. For over a century the region had resisted federal intervention, eschewing even the promise of economic development in order to maintain complete control over southern institutions and affairs. But so desperately did the region require relief that the southern people and the very leaders who had long championed the "state's rights" tradition, demanded federal action in the thirties. From his seat in the House of Representatives, Alabama Congressman William B. Bankhead spoke for his section when he demanded the abandonment of "our preconceived notions of economic policy." The time had come in the history of this Congress," Bankhead admonished his colleagues, when southerners were going to have to burn some of their bridges behind them.[44]

The call for federal action, however, did not imply acquiescence in fundamental economic or political reform. Despite the Roosevelt administration's concern for the South, the federal government hesitated to pursue reform during the first five years of the New Deal. Except in agriculture, where the planter elite eagerly sought and presided over the farm program, concerted national effort to heal the South was postponed. The Roosevelt administration relied on state and local leaders to manage its relief and recovery programs in the South. Southern leaders successfully deflected the efforts of reformers in the agriculture, relief, and industrial recovery programs. In 1934, Dr. Will Alexander, a southern New Dealer, observed the Administration's unwillingness to challenge the southern status quo: "The facts are that the people in Washington feel that the South is just so sensitive that they must not say anything about these matters at all."[45]

The President's political power base—his reliance on southern elites—forced such compromise and stalled any attempt to restructure the southern economy. Roosevelt owed his nomination to the southern Democrats in Congress. More than any other national politician at the time, FDR understood the problems of southern Democrats and relied on their support. Roosevelt feared that inflaming southern politicians would revive the state's-rights tradition, create a recalcitrant Congress, and threaten his own political position. To avert those possibilities, FDR proceeded slowly. His Administration established close ties with southern committee chairmen. Much to the consternation of liberal New Dealers like Interior Secretary Harold Ickes, the President consulted on matters of policy with only three members of Congress—Senate Majority Leader Joseph Robinson of Arkansas, Senator James F. Byrnes of South Carolina, and Senate Finance Committee Chairman Pat Harrison of Mississippi. These men and their southern colleagues shared the Administration's desire for recovery. But the President's goals for the region—a floor under wages, cheap electrical power, and expanded social services—were not theirs. Southern leaders welcomed federal assistance, but opposed intervention that might erode the regional wage differential or threaten the system of white supremacy.[46]

New Deal policymakers, however, generally opposed the objectives of southern leaders. No southerner was among the "Brains Trusters" who devised the programs of the Hundred Days or at the helm of the agencies that directed them. An almost complete divorce prevailed between federal administrators and the southerners who enacted and implemented federal programs. The New Dealers desired reform of the South. At the least, they refused to exempt the region from the national programs then under way. But southern leaders, jealous of their region's prerogatives, demanded special treatment.[47]

Not until 1938 would Roosevelt address this paradox. By then he had grown frustrated with southern politicians and with the region's lingering poverty. The region's peculiar economic maladies and growing political intransigence appeared to subvert all attempts at national economic recovery. Meanwhile, changing political currents eventually emboldened Roosevelt. He decided to commission the *Report on Economic Conditions* and to employ the full power of the national government to restructure the southern economy.

But even if concerted national effort waited until Roosevelt's second term, the early New Deal nonetheless transformed the South. National economic policy effected the greatest upheaval in that poorest of the nation's regions. Moreover, once southern leaders accepted federal intervention in their region, they could not completely control the forces they had unleashed. Like a virus, the New Deal seeped into the region's economic body, infecting traditional political and economic arrangements and preparing the way for more extensive federal action in the future.

AAA AND THE COLLAPSE OF SOUTHERN TENANCY

When Roosevelt took office in March 1933, he confronted a desperate situation in the rural South. The Great Depression had compounded the longstanding problems of southern agriculture. Southern farm income, already the lowest per farm and per acre in the nation in 1929, dropped even further during the first four years of the Depression. Cotton and tobacco prices collapsed. Local agencies failed to boost commodity prices or to mitigate human suffering.[48]

The inadequacy of local efforts inspired the federal agricultural program. New Deal agricultural policies focused on the problems of the South, especially those related to cotton cultivation. They bore the imprimatur of the southerners who controlled and administered the programs. "The Cotton States have found it impossible to act independently or in unison" to restore the rural economy, President Roosevelt asserted in 1934. "They have asked for the use of Federal powers. A democratic government has consented."[49]

Management of the farm program fell to the Agricultural Adjustment Administration (AAA), an agency of the Department of Agriculture. The AAA's first administrator, George N. Peek, had led the fight for farm parity in the 1920s. Peek's deputy and successor at the helm of the AAA, Chester Davis, had served as Peek's lieutenant in that battle. Both men saw the Agricultural Adjustment Act as a precision tool for raising farm prices and not as a multi-purpose vehicle for effecting social reform in rural areas. In this respect, the AAA chiefs differed from liberal Secretary of Agriculture Henry A. Wallace. Nonetheless, the Secretary maintained an amicable relationship with Davis, primarily because Wallace never challenged the AAA's emphasis on raising prices.[50]

Southern planters remained on good terms with both Wallace and AAA leaders. They shared the administrators' view that the federal government ought not interfere with existing agricultural practices. They applauded the agency's idea that higher commodity prices would benefit all southerners. And the landlords had no quarrel with Secretary Wallace because he never pressed his plans for reform and he permitted sufficient decentralization of AAA operations to ensure local control.[51]

Southern landlords and their champions held the positions of power within the AAA's decentralized bureaucracy. Oscar Goodbar Johnston, supervisor of the South's largest cotton plantation, dominated federal agricultural policy for the region. Johnston served as comptroller of the AAA, director of the Cotton Pool, and assistant director of Commodities Credit Corporation. Cully Cobb, a Georgia editor who reportedly hoped to win for southern planters what Alexander Hamilton had achieved for northern industrialists, ran the Cotton Production Section. These men and their allies controlled the cotton program and influenced the formation of all federal farm policy.[52]

Not surprisingly, the cotton production program protected the regional status quo. The growing season was already under way when the Agricultural Adjustment Act became effective in 1933. Production cutbacks therefore required the ploughing up of more than ten million acres of cotton, about one-quarter of the planted acreage. Producers received a cash rental payment (based on estimated yield per acre) plus options to purchase government cotton at 6 cents per pound to compensate for the retired acreage. Participation was open to all farmers, irrespective of land tenure. Legal ownership of the crop determined the distribution of rental payments: cash tenants were entitled to the entire payment, share tenants to three-quarters of it, and sharecroppers to half. The government sent the checks to the landlords, however, and relied on the owners to distribute the benefits.[53]

Before the next growing season, the AAA promulgated a revised contract and enlisted more than one million producers in the cotton limitation program. The 1934–35 contract set the precedent for future arrangements by dividing benefits into separate "rental" and "parity" payments. Landowners were required to turn over the entire rental fee to cash tenants and half of it to "managing" share tenants. The owners kept the parity payments.[54]

In just three years, the adjustment program revolutionized the Cotton Belt and all of southern agriculture. Cotton prices soared from 6.5 cents in 1932 to 10.17 cents in 1933, stabilizing at 11.09 cents in 1935. Tobacco prices doubled in the same period. Meanwhile, cotton acreage plummeted by nearly 30 percent from 1932 to 1935. "King Cotton" would never again account for so great a proportion of the South's cultivated land.[55]

The Department of Agriculture claimed even more extensive success for its "domestic allotment" program. As a typical case, the AAA cited the example of a farmer from Jackson County, Oklahoma. In 1932 the farmer had netted $4 on each of the 48 bales of cotton he produced. In 1933, he cleared $12 for each of 36 bales he harvested and collected an additional $80 in rental benefits. "Last year I couldn't pay a thing," he remembered. "This year, I can live, pay more for my groceries and buy a few things." Such prosperity extended through the entire farming community. In Altus, Oklahoma, the Jackson County seat, bank vice president H. B. Bellinger claimed that "the results of the Goverment's cotton program have been wonderful." Bellinger concluded that the "program just about saved our lives. I never have seen anything work out better in my life."[56]

Many southerners shared the enthusiasm of that Oklahoma banker. Southern cotton producers voted overwhelmingly in favor of every referendum to continue mandatory controls. The representatives of those farmers, in Washington and in the region's state capitals, found that support for the AAA formed a successful campaign strategy. Mississippi Senator Pat Harrison won reelection in 1936 by stressing his association with the New Deal in general and the agricultural program in particular.[57]

The AAA, however, underrated the agricultural program's effects on southern agriculture. It was unprepared for the alterations in economic incentives fostered by the program. The AAA conceded the need for "long term planning on acreage reduction" and acknowledged the "human phase of the problem." Still, it operated on the assumption that planters would cut production proportionally from the wage-labor and tenant sections of their holdings. But since landlords would have to share the rentals for retired acreage from the tenants' plots, AAA policies encouraged owners to sign on fewer tenants before the growing season began and the AAA took force. In so doing, the agricultural program enlarged the asset value of land without securing the tenants' share rights. Or as one luckless tenant put it: "The tenants want this rental money and the landlords don't want to give it." The initial "displacements" furthered the decline of tenancy by increasing the pool of wage labor. That eased the harvest labor constraint, which had long forced planters to maintain tenants on an annual basis. The upshot of these altered incentives was the reallocation of cultivated land from tenancy to wage labor.[58]

The AAA benefit structure was largely responsible for that reallocation of land. The 1934–35 cotton contract required landlords to "endeavor in good faith" to retain "the normal number of tenants." Nonetheless, it extended rental shares only to cash tenants and "managing" share tenants. Moreover, the landlords continued to receive the benefit checks through the 1936 season. One sharecropper in the Mississippi Delta region of eastern Arkansas expressed his frustration with the AAA's failure to limit the power of landlords. "De landlord is landlord," he complained, "de politicians is landlord, de judge is landlord, de shurf is landlord, ever'body is landlord, en we ain' got nothin'."[59]

AAA officials admitted the pro-landlord bias of the crop reduction program. Chester Davis believed that without such concessions to the landlords, the AAA would have failed. "Our problem," he recalled, "was to go to the limit in protecting share croppers on the land while getting the contract signed. If it wasn't signed, we had no program." Under such a policy, tenants received only a small part of the benefits. In 1934, AAA payments accounted for 39 percent of planters' net cash income; for tenants, the benefits provided only 4 percent.[60]

As New Deal farm policy evolved, the tenant sector of southern agriculture shrank while the pool of wage labor swelled. Unlike earlier periods of agricultural recovery, reinstatement of tenants failed to accompany rising prices between 1933 and 1936. According to the WPA's study of 40 southern plantation counties, the number of wage hands grew 20 percent between 1930 and 1934, while the number of tenants held steady, and the proportion of farms operated by tenants declined. This collapse of southern tenancy led a group of southern liberals, the Southern Committee for People's Rights, to charge that "neither the AAA nor the Department of Agriculture dares publish the true story of the South's landless people who, for all intents and purposes, have few, if any civil or democratic rights."[61]

But, in truth, it was not national attention that landless southerners lacked. Nearly all the writers and photographers who searched out the torment of the American people during the Great Depression turned to the rural South, the nation's most impoverished region. Beginning with Erskine Caldwell's *Tobacco Road* in 1932, many documentaries and fictional dramatizations of the plight of sharecroppers thrust the fate of southern tenants before the public eye. The Rockefeller and Rosenwald foundations supported extensive scholarly investigations into the tenant problem. So did the Department of Agriculture. The Roosevelt administration could hardly have ignored the tenant problem.[62]

Nor did the Department of Agriculture desire to do so. In no place was the controversy over the rights of southern tenants fiercer than within the offices of the Agricultural Adjustment Administration. In its public statements, the AAA promised to "see that the hired laborers and tenants do not bear the full brunt of the adjustment program." In private, most officials of the Department of Agriculture were more circumspect. Secretary Wallace himself circulated a confidential report in May 1934 which indicated the government's intention to defend tenants and conceded that "this protection for various reasons now appears to have been inadequate." Secretary Wallace reminded his fellow federal officials that in the South the Roosevelt administration had "inherited an economic and social problem about as mean as the unemployment problem in the industrial sections." The depth of the economic emergency prevented any assault on "this extremely complicated and difficult problem" of tenancy.[63]

Wallace's caution reflected his attempt to mediate between the contending factions in the AAA. On one side stood southern planters and their allies, on the other liberals in the Legal Division who championed tenant's rights. Despite the Secretary's efforts to contain the tension, the two factions erupted into a bitter conflict leading to the celebrated dismissal of the Legal Division liberals in February 1935. This "purge," however, did not end the pressure to improve protections for southern tenants. In 1936, the Comptroller General of the United States prohibited the Agriculture Department from distributing tenants' benefits through landlords. To meet the Comptroller General's objections, the Cotton Section adopted a joint payment system. After 1936, both landlord and tenant had to endorse the tenant's checks.[64]

At the same time, the AAA mandated that tenants deserved a larger share of federal farm subsidies (Table 1-1). In so doing the federal government ironically exacerbated the problem it sought to alleviate. In a report

Table 1-1
AAA Benefits and Tenant Shares, 1933–38
Percentage Shares of AAA Rental Payment (1933–35) and Soil Building Payment (1936–38), by Tenure Type (%)

	1933	*1934*	*1935*	*1936*	*1937*	*1938*
Cash Tenant	100	100	100	100	100	100
Managing Share Tenant	75	50	50	50	50	75
Share Tenant	75	0	0	50	50	75
Sharecropper	50	0	0	25	25	50
Wage Worker	0	0	0	0	0	0

Source: U.S. Department of Agriculture, Agricultural Adjustment Administration, *Reports, 1930–1940* (Washington, 1934–41).

commissioned by the Department of Agriculture, Duke University Economist Calvin Hoover recognized that the more favorable the division of benefits to the tenant "the stronger will be the motive for the landlord to reduce the number of his tenants." The AAA ignored Hoover's warning, and displacement of southern tenants began in earnest after 1935. From 1935 to 1940 the tenant population declined 25 percent. In this same period the number of agricultural wage laborers grew by 14 percent.[65]

Mechanization accompanied the decline of tenancy. "The roar of tractors," agrarian reformer H. C. Nixon observed in 1938, "has been replacing the sound of Negro songs" in the South. Southerners purchased more than 100,000 tractors in the first seven years of the New Deal, doubling the percentage of farms reporting tractors between 1935 and 1940. Most historians have attributed the rapid introduction of tractors in the 1930s to the AAA benefits, which provided cash where it had been previously unavailable. But want of capital had not been the most important constraint on mechanization in the South.

The largest, wealthiest plantations had been the least mechanized; Oscar Johnston's immense plantation, for example, employed little machinery before World War II. The AAA fostered mechanization not so much by injecting cash into southern agriculture as by reducing acreage and shrinking the number of tenants. Thus, the farm program remedied the shortage of wage labor that had so long forced landlords into year-round contracts with tenants. The AAA allowed cotton planters to mechanize pre-harvest operations without worrying about labor needs for the as yet unmechanized cotton harvest. Oscar Johnston later observed that the collapse of southern tenancy had resulted from a fundamental reorganization of southern agriculture rather than "technological unemployment." Mechanization, Johnston asserted, "is not the cause, but the result of economic change in this area."[66]

Johnston's remark came fifteen years after the enactment of the Agriculture Adjustment Act. But by the end of Franklin Roosevelt's first term, the New Deal had prepared the South for its agricultural revolution. The AAA reversed longstanding trends toward smaller units and larger number of tenants. The program also weakened the system of fragmented plantations—large holdings divided in small tenant-operated units. In the process, it eroded the vestigial paternalism of the landlord-tenant relationship.[67]

The New Deal also eroded the partnership between landlords and the operators of the region's low-wage industries. The planters still supported their industrial allies in the fight against federal wage regulation. But, gradually, their economic stake in the alliance diminished. The AAA had strengthened southern planters. Unlike southern manufacturers, they could afford to drop their guard and welcome federal intervention in the region.[68]

NRA AND THE REGIONAL WAGE DIFFERENTIAL

Southern business found the New Deal more nettlesome than the planters did. Southern manufacturers forestalled a direct attack on the region's industrial organization, but unlike the planters, they could not wholly adapt the industrial recovery program into an agent for their interests. The centerpiece of the recovery program was the National Industrial Recovery Act (NIRA). Through the establishment of a National Recovery Administration, the Roosevelt administration sought to increase purchasing power, curtail unemployment, and stimulate business recovery. "The law I have just signed," President Roosevelt said of the NIRA on June 16, 1933, "was passed to put people back to work, to let them buy more of the products of farms and factories and to start out business at a living rate." The NIRA was an emergency measure, unconcerned with the reform and development of the southern economy. Neither the recovery bill nor the brief Congressional debate over its enactment had focused on the NRA's role in the nation's poorest region.[69]

On the very day President Roosevelt signed the Recovery Act, however, he announced that simple business recovery would not suffice. After pledging to restore the nation's idle industrial capacity, Roosevelt asserted that "it seems to me equally plain that no business which depends for its existence on paying less than living wages to its workers has any right to continue in this country." The President's experiences at Warm Springs, his regular journeys past the lumberyards and textile mills of the Southeast, convinced him of the malevolent effects of low wages and inefficient factories. Roosevelt called on the NRA to lead the transformation from "starvation wages and starvation employment to living wages and sustained employment," a change that would not only uplift laboring people but also revive industry by increasing purchasing power.[70]

Southern industrialists realized that most of the businesses that the President asserted had no right to continue were their own factories and mills. A few months after the passage of the NIRA, David Coker, the owner of South Carolina's largest seed company, attended a speech by General Hugh Johnson, the administrator of the National Recovery Administration. According to Coker, Johnson declared that "Southern industry has been built upon the degradation of Southern labor" and promised to eliminate special treatment for the region's industry. "What hope has the South of industrial justice," Coker asked a NRA official, "with a man in charge of its operations who is not only densely ignorant of our Southern situation and problems, but who unlimbers his vitriolic machine in our direction."[71]

By early 1934, the battle lines were drawn between southern businessmen eager to protect their region's low wage position and the proponents of uniform national standards. That controversy grew out of the debate over the code for the cotton textile industry, the first NRA code, adopted on July 9, 1933. The textile industry, the South's largest, had suffered enormously from the Depression and quickly acquiesced in a code that mandated the 40-hour week, the abolition of child labor, production limitations, and a minimum weekly wage of $13 in the North and $12 in the South. The code demanded significant concessions from southern mill operators. Before NRA, northern millhands found on average over $3 more in their weekly pay envelopes than did their southern counterparts, even though the southern work week was longer. During the month before the code went into effect, more than 75 percent of the South's millhands received wages lower than the impending minimum as compared with fewer than 40 percent of the textile workers outside the South. Nonetheless, so desperate for regulation was the textile industry and so insistent on establishing the first NRA code was General Johnson that most southern mill owners shelved their objections and accepted NRA supervision.[72]

That amiability lasted only a few months. Average hourly earnings in southern plants skyrocketed by nearly 70 percent in five months. At the same time, northern plants absorbed only a 51 percent increase. Southern businessmen soon complained about "discrimination." They recognized the code's benefits; it granted price and market advantages to southern textile firms and accepted lower southern wages in principle. Still, many southern leaders, both inside and outside the industry, feared that the NRA would eliminate their region's chief competitive advantage. They worried that the NRA's injunction against "wage chiselers" really referred to cheap southern labor generally. They realized that national policymakers might interpret "cutthroat competition" as labor-intensive southern competition, and "starvation wages" as low southern pay scales. General Johnson's statements threatened a crusade against the region's labor system, and southern industrialists determined to prevent it.[73]

Their anxiety found institutional expression in December 1933 with the formation of the Southern States Industrial Council (SSIC) in Chattanooga. The SSIC, its president John E. Edgerton conceded, was "a child of the NRA." Its purpose was "to protect the South against discrimination," a euphemism for maintaining regional wage differentials. As reasons for low southern wages, the Council's wage differentials committee cited the need to employ "negro labor of subnormal capabilities," and white workers "unable to adapt themselves to close and routine application to work, which the northern worker seems to take for granted." Drastic changes in the region's labor standards, the SSIC warned, would ruin southern industry and displace thousands of workers.[74]

This seemingly arcane controversy over regional differentials embodied in fact a dispute between two conflicting visions of the South's economic future. Southern leaders saw low wages as their region's ticket to prosperity. Low wages would attract industrial plants. These factories would free the region from the shackles of a colonial, extractive economy, one imposed by northern businesses in league with the national government. The NRA's attempt to "standardize everything," the SSIC declared, would impose northern wage scales and northern labor strife on southern industry. It would reverse southern progress. "To some of this section's competitors in other sections," Edgerton maintained, "the opportunity had come to 'reform' the South and to lift it to a higher economical level and, incidentally to destroy its competitive power."[75]

Southern manufacturers viewed labor-intensive production as a rational response to the region's natural endowments and deficiencies. In their opinion, low-wage manufacturing represented the South's best hope. It also offered prosperity and security to downtrodden rural southerners, for southern industrial wages, however pitiful by national standards, far surpassed earnings in southern agriculture. And, according to the region's industrial leadership, what southern laborers lacked in their pay envelopes, they were compensated for in respect and concern from their employers. Southern manufacturers celebrated the "friendliness" of southern labor relations.[76]

The region's liberal newspapers, on the other hand, criticized all regional differentials. The Raleigh *News and Observer,* for instance, lamented "the differentials which . . . keep down the buying power of the South." Enthusiastic supporters of the New Deal, these southern liberals lacked power in the region's state capitals and boardrooms. They nonetheless exerted considerable influence on federal policies concerning the South. Josephus Daniels, FDR's mentor during World War I and one of the few men who called the President "Frank," ran the Raleigh paper. His son and the paper's principal editor, Jonathan Daniels, was one of many young liberal southerners who migrated to Washington during the 1930s and took up positions in the Roosevelt administration. This group included Clark Foreman and George Mitchell—both sons of famous southern liberals—Brooks Hays, Aubrey Williams, Leon Keyserling, and Frank Graham. They all believed that the regional wage differential perpetuated southern poverty.[77]

Many members of the Roosevelt administration shared this view. The NRA endorsed it. It believed higher wages would increase the efficiency of southern workers. An NRA bulletin described one of the agency's purposes as "to increase weekly earnings, both in dollars and goods, particularly of those occupational groups and those sections, geographical and social, where they are below a level making possible maximum product per worker." President Roosevelt himself had long believed that low wages hurt the South more than they helped it. But he was unwilling to challenge openly southern industrialists and their allies in the Congress. The young economists in the Department of Labor felt no such restraints. They prepared a report for the NRA which declared that "there were no economic reasons for wage differences between regions and that these should be eliminated as soon as feasible." These views gained wide currency in the Administration. In fact, a leading government opponent of this view, NRA Director of Research Charles Roos, admonished his NRA colleagues for the "crusading zeal" with which they "faithfully followed the dogmatic advice of the Department of Labor."[78]

Southern manufacturers resented the Administration's position. They attributed the South's reliance on cheap labor, in one's words, to "the fact that much of the native labor cannot learn to operate complicated machinery like that used in the North." That argument exposed one of the issues underlying the debate over geographic differentials—the question of wage rates for black industrial workers. Southern businessmen demanded not only regional differences but explicit racial wage differentials in the NRA codes. Many southern employers insisted that blacks were unproductive. "Colored labor has always been paid less than whites and for good reason," declared the representative of an Alabama paper company, "since these people are not as efficient as white people, needing constant supervision over any work that they perform."[79]

Moreover, southern employers argued that they had a moral obligation to employ these "inefficient laborers." A Georgia businessman informed the NRA of the high concentration of unskilled workers in the region—"laborers who are incapable of performing other than casual or incidental service. This is, of course, due," he added, "to the large Negro population of the South." The same minimum wage for blacks and whites would force the dismissal of black laborers. "We have this class on our hands," he pleaded, "and somebody must take care of them and if industry cannot do so, who will?"[80]

The potential displacement of black workers posed a dilemma for black leaders. They faced the choice between endorsing racial differentials and thus writing black inferiority into the NRA codes, or risking the jobs of thousands of black laborers. Many black southerners feared that the Depression had already set off a Negro displacement movement. In hard times, whites no longer spurned menial labor. Across the South, whites moved into occupations which blacks historically had dominated. "Why, they won't even let us keep our jobs as truckdrivers, let alone anything higher," a black cashier in North Carolina complained. In Virginia, an association of window washers and furnace keepers excluded blacks from jobs they had long held in great numbers. These developments alarmed black leaders like Richmond minister Gordon Blaine Hancock. Hancock feared that high NRA wages without protection for black workers would deny blacks access even to jobs in traditionally black lines of work. Hancock criticized black leaders who placed abstract recognition of racial equality ahead of concrete economic progress for their constituency.[81]

Notwithstanding Hancock's concerns, every national black organization opposed racial differentials. Black leaders in the 1930s envisioned an expanded role for the federal government in American life. They feared that NRA racial differentials might establish the precedent for blacks to receive only substandard federal assistance in years ahead. Such differentials would also validate southern claims of black inferiority. In their view, the danger to the civil rights cause outweighed the benefits of any temporary protections for black southerners. The National Association for the Advancement of Colored People warned in August 1933 that "these codes will serve as standards for decades, if not generations."[82]

The National Recovery Administration cautiously occupied the middle ground. The NRA neither established explicit racial differentials nor affirmed the right of black southerners to equal pay for equal work. It rejected subminimum wages for blacks without denying the rationale for racial differentials. The NRA's Division of Economic Research and Planning conveniently decided that "no conclusion can be drawn" on the alleged incapacities and "leisurely" work habits of southern blacks. It even speculated that syphilis caused black inefficiency. "It is generally known that there is a very high percentage of syphilis among the Negro race," one report concluded. "If allegations of slowness of reaction are true, this factor would account for it." In such a strategic fashion, the agency conciliated civil rights organizations without offending southern industrialists.[83]

The National Recovery Administration deftly resisted southern pressure for racial wage differentials. Nonetheless, it compromised with southern business interests on the region's top priority—separate code provisions on wages and hours for the South. The NRA certainly narrowed regional differentials, but it reaffirmed the principle of lower southern wage rates. The agency declared that ". . . any drastic movement to lower or entirely eliminate the North-South differential would result in a complete disruption of [the] present industrial and social systems of the South." The NRA conceded that uniform national standards might become an object of federal policy, but only "in the future with the aid of education and sociological developments in the South."[84]

Southern industrial interests thus succeeded in embedding a geographic wage structure into NRA policy. Of 578 approved codes, 298 contained regional differentials. These generally delimited a simple division between the South and the rest of the nation, although some codes designated three or more regions. In each case, the rate for the southern region was lowest. A 1935 Brookings Institution study of the recovery program established that "*in fine,* directly or indirectly a geographic wage structure is a characteristic feature of the codes." The Brookings report concluded that "very seldom in terms of employee coverage was there failure to grasp an opportunity to apply differentials."[85]

The manufacturers also controlled the administration of the codes in the South. The NRA devoted few of its resources to its understaffed field units. The furniture industry code authority, for example, delegated nearly all of its administrative functions to the two trade associations that had sponsored the code—one southern, one representing the rest of the nation. The Southern Furniture Manufacturers Association had almost free rein to enforce the NRA in its branch of the industry.[86]

The NRA shrank back from any concerted effort to reform the southern economy. Recovery remained the NRA's and President Roosevelt's top priority and the program required the support of southern politicians and the acquiescence of southern employers. In return for their cooperation, southern leaders extracted from the NRA a rhetorical commitment to the region's low-wage position and local administration of national policy. Southern manufacturers never turned federal policy to their advantage like their erstwhile planter allies. They nonetheless postponed a direct attack on their interests.

Still, the NRA codes fell heaviest on the nation's poorest region. The codes officially sought to shorten hours, to place a floor under wages, and to increase productivity. In the South, where hours were longest and wages and productivity lowest, those principles were bound to effect the greatest change. Southern manufacturers prevented the NRA from adopting an explicit southern policy. But below the Mason-Dixon line, the NRA's national policies achieved their maximum potency.[87]

Every NRA code of fair business practices contained restrictions on maximum hours, regulations which chiefly benefited southern wage earners. Before the NRA introduced the 40-hour week, most southern plants had run 10- to 12-hour shifts without overtime pay. Not surprisingly, average weekly hours, according to the NRA, dropped most precipitously in the South between February 1933 and February 1934. At the same time, the codes introduced into southern industry overtime pay, remuneration for waiting time, and limits on consecutive working hours. Many of these gains outlasted the NRA.[88]

The minimum wage provisions of the NRA even more decisively reshaped southern manufacturing. Skewed wage distributions—a tendency for all workers in an industry to receive the same wage—indicated that NRA regulations, rather than other market factors, drove up wages in southern industry after July 1933. In the southern branch of the cotton textile finishing industry, for example, more than one-third of the millhands earned wages within 2.5 cents of the code minimum. In New England, on the other hand, where the minimum rate was higher, only 2 percent of the workers fell into that category. Similar patterns emerged in the southern baking, lumber, textile, and iron and steel industries. In boot and shoe manufacturing, more than 40 percent of the southern workforce received exactly the minimum wage (Fig. 1-1).[89]

The codes most affected unskilled workers and workers in low-wage factories, thus narrowing the regional wage differential the Southern States Industrial Council and other regional business interests had so jealously guarded (Table 1-2). Before NRA, there had been no appreciable reductions in the North-South differential in textiles since 1889 or in lumber since 1921. But between 1933 and 1935 the average hourly entrance rate for common labor shot up 40 percent in the South. It rose only 20 percent in New England and 23 percent in the maufacturing-rich East North Central States. In 1935 the NRA concluded that in the South "there have been increases in wage rates nothing short of phenomenal."[90]

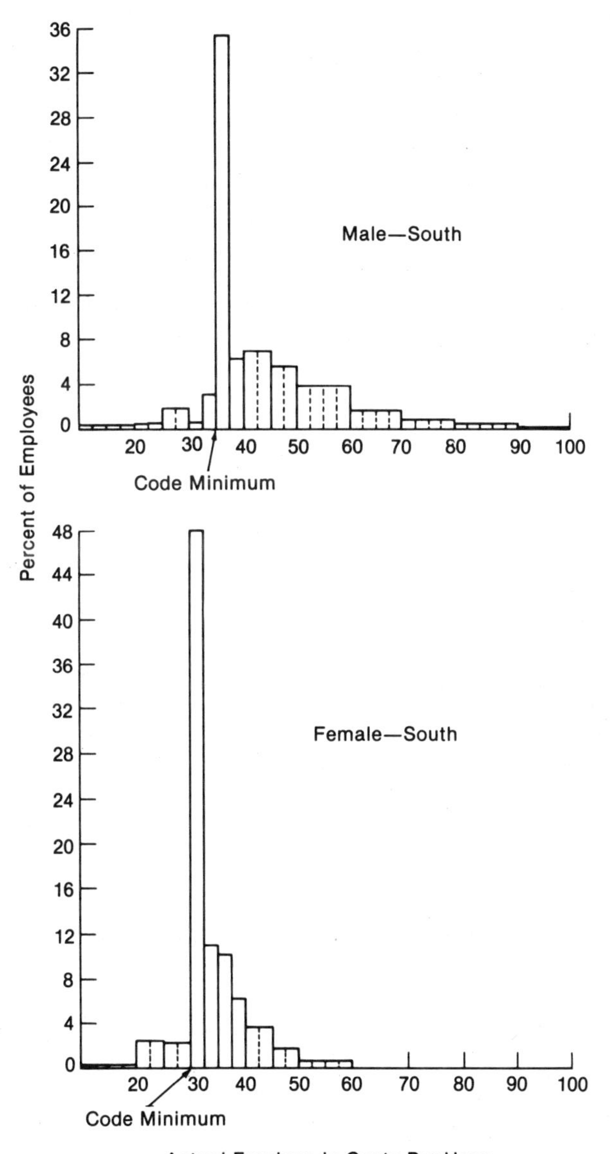

FIGURE 1-1 Earnings Distribution, Southern Boot and Shoe Industry, 1934

TABLE 1-2
The NRA and Regional Wage Differentials, 1931–1934
Southern Average Hourly Earnings as a Percentage of the Non-South* (%)

Industry	*1931*	*1932*	*1933*		*1934*
			(pre NRA)	*(post NRA)*	
Cotton Textile	—	74.0	74.3	82.0	84.4
Furniture	63.2	—	80.5	86.4	—
Hosiery (Seamless)	—	—	74.3	—	96.7
Lumber	—	43.7	40.9	54.7	—

*For hosiery and textiles the conventional designations stand. For the furniture industry, the South is the Census South (the South plus MD, DE, and DC). The non-South contains the mid-Atlantic and East North Central States. For lumber, non-South is Douglas Fir states (OR, WA); the South is made up of the ten Southern Pine states (South, excluding TN, KY, OK).

Sources: U.S. Bureau of Labor Statistics, *Bulletin No. 663: Wages in Cotton-goods Manufacturing* (Washington, 1938), 72. Charles F. Roos, *NRA Economic Planning* (Bloomington: Principia Press, 1937), 159. Report and Recommendations of Industry Committee No. 3 for the Hosiery Industry, Records of the Wages and Hours and Public Contracts Division of the Department of Labor, RG 155, Entry 16, National Archives, Washington, D.C.

Some southern industries easily adjusted to NRA standards. Others sustained devastating increases in labor costs. In the lumber industry, simply raising substandard wages to the code minimum drove up labor costs 40 percent in the South. Western sawmills faced only a 14 percent rise. Net labor costs similarly increased in silk textiles, bituminous coal mining, and woolen and worsted goods manufacturing. In the cotton garment industry, however, productivity gains outpaced the rise in wages so that unit labor costs actually declined.[91]

An industry's capacity to absorb NRA wage increases rested on its ability to adopt labor-saving machinery or more efficient production processes. Unlike the larger, highly mechanized sawmills of the Pacific Northwest, southern lumbermen relied on cheap, mostly black, unskilled labor to cut timber. Logging 1000 board feet of lumber required 13.6 man-hours of labor in the southern hardwood area, but only 4.9 man-hours in the Douglas Fir zones of the Pacific coast. Southern pine and hardwood producers lacked the capital, the knowhow, and the will to alter business practices, especially because the distribution of small, scattered mills throughout the southern forests made the NRA code nearly unenforceable. Evasion of the code was widespread. And after the Supreme Court invalidated the Recovery Act of 1935, wages in the southern branch of the industry declined.[92]

The southern cotton textile and apparel industries, on the other hand, reluctantly modernized their plants. This process had actually begun back in the 1920s. After World War I the prices of textile products collapsed, but mill owners were unable to force equivalent wage concessions from their workers. Sticky wages in the face of falling prices meant lower profits, so the mills introduced the "stretch-out"—attempts to "stretch" more productivity out of their workforces. The stretch-out included faster work paces, longer hours, the installation of new spooling and warping machines, and conversion to the multiple-loom system. Millhands, however, resisted the stretch-out. Labor unrest was widespread.[93]

Mechanization continued in the 1930s and the Cotton Garment Code Authority claimed the credit for increased productivity in the South. It even asserted that the gains justified the displacement of workers. "The code deliberately proposed to reduce drastically prison labor, child labor, home work, and above all, the legion of sweated, underpaid workers," the panel explained. "To register the big gains within four months it was necessary to remove thousands of these substandard workers." Those employees had been "replaced by fewer, but far higher paid and more productive wage earners." The code authority concluded that "surely it is no tragedy that concerns operating 54 hours a week and paying less than ten cents an hour one year ago have recorded losses in employment." "Under the necessity of trebling the wages, these plants were compelled to replace obsolescent machines, submarginal workers and inefficient practices by more modern methods of doing business." A 1935 NRA report suggested that the experience of the southern cotton garments manufacturers may have been representative of the "readjustment in a substantial number of industries."[94]

Those readjustments enhanced the long-run competitiveness of southern industry, but they imposed the costs of modernization onto those least able to afford them. Unskilled workers lost their jobs to the higher wage scales and the introduction of machinery. A rise in the number employed disguised extensive and traumatic shifts in the southern industrial labor force. Together, the 40-hour week and the minimum wage created more, but different jobs. Mechanization also contributed to this high labor substitution.[95]

Black workers bore the brunt of this displacement for two reasons. First, the codes had their greatest impact on unskilled labor and nearly all black workers fell into that category; the small minority of skilled blacks held their own. Second, many southern industrialists refused to pay blacks the wages which the codes demanded. "I will shut down my business," one Georgia employer declared, "before I will pay a nigger a white man's wages." Other southern businessmen claimed that blacks could not operate their newly installed machinery or followed through on the promise to employ "inefficient Negro labor" only if the NRA had sanctioned lower wages for blacks. As businesses reorganized their labor supplies to meet NRA standards, blacks were most often the first fired and the last hired into new, higher paying positions. According to one Cowles Commission study, "directly or indirectly because of the minimum wage provisions of the codes, about 500,000 negro workers were on relief." Black southerners bitterly referred to the NRA as the "Negro Removal Act."[96]

Those employers who did not dismiss blacks often reclassified black industrial workers as sweepers and cleaners, occupations outside the jurisdiction of the codes. Others paid whites a few cents above the minimum, reserving the minimum wage for blacks. Such practices led Walter White of the NAACP to complain that the differentials were "thinly disguised as geographical, but had been placed into effect on a strictly racial basis." Manufacturers fearful of unions most frequently adopted these tactics. Employers could thereby exploit racial tensions to foil the efforts of labor organizers.[97]

Southern manufacturers escaped most of the costs of the NRA and reaped many of its benefits. Still, they remained ambivalent about their experience under the recovery program. In 1935, William L. Mitchell, director of the NRA's Atlanta office, surveyed the region's businesses after the Supreme Court forced the agency to close shop. Mitchell's questionnaire discovered that the "proportion of employers who felt they benefitted by the former codes" ranged from 55 percent in South Carolina to 90 percent in Alabama, Georgia, and Tennessee. In every southern state, a majority of the industrialists desired continued regulation. Almost all of the respondents, however, attached conditions to their support of renewed federal supervision. They approved of codes in general, but feared regulations that would undercut their reliance on cheap labor. The southern industries that experienced the greatest increases in labor costs under the codes or received only small geographic differentials opposed further legislation. The recipients of wide differentials supported a new NRA.[98]

Among the latter group were even a few southern industrialists who were, according to the Florida NRA director, "red hot for the whole program." In May 1935, one such businessman, Robert H. Gamble, wrote the NRA that the Southern Clay Products Industry was disappointed by the Supreme Court and "wants the NRA back." Gamble explained that "working conditions in our industry, as well as in every industry in the South, were at a point in wages and hours where the wages were so low almost to be classified as a form of peonage, and the hours so long as to properly permit the same phrase to be applied." But "under codification . . . nearly every one of these conditions disappeared."[99]

Fear of the return to pre-NRA standards in the South prompted Gamble to contact the NRA office in Washington. "I am reliably, I think, informed that several large manufacturers in the South have already made plans to revert to the old wage scale and hours of work." In the Southwest, Fred C. Rogers, executive assistant in the NRA Houston office, lamented a similar reversion to sweatshop labor conditions. Noting that several large firms had lengthened their work weeks immediately after the invalidation of the NRA, Rogers shuddered "in anticipation of the influence their actions have on the lives of their employees and those capable persons who remain unemployed." Such apprehensions were borne out to some extent by changes in the southern wage structure. In the two years after the NRA disbanded, average hourly earnings in the southern seamless hosiery and cotton textile industries declined in relation to the North. In 1937, the pay rate for common labor in southern sawmills stood below the NRA minimum.[100]

Nonetheless, the regional differential never returned to its former breadth, and the earnings of southern laborers rose enormously. Rupert Vance appreciated the change in 1935 when he referred to "the far-off days, B.C.—Before Codes." The NRA carefully refrained from challenging the southern status quo, but its policies quietly and indirectly eroded the traditional southern labor system. Concessions to southern business tempered resistance to the changes, but the NRA committed the federal government to the reorganization of southern industry and the improvement of southern labor standards. The South's acquiescence in the codes would later embolden federal officials. It would haunt southern leaders.[101]

NEW DEAL, OLD SOUTH: RELIEF AND TVA

No policy more clearly revealed the New Deal's reluctance to challenge the southern status quo than the practice of releasing workers from the federal relief rolls for field labor during the planting and harvest seasons. Southern politicians insisted that federal relief programs not siphon off their supply of cheap seasonal labor. The Roosevelt administration acceded to their demands, even though the President and his relief officials found the pay and working conditions of southern farm hands abhorrent. In 1934, Georgia Governor Eugene Talmadge forwarded to FDR a landowner's complaint about the relief program. "I wouldn't plow nobody's mule from sunrise to sunset for fifty cents per day," the Georgian protested, "when I could get $1.30 for pretending to work on a ditch." The letter raised the President's ire. He calculated that seasonal farm work at fifty cents a day worked out to an annual wage of about $75. Horrified, FDR composed an angry response. "Somehow I cannot get it into my head that wages on such a scale make possible a reasonable American standard of living." The President, however, sent the reply over Harry Hopkins's signature.[102]

Roosevelt's reluctance to sign that letter himself revealed his administration's ambivalence. Federal officials believed that work relief wages represented the minimum necessary for health and decency. Still, they dismissed workers on relief projects for farm labor at wages well below those standards. The planters, according to one assistant to Harry Hopkins, envisioned a system by which the national government would assume their responsibility for carrying tenants through the slack season. The government would then remove laborers from the rolls before the work season "so these niggers will be good and hungry." Actual policy fulfilled the landlord's demands. Federal relief directors in the South dropped any relief client who was offered work in the fields no matter how low the wage. Some unscrupulous planters exploited this situation by offering even lower than customary wages. Relief workers had no choice but to accept them.[103]

Returning public works employees to the fields was but one of many concessions the New Deal relief establishment made to southern interests. Like the NRA, the relief programs adopted a territorial structure in aid and wages. Southerners received the poorest per capita benefits from this arrangement, even though the region most desperately required relief and possessed few resources to meet the needs. In September 1933, more than 20 percent of the families in South Carolina, Florida, and Oklahoma received relief; over 15 percent in Louisiana, Arkansas, and Alabama. Outside the region, only three states reached the 15-percent level. Yet southern state governments offered only meager assistance to those indigents. In all but one southern state, the federal government contributed more than three-quarters of the relief funds. Many southern states operated their programs on the "surplus commodities only" principle; they distributed surplus farm products rather than cash—a policy designed more to sustain farm prices than to help the needy.[104]

Immediately after the inauguration, the Roosevelt administration decided against granting relief in excess of prevailing living standards. The depth of economic emergency and the desire to disburse funds before the arrival of winter forced the first New Deal relief agency—the Federal Emergency Relief Administration (FERA)—to sidestep conflict with local officials. The FERA never attempted to raise living standards in the South by paying high benefits. In May 1933 the average monthly FERA benefit in the United States was $15.59 per family. Below the Mason-Dixon line, it ranged from $13.32 in Louisiana to a paltry $3.86 per in Mississippi. Among southern states, only Louisiana paid more than $9. The North-South gap closed little over the next three years. One federal relief official concluded that FERA policy in the South restricted aid to "the starvation level."[105]

Work relief projects sparked sharper debate than did direct payments. In 1933 the Public Works Administration established a three-tier wage structure. It specified hourly minima of 40 cents in the lower South, 45 cents in the "central region," which included the upper South, and 50 cents for the rest of the nation. The 40-cent southern rate, although the lowest in the nation, nonetheless amounted to almost double the regional average for unskilled labor; it outraged southern interests. Led by governors Eugene Talmadge of Georgia and "Alfalfa Bill" Murray of Oklahoma, southern Democrats pressured relief director Harry Hopkins into reducing southern pay rates. The Civil Works Admiministration (CWA), the successor to the PWA, dropped the southern minimum from 40 to 30 cents in early 1934. By the end of that year, Hopkins bowed further. He ordered that local committees set all work relief wages at prevailing local rates. Southern wages fell to as little as 12.5 cents per hour.[106]

The formation of the Works Progress Administration (WPA) in 1935, an agency that would survive into the next decade, codified the regional benefit structure. Under the leadership of Harry Hopkins, the WPA adopted a "security wage" plan—a formula that set pay rates for work on public projects at an intermediate between direct relief and the prevailing local wage. The plan established a complicated system which divided the nation into four wage regions (three after July 1936). Southern levels remained far below those of the rest of the nation. The WPA wage for unskilled labor, for instance, furnished less than half the cost of a bare bones, "emergency" standard of living in all surveyed southern cities, but more than half in every city outside the region.[107]

The security wage plan marked a compromise between contending factions in the relief establishments. The debate mirrored the disputes within the NRA and AAA. Howard B. Meyers, a WPA official, framed the debate in his study of relief in the rural South: "Shall relief be used to bolster the living standards of a depressed area or to raise abnormally low wage levels?" In the South, Myers concluded, to grant more than the prevailing standard "might operate to keep people on relief, to cut appreciably below it certainly means semi-starvation and increases in disease and mortality rates."[108]

On one side of the controversy stood a small group of reformers. These officials, many of whom served in field positions, conceived of relief as more than a curb on starvation. These New Dealers agreed with southern leaders that high benefits shifted people out of agriculture and forced up industrial wages, but they approved of such changes. The reformers believed that the alleged regional variations in the cost of living reflected differences in living standards rather than tastes or needs. They pressed for a uniform national benefit scale.[109]

The opponents of that view occupied the positions of power. Under their direction the WPA maintained lower southern pay rates throughout the Depression. Even when the WPA narrowed the differentials after 1937, the agency carefully reaffirmed its commitment to flexible treatment of local and regional peculiarities. Colonel F. C. Harrington, one of Hopkins's successors as chief of the WPA, informed the House Appropriations Committee that his agency would not attempt to eliminate regional differentials. Assuaging the fears of southern Congressmen, Harrington asserted that the WPA "should not be a guinea pig in this experiment."[110]

Below the surface of that policy debate lurked the question of black participation in relief programs. Following the NRA lead, the Civil Works Administration insisted on standard wages for blacks. In 1935, President Roosevelt issued an executive order prohibiting racial discrimination in relief work. Hopkins guided enforcement of that decree with two administrative orders of his own. Southerners offered only muted resistance to those directives, primarily because enforcement in the field remained nearly impossible. Few blacks could find relief employment where there was competition with whites; even fewer could qualify for skilled positions. The most blatant abuses prevailed within the Civilian Conservation Corps (CCC), a work relief agency for young men. Under the supervision of the avowed racist Robert Fechner, blacks received only 3 percent of the first 250,000 assignments. CCC details in some southern black majority counties contained no blacks. The expansion of relief programs, however, eventually enabled southern blacks to find work relief jobs. After 1935, WPA employment grew fastest in the states of the deep South with the highest black populations. And WPA workers often labored in black communities; the agency, for example, set up a diet kitchen and financed an athletic program at an all-black South Carolina school.[111]

As southerners black and white participated in federal programs, they came to rely on the national government for the first time. The flood of barely literate letters from poor southerners to the officials of federal programs revealed an awareness of a locus of authority outside the world of the landlord, mill owner, or small town big man. Relief programs served the local elites' interests, but they undermined their authority, inculcating loyalty to the national government. Lorena Hickok, Harry Hopkins's field investigator, reported from New Orleans in 1934: "A lot of these people who used to look up that way to their paternalistic landlords now switched to the President and Mrs. Roosevelt! They just expect them to take care of them!" Slim Jackson, the supervisor of a WPA supply room explained the new way of thinking: "The way I look at it is this, this is a rich country. I figger it ain't going to hurt the government to feed and clothe them that needs it."[112]

Such views generated uneasiness among the southern elite. They feared that the populace would bypass the traditional social and economic leadership and look to Washington to secure its needs. Southern leaders desired aid and welcomed recovery measures, but remained leery of a relief apparatus that drove up wages, siphoned away labor, subverted local institutions, and promoted popular identification with the federal government. Already southerners had started electing a younger generation of politicians—men like Lyndon Johnson of Texas, Lister Hill of Alabama, and Claude Pepper of Florida—who eagerly exploited President Roosevelt's popularity and ran for office as representatives of the New Deal.[113]

The entrenched leadership, on the other hand, faced a dilemma. They desired federal funds, but resented federal interference. In 1934, Oklahoma governor Bill Murray asserted his state's rights to expend relief funds without regard for federal regulations. The federal government responded by invoking a provision of the Emergency Relief Act which forbade political interference in the distribution of relief money. Federal officials assumed control over the programs in Oklahoma. The Roosevelt administration acted similarly the following year in Georgia, where Governor Talmadge recklessly disregarded federal directives, and in Huey Long's Louisiana, where funds mysteriously disappeared.[114]

The fears of the region's politicians were well founded. Early New Deal policy lacked the force or even the intention to undermine the power of southern interests, but the proliferation of programs, the growing presence of officials, and the charisma of the man in the White House built support for more forceful federal intervention in the South. "I'm proud of our United States and every time I hear the 'Star-Spangled Banner' I feel a lump in my throat," declared John Easton, a North Carolina farm laborer. Easton's familiar patriotism, however, stemmed from novel sources. "There ain't no other nation in the world," he beamed, "that would have sense enough to think of WPA and all the other A's."[115]

No government agency so strongly inspired southern demand for federal economic intervention as did the Tennessee Valley Authority. That ambitious project sought to alter the landscape of the Tennessee River Valley, creating a system of navigable waterways and sources of cheap fertilizer and electric power. With them, it hoped to build a new way of life for the inhabitants of the seven-state TVA region. Even Chattanooga *News* editor George Fort Milton, an opponent of many New Deal measures, praised the Roosevelt administration's attempt to develop the Tennessee Valley. In the TVA, Milton declared, the South and the federal government "recognized one another as partners in a common enterprise; as the cooperators in what many hail as the most intellectually stimulating, economically intelligent and socially satisfactory single enterprise under way in the United States."[116]

In its first five years, however, the Tennessee Valley Authority failed to fulfill the promise of its mandate. Like the other New Deal programs, it limited its efforts to projects that neither upset traditional economic arrangements nor threatened local political prerogatives. The early TVA concentrated on improvement of river navigation and the production of low-cost fertilizer. Such a focus, the Authority's directors believed, would create markets and stimulate commerce. It would restore a potentially sound agricultural economy without undesirable heavy manufacturing, unsightly urban growth, or fundamental change on the farms. The TVA officials interpreted the Depression as a warning against industrialization. At the same time, they saw rural poverty as a land-use problem rather than an economic one. Based upon those assumptions, the river navigation and fertilizer projects formed a comprehensive development program.[117]

The TVA's caution reflected the political constraints on a national agency operating in a depressed region that had long resisted federal intervention. But the reluctance to press economic development also revealed the directors' belief in "decentralization." Decentralization implied small-scale industries spread among rural areas. They would provide employment for surplus agricultural population while preserving the rural way of life. These beliefs became early TVA policy. In a September 1933 "Statement of the Present Attitude of TVA Concerning Industrial Development," the TVA opposed the relocation of northern manufacturing concerns to the Valley. It suggested instead the location of small industries "in smaller communities" to "avoid the unfortunate social consequences of excessive urbanization."[118]

TVA's commitment to decentralization was both sincere and politically expedient. The development strategy called for "industries suited to the region." It advocated small industry "as being best adapted to preserving the spirit of individuality which is characteristic of the people of the Valley." Such a philosophy appealed to the owners of southern textile, hosiery, and lumber mills. Yet it offered little promise of industrial diversification. The TVA desired to maintain the area's tradition of self-reliance. That goal, however noble, precluded the arrival of large firms that would bring to the Southeast unions, high wages, and demand for labor.[119]

Decentralization also stressed cooperation with local and state governments. Whenever the Authority could rely on a local institution, TVA board member David E. Lilienthal remembered, it "sought to have that non-federal agency complete the task." In fact, another TVA director added, "TVA has chosen almost uniformly to work through existing governments instead of through its own organization in discharging its manifold responsibilities." According to Lilienthal, "legalistic argumentation about 'state's rights' or 'federal supremacy' . . . faded into irrelevance." Alone among the New Deal agencies of Roosevelt's first term, TVA won the support of southern governors and legislators.[120]

Decentralization achieved only limited success. The TVA constructed dams, harnessed hydroelectric power, and manufactured fertilizer, but it hardly altered the economic organization of the region. To be sure, the Tennessee Valley experienced advances in income, wages, and manufacturing activity that outstripped the South and the nation. Despite the Authority's claims of success, these gains proved illusory. The statistics were not so much evidence of economic growth as they were indicators of how early and how deeply the Depression had bottomed out in the South. The dramatic surge in wages in the Valley from 1933 to 1937, for instance, barely reflected a return to 1929 levels.[121]

The opening of the Tennessee Valley waterways failed to stimulate commerce or encourage economic diversification. The relative contributions of the various sectors to regional income payments showed no change in the TVA's first six years (Table 1-3). Industry entered the overwhelmingly agricultural Valley, but only traditionally southern, low-wage manufacturers. Of 125 firms commencing operations between January 1933 and October 1938, more than half were textile, hosiery, and food processing plants. According to the TVA's Commerce Department, only 12 of the 125 had beneficial effects on the wages, purchasing power, or living standards for their communities. And the Tennessee Valley remained without a major urban center.[122]

In its lack of diversification and continued poverty, the TVA region served as a microcosm for the entire South. As the first term of Roosevelt's presidency closed in 1937, average annual income in the South stood at $314, barely exceeding half of the total for all other regions. The average tenant family in the Cotton Belt fended off starvation on just $73 per person per year. After NRA wage increases, industrial workers fared slightly

TABLE 1-3
Components of Income in the Tennessee Valley Region, Percentage Distribution, 1929–39, Selected Years (%)

Component	*1929*	*1933*	*1939*
Agriculture	23	19	19
Manufacturing	15	15	17
Trade and Service	37	37	34
Mining	2	2	2
Construction	3	1	3
Property Income	12	11	9
Government	8	16	17
Total	100%	100%	100%

Source: John V. Krutilla, "Economic Development: An Appraisal," in Roscoe C. Martin, ed., *TVA: The First Twenty Years: A Staff Report* (University and Knoxville: University of Alabama Press and University of Tennessee Press, 1956), 225.

better; their annual pay made up 71 percent of the non-South standard. Still, those workers toiled in a regional economy that contained 28 percent of the nation's population, but accounted for less than 13 percent of its value of manufactures.[123]

David E. Lilienthal presented a similar set of statistics in October 1936. The TVA director told a Georgia audience that "no one need point out to you the tragedy behind those figures. Our job," Lilienthal declared, was "not to wring our hands but to face these facts, and set about to do something to change them." Lilienthal identified chronic poverty as the "central problem" of the South. He conceded that the TVA alone could not solve the problem and called for concerted national attention to the economic development and idustrialization of the South. "We cannot have a sound *national* prosperity," the director concluded, "if any region of the country suffers under a low income."[124]

Lilenthal's speech found echoes among the growing ranks of liberals in the universities and editorial offices of the South, in the halls of federal agencies in Washington and in the White House itself. They too realized the insufficiency of TVA. In his second inaugural address, President Roosevelt pledged to uplift the one-third of a nation which remained ill-housed, ill-clad, ill-nourished. Early in his second term, F.D.R. left no doubt that "attacking the basic problem of getting a better standard of living for the one-third at the bottom" required increased federal intervention in the nation's poorest region. Marshalling support within and outside the South to confront the altered political and economic landscape of his second term, the President reaffirmed the South's status as the nation's number one economic problem. And in the same sentence, he made southern economic development, "the Nation's problem, not merely the South's." With those words the federal government embarked on a long-term effort to restructure the southern economy. The government adopted a set of policies which would shape the explosive economic changes that revolutionized the South in the next two decades.[125]

Chapter 2 ◎ "Wild Cards and Innovations"

"In poker," recalled Houston magnate Jesse H. Jones in his reminiscence of the Roosevelt administration, "the President preferred wild cards and innovations to the kind of straight five-card draw poker most Texans were raised on." As the New Deal entered its second term, Jones and his fellow southern leaders, in the Congress and in the boardrooms and state houses of the South, were to discover that the President's taste for "wild cards and innovations" ran to politics and policy as well as poker. Over the next three years, the Roosevelt administration, in league with an influential faction of liberal southerners, would abandon its powerful southern allies and embark on an effort to transform the Democratic party and reform the southern economy.[1]

The Administration had yet to deal such a hand, however, as the first term reached its conclusion. The federal government and southern leaders like Jones presided together over a turbulent South. The agricultural program had transformed the rural South, restoring commodity prices without reinstating tenants. One-quarter of the great cotton belt lay fallow on account of AAA crop reduction contracts, and thousands of ex-sharecroppers roamed the countryside in search of land to work. Meanwhile, southern industry had rebounded from the dire conditions of 1932 with the aid of the NRA and the general economic upturn. In 1935 the region witnessed the groundbreaking for a dozen large plants and a business recovery so dramatic that *Business Week* reported enthusiastically that "Depression is forgotten in the South."[2]

In the Southeast the land itself was changing. TVA dams created new lakes, tamed flood waters, and carved a navigable channel that would ultimately stretch from Knoxville in the Great Smokey Mountains to Paducah on the Ohio River. By manipulating water levels, TVA medical authorities destroyed the breeding grounds of the anopheles mosquito and all but eliminated malaria from the valley. The Authority's labor relations policies maintained high worker morale and encouraged union growth in the region. And by 1934, the city of Tupelo, Mississippi had become the first recipient of cheap TVA electricity. Lorena Hickok, Harry Hopkins's field investigator, was flabbergasted when she visited the TVA area in 1934. "A Promised Land, bathed in golden sunlight," Hickok exclaimed to her chief, "is rising out of the grey shadows of want and squalor and wretchedness down here in the Tennessee Valley these days."[3]

Yet despite the upheaval, those gray shadows of want and squalor continued to hover over the South. Indeed, the 4000 families the TVA relocated from the Norris reservoir site in Tennessee earned an average annual income of less than $100, less than one-fifth of the national standard. Beyond the tamed waters of the TVA zone, malaria raged on, accounting for one-sixth of all physician's calls in the South in 1935. Tuberculosis, pneumonia, and pellagra also tormented the southern people with peculiar force, while the region's health care facilities remained the worst in the nation. One North Carolinian told a WPA interviewer that his family stole wood just to keep from freezing during the winter of 1934–35; when they could scrounge up money for food, they ate until they were sick, unsure about when they would next enjoy the sights and smells of food. The southern people, one government source concluded, lacked food, lacked clothes, and lacked decent homes.[4]

Such hardship failed to extinguish the pride and hope of the region's legion of impoverished. "I'm not ashamed of our house although its the worst we've stayed in since we was married thirty-two years ago," a southern farm laborer told a WPA interviewer. "The main thing is," she explained, "we hope to do better after next year." Faith in President Roosevelt and his New Deal supplied much of that optimism. One poor North Carolinian affirmed his support for the AAA, WPA, NRA, and NYA. "I'm a Democrat," John Easton declared. "I stand for the New Deal and Roosevelt." The President's comforting radio messages and expressions of concern relieved the distress of another struggling southerner. "I do think that Roosevelt is the biggest-hearted man we ever had in the White House," George Dobbin declared. "It's the first time in my recollection that a president ever got up and said, 'I'm interested in and aim to do somthin' for the workin' man.'"[5]

This optimism infected a fledgling group of southern liberals. Drawn mainly from the middle-class bastions of the urban South, these journalists, educators, and labor organizers had little power to challenge the region's entrenched economic interests and still less influence within the closed political systems of the South. The Roosevelt administration galvanized these liberal southerners, granting them new-found popularity and supplying them with a unifying focus in their common endorsement of the New Deal. Like-minded reformers across the South first organized in 1934 when Francis Pickens Miller, a Virginia lawyer and opponent of that state's political machine, formed the Southern Policy Committee (SPC). Unlike earlier southern liberal organizations, the SPC did not focus narrowly on racial issues, but sought full-scale economic reform of the South and an end to the domination of the region by "exclusive economic interests and political sensationalists." Rejecting the southern elite's claims for exemptions from federal standards and regulations, the SPC's "ultimate goal" was the integration of the South into the economic life of the nation through "the total elimination of regional discriminations."[6]

Although never an active member of the SPC or its successor organizations, the intellectual mentor of these southern liberals was Howard Washington Odum. The South's first modern sociologist, Odum ruled a veritable fiefdom of research projects, academic programs, and scholarly journals at the University of North Carolina in the 1930s. Odum headed the *North Georgia Review*'s 1940 list of the one hundred most influential southerners. As a youth, Odum had hardly aspired to such authority. Born in rural Bethlehem, Georgia, in 1884, the young Odum lacked confidence. During his undergraduate training at Emory College, he remembered describing himself as "A.U.I.P. or Awkward, Ugly, Ignorant, and Poor." Education stoked his ambition. After taking doctorates at both Columbia and Clark, Odum determined to build southern social science from scratch. According to one of his Chapel Hill colleagues, Odum "was an agent of social change, and at times the area around him seemed charged with ozone." At Chapel Hill, Odum trained the South's first generation of social scientists, including Rupert Vance, Arthur Raper, Guy Johnson, and George Mitchell. He inspired many other young southerners to political action. In the words of one student of the generation of liberal southerners who reached maturity during the Depression, Odum was "their beacon of light, their most reliable guide to a genuinely new South."[7]

Like many observers, Odum catalogued the distinctive features of southern life. But alone among his contemporaries, Odum was a southerner unafraid to criticize the South, and one who looked forward to an industrial future rather than back to the agrarian past. With painstaking detail, he charted the region's social defects and labeled as intolerable some passionately defended southern practices. To ameliorate those ills, Odum proposed "regionalism," a program of scientific investigation and social planning under the direction of a Council on Southern Regional Development. For Odum, such an effort offered an antidote to the divisive sectionalism that enforced southern backwardness and a path toward a "balanced democracy and the redistribution of wealth." In Odum's conception, regionalism would tame sectional conflict by blending the South with the national culture, making national welfare the lodestar in policy determinations. Odum believed that regionalism, unlike sectionalism, would "point toward a continuously more effective reintegration of the southern regions into the national picture."[8]

Odum's investigations also contributed a succinct diagnosis of the region's maladies and suggested a radical treatment. It was Odum who established the portrait of the South as a land rich in resources, yet impoverished in its economic and institutional development. "Until recently," Odum declared in his 1936 opus, *Southern Regions of the United States,* "the south was a furnisher of raw materials to the manufacturing regions, essentially colonial in its economy." "Uneven technology," Odum argued, deprived the South of its share of national wealth. But that problem need not prove insoluble. Odum's very juxtaposition of the potential richness of the region against its current poverty implied that proper planning and resource development could lift the region out of its colonial degradation. "The chasm between abundance possibilities and deficiency actualities," Odum concluded in his characteristically turgid prose, "lies at the heart of southern problems."[9]

The young southern liberals around the Southern Policy Committee embraced Odum's ideas and built upon them. Odum taught them to break the traditional strictures against criticism of the South. But whereas Odum recognized southern problems and advocated scientific planning to solve them, these liberals, emboldened by the New Deal, sought massive federal interference to reform their native region. These young southerners shared Odum's faith in planned regional development, but not his ambivalence about the New Deal. They also lacked Odum's caution. Odum's empire relied ultimately on the benificence of the North Carolina legislature and the support of southern public officials. He developed the habit of conciliation, of gently and cautiously criticizing southern conditions without challenging official authority.

These young southerners had no such compunctions. They sprang from the region's educated, professional classes. Clark Foreman and Jonathan Daniels, for example, came from journalist stock. Their fathers edited the region's most progressive and influential newspapers. All the young liberals, even those from more modest backgrounds, had university educations. Few had connections to the region's mills and plantations or to the county-seat political elites. Most of the young liberals hailed from the region's cities, areas of little influence in the region's rural-dominated economies and political systems. Others, bred in rural areas, initially challenged local elites before looking to the national government. Brooks Hays, an Arkansas country lawyer, lost a race for a congressional seat before joining the Roosevelt administration.[10]

With no opportunity to gain influence in the state governments of the South, the *loci operandi* of the entrenched interests they opposed, young southern liberals migrated to Washington in the early days of the New Deal. Blocked at the state level, they sought power through an expanded federal establishment. Clark Foreman and George S. Mitchell, scions of progressive and ineffectual southerners, found positions, respectively, in the Interior Department and the Farm Security Administration. Aubrey Williams left his Alabama birthplace and eventually took the reins of the National Youth Administration. Leon Keyserling of South Carolina became the principal draftsman of the National Labor Relations Act. At the same time, a small bloc of liberal southerners developed in the Congress. Led by two Alabamians, Senator Hugo L. Black and Representative Lister Hill, the group included Florida senator Claude Pepper, Texas congressman Maury Maverick, and after his election in 1937, Texas representative Lyndon B. Johnson. Southern Policy Committee members met regularly with these congressmen in Hill's Restaurant, and the nation's capital became the headquarters from which southern liberals sought to implement their "new regional plan."[11]

Southern liberalism in the 1930s, though largely a movement of young professionals, was hardly a class movement. These reformers aimed not only for the economic rehabilitation of the region, an objective long shared by the South's powerful boosters and business progressives, but for the uplift of the South's poorest people. They focused not so much on the problems of southern agriculture as on the struggles of southern tenant farmers. They cared little about luring industry to the region, unless it was modern industry paying high wages to common laborers. But they were more than humanitarians or champions of struggling southerners. They hoped to expand the activities of the national government and, at the same time, to remove control of federal programs from local elites and concentrate that power in the hands of federal officials like themselves. They wanted the national government to become a major force in the South, overthrowing traditional political and economic arrangements.[12]

During the course of the 1930s, the young southern liberals did what their mentors and predecessors, the liberal social scientists and journalists of the pre-Depression era, had never dared. They forged close ties with national labor unions, expecially the CIO, believing that unionization would break the power of the repressive, inefficient southern businesses and guarantee the region's downtrodden a healthy standard of living. And in the programs of the Roosevelt administration, they found a detour around the "exclusive economic interests and political sensationalists" who obstructed reform. United by their desire to import the New Deal below the Mason-Dixon line, southern liberals proposed an agenda of federal relief and unemployment insurance, housing, health facilities, labor standards, and assistance to public education. Surveying the region in 1936, Jonathan Daniels noted that for generations "do without" had served as the South's unofficial motto. "In more ways than one that has been the regional plan of the American South," Daniels declared for himself and his fellow southern liberals, "and I for one Southerner, speaking also without fear of contradiction of 25,000,000 others, am ready to find another."[13]

President Roosevelt himself recognized this cohort of young southerners and sought to nurture it. He made efforts to include liberal southerners in progressive circles. And, in 1936, FDR confided to TVA director David Lilienthal his hopes that the efforts of young southerners pointed toward a new balance of power in the region: "They say that the Democratic Party can't be liberal because of the South but things are moving toward progression in the South." Roosevelt, according to Lilienthal, conceded that the Old Guard—men like Mississippi senator Pat Harrison—would prove recalcitrant. "But the young people in the South, and the women, they are thinking about economic problems and will be part of a liberal group in the South."[14]

Neither Roosevelt nor his southern supporters anticipated the coming conflicts of the second term. Back in the home states of the southern liberals, the region's leaders generally joined in the chorus of support for the Roosevelt administration as the first term ended. With one finger on the pulse of the southern electorate, southern political leaders recognized President Roosevelt's personal popularity and greedily pursued his electoral coattails. With the other hand on the reins of power, the region's entrenched interests recognized that for all the Roosevelt administration's rhetoric, the progress of the early New Deal had been to their liking. Planters controlled the AAA and employed the agricultural program to secure their control over land and labor, without offering protection to tenants. The Relief Administration maintained steep regional differentials in benefits and acceded to the southern demands to release public works employees from the rolls during the harvest season. And the NRA, the most nettlesome of the New Deal agencies, had been disbanded by the Supreme Court.

Moreover, despite the Administration's overtures to blacks, white supremacy remained unchallenged in the South. Even the ambitious Tennessee Valley Authority welshed on its promise to employ blacks in proportion to their percentage of the population. The Authority drew severe criticism from the NAACP for practicing "Lily-white Reconstruction" by extending most services to whites only and limiting blacks to unskilled jobs. The most notable example of the Roosevelt administration's unwillingness to tamper with southern racial practices, however, was FDR's refusal to endorse a federal anti-lynching statute. Noting that southerners controlled the leadership positions in the national legislature, the President explained his timidity to NAACP Secretary Walter White. "If I come out for an anti-lynching bill now," Roosevelt argued, "they will block every bill I ask Congress to pass to keep America from collapsing. I just can't take that risk." Certainly, Roosevelt strongly criticized the practice of lynching, as did every one of the southern opponents of a federal statute, but the President's refusal to support the Wagner-Costigan Anti-Lynching Law in 1935 indicated his accommodation with white supremacy.[15]

Nonetheless, some tension had developed between the Administration and southerners in Congress. Dedicated conservatives like Virginia senator Carter Glass and North Carolina senator Josiah Bailey opposed most New Deal initiatives. And Democratic party stalwarts like Senate Majority Leader Joseph Robinson (D-Ark) and Senate Finance Committee chairman Pat Harrison (D-Miss) had a run-in with the President over the handling of the 1935 tax bill. Still, these southerners concealed their misgivings, and when Harrison came up for reelection in 1936, the "Grey Fox of the Delta" forgot his private objections and ran as a New Deal candidate. Politicians throughout the South followed suit, confident of their electoral prospects and of winning control of national policy in a second Roosevelt term. So confident were the southern Democrats that they acceded with little fight to the repeal of the two-thirds rule at the 1936 Democratic National Convention. It had required a candidate to garner two-thirds of the delegate votes to secure the Democratic presidential nomination and it had therefore long functioned as the South's *de facto* veto over any potential candidate. With Missouri senator Bennet "Champ" Clark holding the gavel, repeal of the hallowed rule passed the convention by simple voice vote. "The Democratic Party is no longer a sectional party," intoned Clark, the son of a man once denied the presidential nomination by the two-thirds rule, "it has become a great national party."[16]

The Roosevelt landslide in the national elections marked the advent of the party of Jackson as a "great national party"; it also augured ill for the southern leadership. Abrogation of the two-thirds rule, a policy Roosevelt's campaign chief Jim Farley claimed FDR personally engineered, deprived the region of control over the nominating process. And Roosevelt's overwhelming victory, capturing every state but Maine and Vermont, dramatized that he was the first Democratic nominee who could have won the presidency without a single southern electoral vote. The convention itself offered the first sign that the Administration recognized and was prepared to act on this shift in its constituency from South to North, or in Raymond Moley's words, from "acreage to population." For the first time in its annals, the Democratic party seated black delegates at its national convention and integrated the regular press box. Some southern leaders bridled at these gestures. When a black minister rose to offer the convocation at one session, South Carolina senator "Cotton Ed" Smith led a small group of southerners out of the convention. That walk-out foreshadowed the full-scale defection of southern delegates twelve years later, and it suggested the imminent decline of southern influence in the party's councils.[17]

That decline would continue as the nascent Roosevelt coalition congealed. The strength of Roosevelt's candidacy outside the South revealed potent Democratic strongholds in the nation's cities, union halls, and especially, in its black communities. In 1936, the national Democratic party had made its first effort to court the black vote, and with stunning success. Black voters registered in unprecedented numbers in northern cities, and black neighborhoods gave the Democratic ticket large majorities. For Democratic politicians the potential black vote in future elections took on enormous, even exaggerated, significance in their calculations. Together with labor and liberals, northern blacks represented millions of votes and interests inimical to the southern establishment. The national political game had become, in the words of *Business Week,* "New Dealer's choice." Roosevelt possessed the support outside the South to indulge his taste in "wild cards and innovations." And he did just that in the 1937 session of Congress, pursuing reorganization of the Supreme Court and the Executive Branch, new appropriations for relief, and national labor standards.[18]

Disaffection among the southern elite mounted gradually and sprang from diverse roots. Distribution of federal patronage to reformers, circumventing the traditional state machines upset politicos like Virginia boss Harry Byrd and Vice President John Nance Garner. The Administration's courtship of organized labor and its hostility to business, especially evident in its call for a national minimum wage law, alienated southern industry and its Congressional patrons like Pat Harrison, Georgia senator Walter George, and Texas senator Tom Connally. South Carolina's leading politician, Senator James F. Byrnes, first broke with the White House over the Administration's failure to condemn the CIO sitdown strikes in the winter of 1937. He did so because John L. Lewis and the CIO's Textile Workers Organizing Committee were beginning to organize South Carolina's cotton mills. Southern leaders also dreaded the racial implications of the Administration's initiatives. The expansion of federal relief cemented the connection between the Democratic party and black voters in the North, while elevating the economic status of blacks in the South. And Pat Harrison feared that an enlarged, more liberal Supreme Court might tamper with the legal edifice of Jim Crow.[19]

In January 1938 the Senate Majority Leader, Alben Barkley (D-Ky), refused James Byrnes's request to rearrange the Senate calendar. Barkley had promised the NAACP that an anti-lynching bill would reach the floor first and he could not make the change unless the NAACP's leaders released him from his promise. That episode only confirmed Byrnes's fear that blacks were overrunning the party of the fathers. "Barkley can't do anything," Byrnes fumed, "without talking to that nigger first." For Jimmy Byrnes and other conservative southern leaders, a Democratic party under the growing sway of labor and blacks no longer seemed a safe home and no longer commanded unswerving loyalty.[20]

If southern politicians rued growing black influence in the North, they were horrified by the politicization of blacks in the South. Across the urban South, blacks revived old organizations or formed new ones aimed at securing voting rights or access to segregated facilities.[21]

Such efforts, of course, made little progress. The New Deal posed no real threat to white supremacy. But it did threaten the longstanding relationship between state and society in the South. Despite the efforts of early new Deal programs to conciliate southern elites, federal programs could not help but change southerners', and especially poor black southerners', relationship with their governments. Whatever the AAA's injustices to sharecroppers, one black tenant insisted with justification that "the government was helping the poor colored people more than anybody else." He understood that even partial federal supervision promised fairer treatment than control by local elites. "It's just a better deal for the colored race all the way through than to let these people here preside, the Watsons and the Graces and the other moneyed men." And, although it offered few concrete benefits for southern tenants, the reestablished AAA included black farmers in its elections during Roosevelt's second term. Relief recipients also realized that the government in Washington affected their lives, that they were no longer completely dependent on the good will of wealthy local whites. At the same time, the Rural Electrification Administration brought light and comfort to the homes of many poor southerners, while the Tennessee Valley Authority organized local cooperative associations, provided cheap power and fertilizer, and challenged powerful utility companies.[22]

These initiatives brought little real political or economic reward to black southerners, or poor southerners generally, but they did challenge the foundations of southern political economy. Government benefits went directly into the hands of poor people, and public funds provided for relief and health care. Such aid bypassed the state and local leaders who controlled the pursestrings in the New South regimes. It also allocated resources for relief and public welfare, a departure from the business progressives' emphasis on public investment for the benefit of business. Finally, since the end of Reconstruction, southern governments had directed their programs and expenditures toward whites. New Deal programs, despite the persistence of local control and segregation in many of them, included blacks. This threatened the *Herrenvolk,* whites-only nature of state action in the South.[23]

Even if the early New Deal had not upset the political economy of the South, Roosevelt had already gone too far for southern politicians. For them, the fundamental issue, whether in its racial, economic, or political manifestation, was the federal presence in the South and its disruptive impact on existing socio-economic arrangements. The general effect of the New Deal was the erosion of the authority of the county-seat elites, the "small town big men," whose control of labor, credit, and municipal government guaranteed their power. Relief interposed federal officials between the laboring people and the local "bosses." Labor standards eliminated the low wages that forced entire families to work in one man's textile mill. Government credit skirted bankers. So deep and widespread were the marks of increased federal intervention in Dixie that southerners saw it as a second Reconstruction. In such terms, Carter Glass starkly cast the prospects of the southern establishment in 1938: "The South would better begin thinking whether it will continue to cast its 152 electoral votes according to the memories of the Reconstruction era of 1865 and thereafter, or will have spirit and courage enough to face the new Reconstruction era that northern so-called Democrats are menacing us with."[24]

The prospect of a second Reconstruction was conjured up not only by the opponents of the New Deal in the South but by its supporters as well. Liberals embraced that notion, contending that the second time around the federal government would rescue rather than plunder the South. Claude Pepper's "A New Deal in Reconstruction" instructed southerners that while the federal government had created southern poverty after the Civil War, that same agency was building prosperity under the beneficent New Deal. "If we go along joyfully," Pepper advised, "we shall enter the councils of the common course, have our just rewards, and take our place as the born leaders we are, in the making of the bright future which a humane and liberal government has begun to open up for the South." Jonathan Daniels went so far as to compare the TVA to the "last earlier and significant representative of the USA" in the region, General William Tecumseh Sherman. "The movement in occupation in 1933," Daniels concluded, "was perhaps no less an invasion but a strange, new blessed one for the South."[25]

The President, in his public statements, more and more stridently promised the economic and political "Reconstruction" that southern leaders feared and southern liberals cheered. In Gainesville, Georgia, Roosevelt asserted that the South remained a feudal economy and declared that "there is little difference between the feudal system and the Fascist system." Roosevelt warned that "Georgia and the lower South may just as well face facts—simple facts presented in the lower South by the President of the United States. The purchasing power of millions of Americans in this whole area is far too low." "And let us remember," he reminded the crowd, "that buying power means many other kinds of better things—better schools, better health, better hospitals, better highways." Roosevelt warned that "Those things will not come to us in the South if we oppose progress—if we believe in our hearts that the feudal system is still the best system."[26]

In that and other pronouncements between 1936 and 1939, FDR sounded three themes he had adopted from southern liberals. First, the President stressed the necessity of regional economic development. The South must industrialize, he contended, but not by wooing still more low wage plants. The region instead needed factories which employed the most sophisticated technology and paid wages on the scale of the North and the West. In Roosevelt's view, higher wages were fundamental to southern economic growth; they would create purchasing power, finance public services, and attract market-oriented businesses. "On the present scale of wages and therefore the present scale of buying power," Roosevelt asserted, "the South can not and will not succeed in establishing successful new industries." Second, the President called for the integration of the South into the national economy. Crediting his approach to the "new school of thought" in the South, "a group principally recruited from young men and women who understood that the economy of the South was vitally and inexorably linked to that of the nation," the President urged southern audiences to abandon sectional particularism, accept the direction of the national Democratic party, and, with federal assistance, rebuild the regional economy along the lines of the rest of the nation. "Nationwide thinking, nationwide planning and nationwide action," Roosevelt declared, "are the three great essentials to prevent nationwide crises for future generations to struggle through."[27]

Finally, Roosevelt asked for a new Democratic party in the South, one that would unhesitatingly embrace the New Deal. At one 1938 press conference, Roosevelt speculated on the future of southern political allegiances: "I think the South is going to remain Democratic, but I think it is going to be a more intelligent form of democracy than has kept the South for other reasons, in the Democratic columm all these years." "It will be intelligent thinking, the President added, "and in my judgment, because the South is learning, it is going to be a liberal democracy."[28]

Ample evidence seemed to confirm the President's conviction that the South was learning liberalism. At least the region's voters embraced FDR and his supporters. A Gallup poll on the "Court-pack" revealed that a majority of southerners approved the reorganization, while most Americans opposed the plan. And early in 1938, Lister Hill and Claude Pepper won smashing victories in senatorial primaries. Pepper, whose lieutenants had secured repeal of the Florida poll tax in 1937, rode his support of New Deal legislation to reelection in the Sunshine State. Hill, a leading southern liberal, won the seat vacated by the elevation of Hugo Black to the Supreme Court. Hill triumphed by nearly two to one over "Cotton Tom" Heflin, a bitter foe of all New Deal programs. One northern analyst viewed those elections as "proof positive that progressives are gaining ground" in the South, a sign that the Solid South had "cracked" and that the "Bourbons in the Senate" stood "on pretty soggy ground."[29]

By 1937, New Deal sentiments had burrowed into southern political rhetoric. In Mississippi, governor and business booster Hugh Lawson White won support for his industrial subsidy program—"Balance Agriculture with Industry"—by invoking the rhetoric of the New Deal; he alluded to FDR's willingness to "spend millions to help millions," to secure approval for his plan from the senate legislature. Ordinary southerners echoed the calls for application of the New Deal to their region. One mill worker told a W.P.A. interviewer that the workplace required stricter regulation by a committee of "men hired by the governmint and not the comp'ny you understand." That same millhand believed it imperative that southerners replace their political establishment with people more sympathetic to the New Deal. "Roosevelt picked us up out of the mud and stood us up but whenever he turns us loose I'm afraid we're goin' to fall deeper in the mud then we was before." "That's because, George Dobbin explained, "so many of his own party has turned against him and brought defeat to lots of his thinkin' and plannin'." The prospect of the southern Democrats beating off the New Deal frightened "Old Man Dobbin." "If they keep abuckin' against him," he said, "desolation will follow in this country."[30]

Emboldened by the growth of a southern liberal brain trust in Washington and New Deal constituencies in Alabama and Florida, Roosevelt decided to pursue his desire to reform the South in 1938. He also resolved to challenge the southern obstructionists who frustrated his plans for the region and the nation. On the top of FDR's hit list sat Georgia senator Walter George, an opponent of the Court plan, the minimum wage bill, and relief appropriations. Roosevelt summoned Clark Foreman, a native

Georgian, to the White House and asked him to suggest an opponent to George for the 1938 Democratic primary. Foreman named no candidates, but used the opportunity to press his idea that the Administration distribute a pamphlet detailing the benefits of the New Deal to the South. The President approved the suggestion, but insisted that the pamphlet describe the region's problems, without prescribing solutions. Foreman recalled the President remarking that, "If the people understand the facts, they will find their own remedies."[31]

Roosevelt assigned preparation of the report to Lowell Mellett, a trusted personal adviser and Executive Director of the National Emergency Council. To ensure a positive reception for the draft Mellet prepared in June 1938, the Administration convened a conference of prominent southerners to examine, revise, and approve it. Mellett and Foreman believed that a Conference on the Economic Conditions of the South, composed entirely of southerners, would forestall critics of Yankee interference in southern affairs and mobilize supporters of the New Deal throughout the region. The panel carefully assembled representatives of all sectors of the southern economy—labor, industry, farmers, government, and the press. Mellet selected University of North Carolina president Frank Graham, a leading southern liberal, as chairman of the Conference. Graham's service to the Conference would lead him to the chairmanship of the Southern Conference for Human Welfare, a seat on several Labor Department investigations of southern industry, and eventually, a position on the National War Labor Board. As the Conference convened, Graham received a stirring message from FDR. "No purpose is closer to my heart at this moment than that which caused me to call you to Washington," the President instructed the Conference. "My intimate interest in all that concerns the South is, I believe known to all of you," he wrote, adding that "this interest is far more than a sentimental attachment born of a considerable residence in your section and of a close personal friendship with so many of your people. It proceeds even more," he asserted, "from my feeling that the South presents right now the Nation's No. 1 economic problem—the nation's problem, not merely the South's." Southern underdevelopment, had created an economic "unbalance" in the nation, an unbalance, Roosevelt concluded, "that can and must be righted, for the sake of the South and of the Nation."[32]

The *Report* distilled the condition of the South into fifteen concise, brutally descriptive sections on topics from soil depletion to shabby housing, from deficient health care to meager sources of credit, each section delineating the poverty, backwardness, and misery of the region. Nonetheless, the *Report on Economic Conditions of the South* was not the straightforward presentation of facts it purported to be. It embodied the diagnosis of the southern liberals, adopting the conception of the colonial economy and reiterating Odum's theme of coexistent abundance and waste. The pamphlet also recommended the liberals' preferred treatment for the South, national action that would develop the region's natural and human resources, eliminate barriers to southern participation in the national economy, and restructure the region along the lines of the rest of the nation. In the *Report*'s terms, the South formed "the Nation's greatest untapped market," and its problems, "the most pressing of any America must face," required the intervention of the federal government. Roosevelt himself depicted the *Report* as a prelude to extensive reform. He asked the Conference for "a picture of the South in relation to the rest of the country, in order that we may do something about it: in order that we may not only carry forward the work that has begun toward the rehabilitation of the South, but that the program of such work may be expanded in the directions that this new presentation will indicate."[33]

The NEC report signaled a shift in the direction of federal policy toward the South. It marked the onset of a concerted effort to restructure the regional economy, and the end to the national administration's conciliation of southern interests. The federal government would embark on a long-term sponsorship of southern economic growth, pursuing development along the lines favored by southern liberals—the elimination of low wage employment, nationally financed improvements in education and public services, and encouragement to industrialization.[34]

Federal intervention, however, would look beyond mere regional development. It would mobilize the instruments of the national government to aid the people of the South. Southern New Dealers envisioned a Farm Security Administration to rehabilitate poor farmers and settle sharecroppers on their own land, and an expanded National Youth Administration to equip young southerners with skills for employment in new, high wage factories. They supported also the abolition of the poll tax, not so much as a means of enfranchising blacks but as a way to build a coalition of white New Deal supporters in the South.[35]

The impassioned reactions to the *Report on Economic Conditions of the South* revealed its import. Northern magazines generally endorsed the *Report*'s findings and the prospect of increased federal involvement in the region. "The South cannot pull herself out of the morass," the *New Republic* declared, "without aid from other, more fortunate parts of the country." Southern liberals joined in the chorus of approval. Maury Maverick appealed to his fellow southerners, "Let's Join the United States." Lister Hill praised the *Report,* and a committee of Georgians recommended a five-point program of federal intervention in their state. Jonathan Daniels accepted the general tenor of the *Report,* but wondered if the President had not underestimated the strength of the southern elite. "No other President," the Raleigh editor reflected, "has ever done so much for the South and tried to do so much more against a reaction in the South which conceives of the region in terms of a rigid order that must break if it is altered." Daniels worried that without the initial emergence of a truly democratic political leadership in the South, efforts to improve the southern economy would falter.[36]

Daniels's warning was borne out by the hostile reaction of southern leaders, the men and interests Frank Graham labeled "southern patriots." Some southerners blamed the region's troubles on the size of its black population, others on the curse of the boll weevil. Industrial interests denied the label of No. 1 economic problem altogether, citing the region's "incontrovertible data of progress." Invoking the ghosts of Reconstruction, *Manufacturers' Record* conceded that "the South was a problem in 1865 when it was stripped of its wealth," but since then, "the South's upward climb from poverty has amazed the world." The Southern States Industrial Council, the lobby for low wage southern manufacturing, contended that the *Report* "did the South a grave injustice." "Southern businessmen," added Council president Fitzgerald Hall, considered "the South the number one economic hope of the nation" and found the *Report* counterproductive. Southern patriotism emerged from other quarters as well. In Congress, Josiah Bailey and Carter Glass vilified the *Report,* an attack echoed by many of the region's proud impoverished. "I'm proud of North Carolina too," one Tar Heel farm worker asserted, "and I don't like for our President to call it no problem."[37]

Disguised as an objective analysis of the regional economy, the *Report on Economic Conditions of the South* was a manifesto for the southern liberal program for regional development. It was also, in Clark Foreman's words, "a part of the President's program to liberalize the Democratic Party." FDR's so-called purge, a general effort to unseat Congressional opponents of the New Deal in the 1938 mid-term elections, concentrated on defeating the recalcitrant southerners in the President's own party. After returning from a tour of the Pacific in August of 1938, the President campaigned vigorously through the South, repeatedly citing the *Report*'s findings from the stump. Hostility to the President's interference in the southern primaries soon translated into rejection of the *Report.* The Atlanta *Constitution,* for example, initially endorsed the pamphlet, but later warned schoolteachers against its biases, and the federal government against "meddling" in southern affairs. "No new reconstruction government," the paper declared, "is desired or will be tolerated." Birmingham columnist John Temple Graves concluded that by introducing the economic report into the election, "the President has called the South a No. 1 economic problem and he has made it a No. 1 political problem."[38]

Roosevelt thought southern support for him and his program sufficient to defeat recalcitrant southern legislators. While the Administration worked behind the scenes against the Byrd organization in Virginia, the President publicly attacked Georgia senator Walter George and South Carolina senator "Cotton Ed" Smith. According to Jim Farley, Roosevelt "wanted to make an object lesson of George because he thought such a defeat would furnish a lasting lesson to the southern bloc in Congress." In South Carolina, popular governor Olin Johnston, a New Deal supporter who had announced his candidacy from the White House steps, faced the intransigent Smith. Without mentioning either candidate's name, the President heaped criticism on Smith at a whistle stop in Greenville, S.C. The President's resentment of George was so strong (he reportedly told aides that he would back someone against the Senator, even if it had to be the tenant on his Warm Springs farm) that Roosevelt endorsed George's opponent Laurence Camp at an address at Barnesville. FDR mentioned the NEC report's description of the South to the Barnesville crowd and asserted that "It is not an attack on state sovereignty to point out that the national aspect of all these problems requires action by the Federal Government in Washington." To achieve his goals for the region, he needed sympathetic legislators in Congress. FDR assured the voters of his "other state" that Walter George was his "friend," but if he were voting in the Georgia primary, he would "most assuredly cast his ballot for Laurence Camp."[39]

The President, however, badly overrated the impact of his direct involvement in southern affairs and miscalculated the vulnerability of the southern conservatives. Black southerners, after all, were generally denied the suffrage, as were many southern whites, and the specter of "outside interference" had lasting emotional resonance in the land of Dixie. The usually shrewd political tactician ignored the advice of Jim Farley and Harry Hopkins and as a result "outside interference" replaced support for the New Deal as the major campaign issue. Walter George termed the President's appearance "a second march through Georgia." Before a statue of Confederate hero Wade Hampton, "Cotton Ed" Smith declared that "no man dares to come into South Carolina and try to dictate to the sons of those men who held high the hands of Lee and Hampton." The President's statements also forced liberal state officials to remain aloof from the contests. Georgia governor Eurith Rivers, architect of the state's "Little New Deal," remained neutral, a posture which so infuriated FDR that he halted PWA aid to Georgia for nearly a year. On primary day, Smith held onto his seat. George won easily, with Camp a distant third. The attempt to remove the obstructionist southerners in Congress backfired completely. When "Cotton Ed" Smith suggested after the elections that FDR was his own worst enemy, George replied, "Not as long as I am alive." Not only did the failed purge win the Administration intractable enemies in George and Smith, it also alienated the remaining southern leaders in Washington.[40]

Recognizing his overblown expectations, Roosevelt conceded to Jim Farley: "It takes a long, long time to bring the past up to the present." In Farley's mind, that statement and the President's behavior during the campaign evinced the irrational hatred that FDR felt for his opponents. But although Roosevelt was a tough politician with little affection for his "friend" Walter George, it was not blind passion that accounted for FDR's poor judgment. The President mistakenly interpreted his personal popularity in the South and the region's new-found penchant for New Deal rhetoric as a commitment to Administration policies and a mandate to replace southern political leaders. The Administration and its liberal southern allies proved incapable of reshaping southern politics between 1936 and 1938. They also failed to realize that populist resentment to the conservative southern leadership did not necessarily translate into a desire for federal involvement in the region.[41]

But if the South of 1938 was not yet vulnerable to political realignment, it was susceptible to economic reform. Central to the southern liberal program and to the purge campaign was the struggle over federal wages and hours legislation. The *Report on Economic Conditions of the South* contended that low industrial wages caused poor public services. And "wild wage differentials" between the South and the rest of the nation perpetuated poverty and malnutrition, forced mothers and children into the mills, and forestalled the development of market-oriented, high grade industry. Raising wages was an important component of the liberals' plan for the region. "A low wage scale," the pamphlet concluded, "means low living standards, insufficient food for many, a great amount of illness, and, in general, unhealthful and undesirable conditions of life."[42]

The President himself sharpened the message that improving southern wages formed the first step in regional development. In June 1937, FDR informed reporters that his promise to secure a better lot for the one-third of the nation at the bottom of the economic ladder required legislation correcting wages in the areas where they remained "far below any decent standards." In his 1938 State of the Union address, Roosevelt reasserted "the immediate desirability of increasing the wages of the lowest paid groups in all industry." A few weeks later in his Gainesville speech, the President emphasized the special significance of this for the South. "Most men and women who work for wages in this whole area get wages which are far too low," FDR complained. Such low pay rates maintained southern backwardness, not only by reducing purchasing power and the tax base, but also because "efficiency in operating industries goes hand in hand with good pay."[43]

The movement for federal regulation became a southern issue, not only because the South was the low wage region and would be most dramatically affected by any such law, but also because in 1937 the Southeast remained the only region in the country without any such state legislation in force. The demand for a federal statute first emerged out of the need to replace the defunct NRA codes and the ineffective Walsh-Healy Public Contracts Act. But after 1937, proponents sought the regulations as a device to restructure southern industry. Indeed, if any doubt existed about the Fair Labor Standards Act's intent to reform the South, the course of the two-year long debate over the measure dispelled it so dramatically that Walter Lippmann would tag the final version "a sectional bill thinly disguised as a humanitarian reform."[44]

The Administration initiated the battle for a national minimum wage law in 1937 by enlisting Senator Hugo L. Black, the premier southern liberal in the Congress, as a co-sponsor of the bill. The bill proposed the appointment of a special panel in the Labor Department that would prescribe "non-oppressive" wages and hours after considering the economic conditions of the various industries and areas. With the South providing most of the Democratic opposition, a coalition of Republicans and conservative Democrats had the bill recommitted without a vote during the 1937 session. The plan never reached the floor of the Congress both because southern opponents controlled the House Rules Committee and because even supporters of a minimum wage law disapproved of the specifics of the Black-Connery bill. Early in the 1938 session, Robert Ramspeck, a moderate southern Congressman representing Atlanta, proposed a bill that established an independent five-person wage commission and incorporated formal regional differentials into the law. And Texas Congressman Martin Dies suggested to FDR that the law grant state governments the authority to roll back wage and hour standards. The President rejected such regional variations. "Call up Martin Dies," he instructed his personal secretary, "and tell him having an individual state vary a national Wage and Hour bill is not only unsound but would destroy the effectiveness of building up purchasing power in those sections worst needing it."[45]

Dies's proposal was never considered, and the House Labor Committee quickly rejected the Ramspeck plan with its regional standards. The committee endorsed instead a substitute proposed by AFL president William Green which fixed national labor standards, eliminated the discretionary panel, and forbade geographic differentials. The southerners on the Rules Committee again prevented a vote, and the measure appeared dead until Claude Pepper, a supporter of the law who had campaigned on its behalf, won his smashing primary victory. A few days after the election, House members flocked to the House podium to endorse the necessary discharge petition, although only 22 of the 183 Democratic signers were from the South.[46]

The bill passed the House 314–97 with southerners casting 52 of the 56 Democratic nay votes. In the Conference Committee, Administration partisans beat off a last ditch southern effort to reinsert regional differentials. They reported a "compromise bill" that established fixed rates, but escalated them gradually over a seven-year period. The compromise package also included numerous exemptions, especially for agricultural labor, and established Industry Committees that could adjust the timing of the escalations to consider "competitive conditions as affected by transportation, living and production costs."[47]

The progress of the Act displayed the shift in federal policy toward the South. The bill's legislative history followed a path making it more and more satisfactory to organized labor and black rights organizations. Outlining the position of organized labor, John L. Lewis of the CIO indicated unions' general support for a national wages and hours law. "It will mean at least a glimmer of sunlight," Lewis maintained, "to millions of submerged American workers who live in economic darkness and despair." Labor demanded, however, that the law specify fixed standards, a 40-cent hourly minimum wage and a 35-hour work week, and expressly reject regional differentials. "I am firmly opposed to wage differentials based on geography," Lewis thundered. "Usually this is no more than a plea for the continuance of low living standards in the Southern States."[48]

Miffed at the operation of the NRA codes, black leaders similarly opposed flexible labor standards. They believed discretionary powers would serve to discriminate against black workers. "The bill, rather than permitting differentials," National Negro Congress spokesman John P. Davis demanded, "should expressly declare as a part of its policy its opposition to any type of differential treatment so far as minimum wages are concerned." Asked whether "differentials had always proved injurious to Negro labor," Davis responded emphatically, "Yes."[49]

The black leadership's unflinching position had far-reaching consequences for their constituency. In Washington, black spokesmen demanded fidelity to the civil rights position. They tolerated no racial distinctions in federal legislation, no protections for black workers that might imply their inferiority. The long-run benefits of that strategy—the integrity and political power of the civil rights movement, accelerated black outmigration from the South—might have outweighed the short-term costs. But in so doing, the national black rights organizations ignored the pleas of some black southerners. Richmond minister Gordon Hancock, for example, believed that the black intelligentsia ignored the concerns of their poorer, southern brethren. In any case, the proposed upgrading of southern employment, in wages and skills, would ease exclusion of blacks from the region's workforce. As it did child labor, the proposed legislation targeted for elimination the low paying, unskilled jobs traditionally held by southern blacks. And it offered no guarantees that blacks would receive the new jobs at higher pay. Nonetheless, from 1938 to 1980, black groups never wavered in their commitment to higher minimum wages without racial or regional distinctions. During the battle for the Fair Labor Standards Act, they helped lead the struggle against regional differentials.[50]

The evolution of the FSLA revealed the Administration's desire to conciliate blacks and labor—two elements of the new "Roosevelt coalition." It also betrayed a willingness to alienate the established white South, for southern economic interests took a united stand against the bill. Industrialists feared the loss of their low wage position. Planters also opposed the measure, albeit with little enthusiasm (landlords felt less pressure than mill owners to safeguard their cheap labor because the law exempted agricultural labor from national labor standards, and also, because of the AAA, southern landlords no longer faced a shortage of unskilled labor).[51]

The South's representatives in the national legislature resisted the law. Pat Harrison broke with the Administration over this measure and lined up against it all but one member of Mississippi's Congressional delegation. Harrison recalled the conciliatory spirit that had dominated FDR's first term, intoning that "It is a beautiful realization for the present generation, to have experienced the almost complete disappearance of sectional feeling." Not known as the Grey Fox of the Delta for nothing, the crafty Harrison added: "In these circumstances, it is unfortunate that any issue be precipitated that might arouse sectional antagonisms or sectional discussions." The shift in federal policy threatened to do just that. "Within the past year," he claimed, "the South has been the object of unjustified criticism, receiving from the invective tongue of the social uplifter and prejudiced journalists a castigation of misrepresentation." "And why are these cruel aspersions cast at the South?" Harrison asked with mock bewilderment, "simply because that section is opposed to a federal minimum wage law."[52]

Southern business interests joined Harrison in his denunciation of the labor standards legislation. An Alabama peanut packager warned that "passage of the Black-Connery Wage and Hour Bill will be ruinous to this section and the South as a whole and shows definite discrimination against the South." A representative of the Southern pine industry asserted simply that the law meant "the annihilation of the Lumber Industry of the South."[53]

Southerners raised two sets of objections to the minimum wage law. First, opponents of the legislation resuscitated all the objections to NRA code provisions. Southern States Industrial Council president John Edgerton sneered that "the bill as a whole presents itself to us as a reincarnation of the NRA in more virulent form." As in the campaign against higher NRA wage minima, southern businessmen trotted out arguments that living costs were lower in Dixie, that southern industry was undermechanized, and that the climate reduced labor productivity. Pat Harrison went so far as to claim that southern workers were content with lower wages because labor received more respect in the South than in other parts of the country. Most common was the claim that black labor was inefficient; blacks allegedly required greater supervision, worked more slowly, or were unequipped to perform work at the level of even the lowest proposed federal standard. Some southern employers simply refused to pay decent wages for "sorry negro labor." Contemplating the impact of a minimum wage requirement on his black laborers, one businessman concluded: "Those that have jobs will have too much leisure and the extra money they get will go for liquor."[54]

The second argument held that a national wage would enforce the economic domination of the North and cripple southern efforts to industrialize. Such traditional southern resentments had lain dormant in FDR's first term, but reemerged during this debate. "One of the reasons that industry in many of the highly industrialized centers is not opposed to this bill," the Southern States Industrial Council averred, "is because it will provide a means by law to do away with the competition from the immature rural South." "If the object of this bill is to prevent the migration of industry from one area to another, especially from the North to the South," SSIC president John Edgerton insisted, "there is scarcely any doubt that the object would be attained by the enactment of this law." The mayor of

Atwood, Alabama, articulated this fear most vividly. "If a minimum wage scale should be set," W. R. Holley predicted, "the Yankees would again whip us worse than in '61–'65."[55]

As in '61–'65, the agents of this invasion were representatives of the national government, not the "blue-bellied devils" of the Union Army, but the besuited administrators of the Department of Labor, come to raze the southern economy. Southern industry welcomed any fair scheme of labor standards, John Edgerton assured Congressional investigators. "What southern industry is mortally afraid of," he explained, "is the result of domination of all industry in the United States by a board with headquarters in Washington. . . . We of the South know that we have never profited from the sacrifices of any of the rights of the states or individuals to Federal discretion."[56]

Such southern patriotism was most pronounced in the controversy over explicit regional differentials, a concern that came to occupy the center stage after it became clear that some sort of minimum wage law would pass during the 1938 Congressional session. Some conservative southerners simply opposed higher wages. But as John Edgerton conceded, many southern industrialists could accept some wage increase as long as the region's relative position was protected. The growth strategy of southern business depended on relatively cheap labor to attract investment. Higher wages would not threaten this strategy so much as the elimination of differentials. To defend low southern pay rates, representatives of southern industry argued over and over again that all earlier Roosevelt administration policy had maintained southern wage differentials. The NRA had sanctioned a plethora of differential formulae, and the WPA continued to pay wages in the South at roughly half the northern rate. Not only did southern businessmen plead that differentials made up a sound, established policy, they also cited a Gallup poll that indicated that voters in all sections of the nation favored geographic differentials in minimum wages.[57]

Aligned against the southern elite in the wage differential debate were organized labor, black rights groups, and, with increasing fervor, the Roosevelt administration. The Labor Department opposed regional discriminations from the outset. The Democratic majority on the House Labor Committee followed suit, demanding that "the wages and hours prescribed apply nationally in each particular industry," without differences between "sections of the United States . . . or between employers." And President Roosevelt's position on the issue slowly hardened. When he first asked for wages and hours legislation in May 1937, FDR allowed: "Even in the treatment of national problems there are geographic and industrial diversities which practical statesmanship cannot wholly ignore." By the following autumn, however, he pledged to secure "fully adequate pay for all labor." Farsighted businessmen, the President explained in a fireside chat, understand "that no one section of the country can permanently benefit itself, or the rest of the country, by maintaining standards of wages and hours far inferior to other sections of the country."[58]

The South's allegations of outside interference had some validity. Many northern business-persons supported the labor standards legislation, most notably Robert Johnson of Johnson and Johnson, who operated highly mechanized plants and paid high wages in both regions. Northern manufacturers generally opposed territorial differentials, none more vociferously than New England textile interests, which had lost ground to the South during the Depression. Labor leaders believed that the FLSA would secure their position in the North. For even if high minimum wages failed to spur organization below the Mason-Dixon line, they would reduce competition from southern plants and protect unionized jobs in the North. Congressional voting patterns revealed similar pressures. In one key vote on the FLSA, only two Republican senators cast votes in favor of the legislation; one was a Pennsylvania liberal with ties to organized labor, the other the Massachusetts conservative Henry Cabot Lodge. That dour Bostonian was no friend of the New Deal, but he was a close ally of textile mill owners in his state.[59]

Despite these currents of support for the FLSA, the accusations of a northern industrial conspiracy proved unjustified. All the major national business organizations—the National Association of Manufacturers, the United States Chamber of Commerce, and the American Mining Congress—opposed the bill and resented further federal intrusion into labor-management relations. The Ohio State Chamber of Commerce even cited concern for the South as one of its reasons for opposing the legislation, a rather philanthropic response for a potential beneficiary of a bill to cripple southern industry. George B. Chandler, spokesman for the Ohio Chamber, even invoked the rhetoric of the Lost Cause in his broadside against the Black-Connery bill. "South Carolina," he enjoined, "fired on Fort Sumter for a far less pretext than this bill affords." The pleas of northern business interests like the Ohio Chamber drew acid responses from Senator Hugo Black. Wasn't it amusing, he chided, to see the Ohio Chamber of Commerce posing as "the special and particular defender of the South." While a substantial number of northern businessmen eventually supported the bill, they did not inspire the legislation, nor did their interests provide its central motivation.[60]

The FLSA was simply not the purely protective measure it seemed to southern manufacturers in 1938 and which it appears to be to many contemporary scholars today. It is tempting to view the FLSA as a kind of regional tariff, a ploy by northern businesses to eliminate low wage competition and by northern politicians to stop runaway plants. Whatever the Roosevelt administration's intentions in drafting the law, this argument asserts, the chief concerns of those who voted it into law must have been protectionist. FLSA supporters presumably cared little about uplifting southern labor or eliminating the sweat shop; their primary aim must have been weakening the competitive position of southern business versus their own regions. The FLSA, after all, covered only manufacturing workers engaged in interstate commerce, that is, those competing with northern firms. Certainly, organized labor and Massachusetts textile manufacturers, strong advocates of the FLSA, supported it for precisely these reasons. But this view, logical as it may be, overemphasizes the economic foundation of American public policy. With the exception of the southern Democrats and the two Republicans noted above, Congressional support for the wages and hours bill closely followed party lines. Republicans, self-declared guardians of northern business interests, opposed the bill. Democrats, responding to pressure from the White House and their constituents (public opinion polls indicated as much as 70 percent of the electorate in favor of the legislation), enacted it. Northern business may not have mounted strong opposition to the law, but its interests hardly motivated the law's formulation or accounted for its passage.[61]

Southern liberals, far more than northern capitalists, pushed for national labor standards. They assailed the arguments of Edgerton and other southern business leaders and denied the alleged cost of living differences between the regions. Partisans of the wages and hours law were convinced that only the worst, most exploititive southern employers opposed the bill. Senator Black postulated that "those who are making the loudest noise are slave drivers." President Roosevelt shared this view, singling out for criticism the recalcitrant southern lumber industry. The Administration asserted that national labor legislation would free southern workers from oppressive wages and hours, make child labor and women's work unnecessary, and spread employment. And, however far-fetched their belief might have been, national policymakers also envisioned the minimum wage as the first step in regional economic development. "We cannot sell goods to millions of underpaid in the United States," one manufacturer explained, encapsulating the prevailing theory that purchasing power drove economic growth, "they are out of the market." Claude Pepper laid out the long-run implications of the law for his native region. The minimum wage, in Pepper's view, would stimulate the "accelerated emergence of a skilled labor supply of strong and intelligent people" that "would attract industry requiring skill and responsibility in the worker."[62]

More notably, however, the enactment of the FLSA inaugurated a new era in the federal government's relations with the South. Both the traditional policy of benign neglect and the more recent effort at piecemeal reform and conciliation of southern interests were to be abandoned. The change did not escape the notice of contemporary observers. "Social and economic affairs below Mason and Dixon's line were for a long time little attended," noted Chattanooga editor George Fort Milton in 1938. "In the last year or so, however, they have shared the spotlight." No one appreciated the extent of the shift more than the man ultimately responsible for it. In a letter to the Southern Conference for Human Welfare, FDR shared credit with southern liberals. "The long struggle by liberal leaders of the South for human welfare in your region has been implemented on an unprecedented scale these past five and one-half years by Federal help. Yet we have recognized publicly this year that what has been done is only a beginning and that the South's unbalance is a major concern not merely of the South, but of the whole nation."[63]

As the *Report on Economic Conditions of the South* dramatized, that unbalance had not even been confronted by 1938; farm income, industrial wages, family income, and public services remained low in the South. But Roosevelt's remarks and the broader policy commitment they pledged marked the onset of sustained federal intervention in the southern economy, intervention which would catalyze tremendous economic change over the following two decades. As stated in the NEC *Report,* the principle of entitlement—the southern liberal view of the South as a national problem deserving of federal aid—would translate into preferential treatment for the region in the disbursement of funds for highways, airports, and public education. Along with those expenditures came increased indirect federal investments, which would accelerate the emergence of an industrial economy. Deliberate federal policies to drive up southern wage rates and reshape the southern labor force were more direct and more controversial. That program and a fifteen-year-long struggle over it were only beginning when FDR signed the Fair Labor Standards Act in June 1938.

When the FLSA went into effect, President Roosevelt and the southern New Dealers shared a commitment to economic reform, but possessed no comprehensive plan and no clear popular mandate. Legislation for tenant famers, for federal aid to education, for the TVA, and for federal relief was on the books or being debated in the Congress. The Interior Department planned a major dam construction program in the Southwest. The AAA included blacks in its special elections, and Roosevelt had privately expressed approval for a number of measures dear to the hearts of southern liberals, including abolition of the poll tax. In February 1939, David Lilienthal took stock of recent events. The President, Lilienthal wrote to a friend, "after a period of caution," has now boldly taken up the theme of regional development. "In his speeches and in the *Report on Economic Conditions in the South,* his help on the transportation matter, and on Southern wage levels, education, etc., he has set the thing going nationally."[64]

Not surprisingly, then, this new turn in federal policy toward the South also introduced tensions into the national administration's relationship with southern leaders. As Lilienthal noted, FDR "was induced to use, or permit be used, some of the words that get folks into trouble in the South" and that had "led to some unnecessary crosscurrents." At the end of the 1939 Congressional session, Claude Pepper rose from his seat on the Senate floor to attack the opponents of the New Deal. "I am unwilling to let this session of Congress end," Pepper inveighed, "without lifting my voice to decry the unrighteous partnership of those who had been willing to scuttle the American government and the American people and jeopardize the peace of the world because they hate Roosevelt and what Roosevelt stands for." Pepper, along with his southern liberal allies in the Administration, recognized that the "unrighteous partnership" included many political leaders of their party and many economic interests in their section. The President and his top advisers understood that as well, but they believed that their new strength in the Northeast, the West, and in the South compensated for the estrangement of the southern "Big Mules." And when it considered the South's longstanding dependence on the Democratic party and the sentiments of the South's common people, the Administration evinced a perhaps unjustified confidence. The President and his advisers concurred with the analysis of Birmingham columnist John Temple Graves in 1939: "The South, it may be said, is looking right, left, up, down and over—but it still loves Roosevelt."[65]

Chapter 3 ❂ The Wages of Dixie

"You need more industries in Texas," President Roosevelt reminded a Fort Worth audience soon after signing the Fair Labor Standards Act in 1938, "but I know you know the importance of not trying to get industries by the route of cheap wages for industrial workers." That salvo opened the first front in the Roosevelt administration's battle to reform the southern economy—the attack on low southern wages. Exploiting the opportunities presented by the Depression and later by the war emergency, the federal government sought to align the wage structure of southern industry with the rest of the nation. It hoped thereby to break the hold of the low wage, labor-intensive sectors that isolated the South from the national economy and, in the minds of policymakers, imprisoned it in poverty. Expounding the message which he and many others would carry into wartime federal agencies, Maury Maverick reiterated the Administration's goal. "If any part of the nation needs protection against low wages," Maverick insisted, "it is the South."[1]

By 1938, federal policymakers had lost faith in the ability of flexible wages to alleviate unemployment. They no longer believed that as long as wages could be reduced, involuntary unemployment could not exist. Simultaneous declines in wages and employment during the Great Depression had discredited that theory. "There is a widely held opinion that low wages tend to make possible the employment of a comparatively large number of people and therefore tend to prevent unemployment," the Department of Labor admitted in 1938. But, it insisted, "the experience of the South with low wages affords no basis for this view."[2]

The Depression also convinced policymakers that uncontrolled wage cutting had ensured the severity and the duration of the crisis. Even though the deflation had ended years earlier, the House Labor Committee warned that "the Federal Government cannot by its inaction permit the channels of commerce to be used to set this spiral of deflation in motion."[3]

Federal policymakers believed that by sustaining aggregate demand, what contemporaries termed "purchasing power," they could avert depressions. But the Roosevelt administration only reluctantly translated this analysis into a program of compensatory government spending. Instead, the federal government enthusiastically applied this principle to wage regulation. The South, with its vast undeveloped market and exceedingly low wage rates, offered the most fertile ground for such reform. Higher southern wages would stimulate aggregate demand, thus freeing the poor from their current want and the South from its longstanding poverty. "Cheap wages mean low buying power," President Roosevelt explained in 1938. And "low buying power means low standards of living." Those conditions in turn perpetuated a backward, underdeveloped South. "And let us remember that buying power means many other kinds of better things," Roosevelt told another southern audience, "better schools, better health, better hospitals, better highways."[4]

In the minds of national policymakers, lifting wage rates would not only fortify purchasing power, it would also enhance productivity in the South. The New Dealers' rejection of orthodox wage theory did not end with the concept of involuntary unemployment; it challenged the very definition of the wage rate as the marginal product of labor. To explain southern labor standards, federal officials introduced the concept of "substandard wages," wages below what normal labor productivity would justify. For workers at these substandard levels, higher wages would actually raise output per man-hour. "Experience reveals," the Bureau of Labor Statistics contended, "that a properly established minimum wage tends to increase the productive capacity of workers as a result of the improvements in physical well-being and morale that higher wages make possible." Wage regulation would thus restore "genuine competition," eliminating the southern labor system in which employers swelled profits by exploiting labor rather than improving productivity. "In too many instances," Isador Lubin, one of the architects of the FLSA, explained, "the ability to sweat one's labor has supplanted efficiency as the determinant of business success."[5]

As well as raising productivity within existing plants, higher southern wages would spark a wholesale reorganization of southern industry along more capital intensive and "efficient" lines. A minimum wage, this hypothesis maintained, would induce southern firms to refine their allegedly lackadaisical management and marketing practices and modernize their old-fashioned production processes. "Low wages have helped industry little in the South," the NEC Report lamented, and "they have made possible the occasional survival of inefficient concerns." According to Howard Odum, the South's cheap labor had resulted "in hand work rather than machine work, in roundabout methods of cultivation and processing, in piecemeal organization, and generally speaking, in wasteful use of labor." Federal wage standards would discourage such practices and offer premiums to a modernized southern industry. President Roosevelt himself conceded in one speech that new techniques and machinery might eliminate some unskilled jobs, but he argued that gains in efficiency and purchasing power would more than compensate for such losses.[6]

In fact, if the FLSA imperiled any southern jobs, the President and other New Dealers assumed only substandard jobs were at risk and bade them good riddance. Those positions, many of them held by children and women, were undesirable in any reformed southern economy. The wages and hours law pursued a social as well as an economic objective. Stable family employment and high family wages mattered more to federal authorities than did the total number employed. One of the perceived evils of low southern wages was that they made a man unable to support his family and forced his wife and children to work. Competition from those women and children further depressed southern pay rates while it destroyed the region's young human capital. Even Lorena Hickok, Harry Hopkins's assistant and Eleanor Roosevelt's confidante, believed that many southern women "got jobs that shouldn't have had them."[7]

Many southern employers freely admitted their reliance on child and female labor. They even advanced it as a defense of low wages. Earl Constantine of the Hosiery Manufacturers pleaded that most low wage workers were "girls and their earnings are merely contributing earnings, and frequently there are two or more girls in the same family working in a plant." Labor leaders, however, decried this charity. "This does not mean, of course, that I am opposed to the employment of women, or even of wives, when this is the result of their own free choice," CIO leader John L. Lewis exclaimed. "But I am violently opposed to the system by which degrading the earnings of adult males, makes it economically necessary for wives and children to become supplementary wage earners and then says, 'see the nice income of this family.'" Federal officials concurred with Lewis. The *Report on Economic Conditions of the South* noted the high incidence of female labor and, especially, of child labor in the South. It identified those trends as products of low wages for adult males. The national government was willing to sacrifice such jobs to provide more secure, better paying employment for the men they assumed to be the heads of these desperate southern families.[8]

Black workers, however, could easily fall through this family safety net. Black southerners—women, youths, and adult males—filled many of the region's low wage positions. Like child labor and women's work, these substandard positions, including those of male adults, were slated for elimination. Civil rights groups and federal policymakers assumed that blacks would win a share of the new jobs at higher pay. They rejected southern employers' claims that blacks worked inefficiently or required less money than whites for subsistence. They failed to recognize that southern employers believed those arguments. Moreover, the legacy of segregation furnished white southerners with better education and more experience in skilled positions than their black fellows. After the enactment of the minimum wage, southern firms could hire whites at the same wages the law required for blacks. With considerable prescience, Louisiana governor

Richard Leche portrayed the wages and hours law as a "second emancipation proclamation," for white southerners. Leche declared in 1938 that the act would free "the Southern *white* laborer from economic exploitation."[9]

The region's economic leadership did not share that enthusiasm. Senator Pat Harrison, the leading Congressional opponent of the FLSA, rued the implementation of the act. The law would not "vitally affect the South," Harrison asserted, "if fairly administered." But "if they construe it in a sectional way," he warned, "it can play the devil with us." Harrison, "the Grey Fox of the Delta," expressed the fears of southern businessmen, who believed, with some justification, that the law sought to weaken their economic and political power and remove their region's comparative advantage. During the week before the FLSA took effect, threats of plant closings and massive layoffs pervaded the region. Rumors of the demise of Texas pecan shellers, the shutdown of sawmills in southern pine and hardwood forests, and the firing of thousands of textile workers flooded the nation's capital. President Roosevelt denounced the reports as the exaggerations of "reactionaries." Elmer F. Andrews, the man Roosevelt had appointed to administer the act, similarly scoffed at the tales of a southern industrial apocalypse. Andrews informed reporters that he "was not at all excited" by rumors of plant closings on account of the minimum wage. Asked whether his calm betrayed a desire to see low wage southern firms close shop, the administrator grinned and declined comment.[10]

A week after the implementation of the act, a more exuberant Andrews faced the press. Southern state authorities had reported few closings and no drop in employment. But while reports of imminent disaster proved false, the FLSA had a deep and immediate impact on the South. The law established a 25-cent minimum hourly wage with a 44-hour work week for the first year. In October 1939, a 30-cent and 42-hour standard became effective. Thirteen percent of covered southern employees were due to receive mandatory wage hikes to reach the 25-cent level. Outside the South fewer than one-tenth of one percent of the affected workers earned below the minimum. The maximum hours requirement also applied to a higher proportion of laborers in the South than elsewhere in the nation.[11]

These regulations reorganized the workplace of the industrial South. In its first two years the FLSA compressed the wage structures of most southern industries. As many as half the workers in the southern branches of some industries earned an hourly rate within 2.5 cents of the legal minimum. At the same time, the hours restrictions—by requiring overtime premiums for long work weeks—spread employment among more laborers. During 1938 and 1939, the number of wage earners in southern manufacturing grew faster than the total man-hours of labor.[12]

The employment effects of the FLSA proved more difficult to gauge and provoked much debate. The Department of Labor reported no loss of jobs in the first two years under the act. The Department asserted that a growing workforce in better paying firms more than compensated for the declines in low-wage establishments. Plant mortality was no more frequent than usual in the sharply competitive southern industrial environment. But this rosy portrait, based on aggregate figures for the region, disguised the dramatic alterations within certain southern industries. The wages and hours law hit hardest in the South not only because southern firms paid lower wages than their competitors within national industries, but also because southern manufacturing was dominated by industries like cotton textiles, which paid low wages everywhere. In some southern industries the law had pronounced effects. In others, it was hardly noticeable.[13]

In the predominately northern iron and steel and cement industries, no employees, north or south, fell under the provisions of the minimum wage law. The act exerted no influence on employment or plant organization in those sectors. Pecan shelling offered a contrasting example. During the 1920s, packers of paper shell pecans in San Antonio, Texas, had closed semi-mechanized facilities and converted to hand shelling on a contract basis. An average sheller cracked eight pounds a day at 5 to 6 cents per pound for a weekly wage of around $2.50. Compliance with the FLSA would have tripled weekly pay rates. Instead, the major pecan processors installed shelling equipment. A catastrophic decline in employment accompanied the rapid mechanization.[14]

The region's principal industries escaped such severe tribulations, but they nonetheless felt the FLSA's impact. In seamless hosiery, the 25-cent standard drove up southern payrolls at a rate ten times that of northern mills. Employment in southern plants dropped 5.52 percent from 1938 to 1940 while rising slightly in the North. These layoffs were limited to low wage plants, those where hourly earnings had stood below 32.5 cents in 1938. The wages and hours law also encouraged the replacement of hand transfer-top equipment by automatic machinery and a corresponding reduction in the number of knitters. One employer, a loyal South Carolina Democrat, expressed bewilderment at the pressure to mechanize. "With the President of the United States calling the South 'Economic Problem No. 1,' with the President urging the South to get out of hock to the North," H. M. Arthur of Excelsior Hosiery Mills protested, "why are they forcing you to buy automatic machinery and get back in hock to the North."[15]

Sawmill owners also felt the act's pressure to mechanize, but the manpower requirements of the region's mills often made such modernization impractical. The South's small, scattered timber stands and heterogeneous lumber crops defied easy automation, so that southern sawmills relied on large quantities of cheap labor. Thirty-five percent of southern sawyers earned less than the legal minimum when the FLSA took effect. At the same time, only 19 percent of the industry's northern labor force and a meager 3 percent of the lumbermen in the highly mechanized Pacific Northwest fell below that standard. By the winter of 1939–40 the new law had skewed the industry's wage structure in the South. Half of the southern labor force earned exactly the minimum wage. Unable to mechanize, many southern sawmill operators simply evaded the minimum wage regulations by selling all of their output within the boundaries of their home states. According to the Southern Pine Industry Committee (SPIC), the FLSA had "balkanized" lumber production in the South. In 1938, of the 500 Texas sawmills, three hundred participated in interstate commerce; in 1941 twenty continued to do so. Even so, industry officials claimed that only wartime demand for timber products forestalled closings and layoffs.[16]

Defense production also buoyed the cotton textile industry during its adjustment to federal regulations. In August 1938, 14.6 percent of southern millhands earned wages below the impending 30-cent minimum; hardly any (less than 1 percent) of the employees in other regions did so. By September 1940 every textile worker's pay rate reached that level. The pace of technological change quickened as well, reflecting both the stimulus of war demand and the substitution of machinery for suddenly costlier labor. Sales of textile machinery hit an all-time high in 1940. The return of prosperity muted resistance to these changes, but southern employers nonetheless resented the federal government's interference with the wage differential. "The South has struggled for 75 [*sic*] years against various obstacles consciously erected to hinder and retard its industrial and commercial development," a group of southern mill owners maintained. Then, the FLSA arrived to "perpetrate the punitive program of the carpetbaggers under the guise of equalizing conditions."[17]

The FLSA became a lightning rod for debate on the southern economy—both for southern opponents of the New Deal and for the southern New Dealers themselves. Although it was a modest piece of legislation which covered only a minority of workers in the heavily agricultural South, the law stirred the hopes and fears of many. Southern manufacturers blamed it for economic pressures—such as changes in demand and the need to mechanize—that had been operating for two decades. Reformers, on the other hand, hoped to extend the purview of the act and thus widen their attack on the region's conservative leadership. They also saw the wages and hours law as the foundation for a labor-liberal political coalition in the South. After all, Claude Pepper's support for the minimum wage had catapulted him into the Senate.[18]

REFORM BEGINS: THE INDUSTRY COMMITTEES

Reformers' hopes and manufacturers' discontent were slated to clash in the Industry Committees. The FLSA established these tripartite bodies, composed equally of representatives of management, labor, and the public, to investigate business conditions in particular industries. They reported their findings and recommended new regulations to the Wage and Hour Division of the Department of Labor. The Industry Committees laid the foundation for federal reform of the southern economy over the next fifteen years. They granted organized labor influence in the South; all employee representatives were union officials, even though only a tenth of the South's nonagricultural laborers carried union cards. They jolted southern manufacturers into concerted, united action. But most important, the committees inaugurated a sustained period of direct federal government involvement in the day-to-day business practices of southern firms and recruited the personnel who would oversee that intervention.[19]

Like the FLSA itself, the Industry Committees concerned themselves primarily with the South. The first three committees investigated regional industries—textiles, tobacco, and hosiery—and their appointment filled southern leaders with apprehension. "Too much depends on the personnel and upon the administrator to be appointed," Birmingham columnist John Temple Graves asserted in a summary of southern opinion. In the wake of the Roosevelt administration's changed attitude toward the region, Graves reported, "there is fear that the South is not politically strong enough at Washington to be sure of a proper representation."[20]

Southern manufacturers, however, mobilized quickly to prevent any such under-representation of their interests. The cotton textile industry offered a case in point. During the 1930s, southern mill owners had been slow to organize opposition to the NRA and the FLSA, partially because their longstanding involvement in national trade associations muted the expression of explicitly regional interests. But under the aggressive leadership of Claudius Murchison, the Cotton-Textiles Institute orchestrated the employers' testimony before the Industry Committees and submitted a brief signed by ten smaller associations. At the same time, industry leaders squelched the voices of those few southern mill owners who favored increasing the minimum wage.[21]

Once mobilized, southern manufacturing groups challenged the basic structure of the Industry Committees. Section 5(b) of the FLSA required that the composition of the committees reflect the geographic distribution of the industry. Although ten of the textiles committee's twenty-one members represented the region, southern employers remained unsatisfied. "While seventy-five to eighty percent of the employees *who would be directly affected by a wage order* is [sic] located in the South," the textiles committee's southern minority complained, the majority of the committee as a whole and the overwhelming majority of the labor members hailed from other sections (emphasis in original). Furthermore, southern operators argued that it was unrepresentative to draw every employee representative from the unions, since the committee was investigating a largely unorganized labor force. Regional interests offered similar opposition to the hosiery and lumber committees. They launched a massive letter-writing campaign to lobby the administrator to reject the committees' recommendations. Administrator Andrews denied the appeals and upheld the wage orders.[22]

For southern businessmen, the composition of the Industry Committees confirmed their deepest fears. Southern manufacturers saw the minimum wage as a kind of wolf in sheep's clothing—an assault on their region's comparative advantage disguised as humanitarian and economic reform. "Is there any constitutional authority," the minority of the textiles committee demanded, "under which Congress or an Administrative agency operating under a law passed by Congress, may penalize one section for the continued economic aggrandizement of another?"[23]

Their objections were not unfounded. As we have seen, northern branches of some prominent southern industries such as textiles favored the national minimum, and organized labor certainly pursued its own interests. Even the administrator of the FLSA asserted that "One of the declared objectives" of the act "was to bring to an end this migration of plants solely to obtain a source of cheap labor."[24]

By the time the Industry Committees convened, southern trade associations recognized that it was too late to escape regulation altogether, so they lobbied for regional differentials in minimum wage standards. Accordingly, management took pains to demonstrate the act's discriminatory effects on the region. Before the railroad committee, southern carriers showed that the impending 30-cent minimum led to wage increases seven times greater in the South than in the rest of the country. Seamless hosiery industry representatives presented evidence that the proposed standards would affect three times the proportion of southern as northern workers and require a six times greater average wage increase.[25]

Southern manufacturers also rested their demand for preferential treatment on the alleged lower living costs in the South and the peculiar characteristics of southern industry. Operators of mill villages, for instance, claimed that the benefits they provided their workers compensated for lower pay scales. They maintained that the minimum wage law would upset the cherished bonds of obligation between southern workers and their bosses. One witness even cited the "factor of obsolescence" as a rationale for special treatment. Most facilities in the region were old, Burt C. Blanton of the Southwest Textiles Manufacturers explained. "They can't compete with new mills with new machinery."[26]

Blanton's appeal raised the most contentious of the southern industrialists' arguments—the assertion that increased minimum wages would force mechanization and cause massive technological unemployment. Low-wage rail carriers, including all southern railroads, were labor-intensive. They operated less than $250 worth of machinery per mile of track. Northern high-wage lines, on the other hand, operated $350 worth of equipment per mile. Southern carriers claimed that a higher federal minimum would require a complete rearrangement of their labor force. The ensuing layoffs would not be limited to workers currently earning wages below the minimum; all classes of labor would suffer. For the textile industry, Claudius Murchison offered an even more dire prognosis. A national minimum of 35 cents, he declared, would place a premium on machinery and lead to a one-third cutback in employment within five years.[27]

Southern trade associations met strong opposition in their effort to preserve the low wage southern economy. To be sure, most northern business groups supported their protests against the Industry Committees. Still, southern interests clashed with some northern employers, principally the New England textiles industry, organized labor, and the southern liberals in Washington. They also combated a more formidable rival—the prevailing notion that labor-intensive production was inherently inefficient and backward. In so doing, southern industrialists challenged a deep-rooted American tendency to equate economic progress with technological advance and capital intensity. Southern manufacturers explained their low wage, labor-intensive methods as a rational allocation of resources and a generator of employment opportunities. Their opponents, however, charged that low wages merely protected inefficient management. "I do not want to be in a position of defending obsolete plants," declared George Taylor, the chairman of the hosiery committee and vice chairman of the textiles committee, "and having a wage rate so low it would have a little tariff wall around itself to keep those firms in business."[28]

Organized labor stressed this presumed connection between low wages and inefficiency. In one hearing, a labor representative pounced on a southern manufacturer who pleaded that a low wage mill was not necessarily a low cost mill. "Then a low wage mill," he chided, "means protection of inefficient management, does it not." The Textile Workers Organizing Committee went so far as to describe a 40-cent federal minimum as the only path to modernized southern industry: "Concerns which have not kept up to the times will be brought up to current technological standards by such demands."[29]

Union representatives also dismissed warnings of technological unemployment. They advocated mechanization and skill upgrading in southern factories. They echoed the Roosevelt administration, claiming that better-paying jobs for male breadwinners would replace the South's many unskilled positions at substandard wages. Labor leaders, after all, realized that skilled workers in automated plants offered more fertile ground for organization than unskilled southerners just off the land. The textile unions, for instance, had learned that unskilled southern millhands looked on unions with suspicion.[30]

The federal government shared organized labor's views. In the late 1930s, the Department of Labor was committed to erecting national labor standards. Charged with administering the FLSA, the Labor Department quite naturally supported strict enforcement of the act. At Industry Committees hearings, witnesses from the Bureau of Labor Statistics testified that the South most desperately required the benefits of higher wages and that the federal government had no desire to save low wage firms. "Congress did not intend that a committee's recommendation be disapproved," Administrator Andrews announced in 1939, "because sub-marginal plants are unable to maintain modern standards of efficiency and consequently may not survive the increased minimum."[31]

The Labor Department implemented the economic policies which southern liberals and senior federal officials had enunciated since 1937. "I am a firm believer in adequate pay for all labor," the President declared the year before the FLSA took effect. "But right now I am most greatly concerned in increasing the pay of the lowest-paid labor." Industry Committees wage orders reflected that priority. In every case involving a major southern industry, including textiles, lumber, hosiery, and tobacco, the administrator approved general increases in minimum wage levels without geographic differentials. A year after the FLSA took effect, Roosevelt in commenting on a southern mill with cheap labor and outmoded equipment, emphatically asserted that "that type of factory ought not to be in existence."[32]

Federal wage policy shoved those plants toward oblivion. But what of the laborers who manned those assembly lines? As policymakers desired, children and women could leave the labor market. Unskilled males, however, lost a traditional source of employment. Many lacked the education or training for other positions. Blacks, deprived of education and of skilled positions even when they were qualified, bore the brunt of these dislocations. As the nation's economy mobilized for war, many unskilled southerners headed north.

THE WAGES OF WAR

In fact, war mobilization soon made the Fair Labor Standards Act obsolete. The war expanded manufacturing employment in the South by fully 50 percent. Average annual wages rose 40 percent between 1939 and 1942. All of a sudden, the competition for labor grew so fierce that most industries reached the 40-cent minimum long before the FLSA's target date. The wage hikes derived partly from government directives, but more frequently from voluntary decisions.[33]

Few observers realized that the defense boom might have averted an economic catastrophe for southern industry and a political disaster for the architects of the FLSA. Southern New Dealers expected the minimum wage, a measure formulated in the midst of economic depression, to eliminate "the sweatshop"—a major sector of the regional economy—without causing massive structural unemployment. The law, they believed, would force southern manufacturers to modernize; it would stimulate mechanization, upgrade the skill level and wages of workers, ban child labor, and even promote unionization. But despite their hopes for large-scale federal investment in the region's economic infrastructure and educational facilities, southern liberals never considered who would pay for the restructuring of southern industry. Nor did they ponder the effects of such modernization on the region's competitive position.

Reformers assumed that modern industry would prove more efficient than "sweatshops," but the South, with its shortage of skilled labor and lack of managerial experience with capital-intensive plants, might not have survived the transition. Eliminating the sweatshop could only succeed if the federal government built up other sectors of the southern economy. The defense program, as Chapter 4 details, accomplished just that. But the historian cannot help but wonder about the consequences of the FLSA had war not intervened and had the Depression continued. Analyses by contemporary economists suggest that the law would only have exacerbated broader economic forces already favoring capital-intensive industries that needed relatively few workers. It would further dry up employment in labor-intensive lines, throwing thousands more southerners out of work. But that never took place. As the nation pulled out of the Depression in 1940, the Wage and Hour Administrator could testify without hesitation that "the South is better off because of the enactment of the wage-and-hour law. The enactment and enforcement of that law has raised the standard of living in the South."[34]

The Industry Committees eased the transition to a war footing by providing continuity in labor policymaking. George Taylor, vice chairman and later chairman of the National War Labor Board, had chaired two of the Industry Committees. Donald Nelson, the chairman of Industry Committee No. 1, presided over the War Production Board. Frank P. Graham, who had supervised the Conference on the Economic Conditions of the South, chaired the railroad carriers committee. Graham served as a public member of the National War Labor Board and the primary liaison between Washington and the two southern regional panels. Labor and management similarly drew on Industry Committee experience to select their representatives to the war agencies.[35]

This continuity in personnel allowed the federal government to escalate its assault on the low wage sectors of the southern economy. Policymakers who entered government service to shape regional economic development maintained that concern during the war. The Roosevelt administration consistently excepted the South from its efforts to "hold the line" on economic conditions. Federal policy continued to "correct substandards of living"; it established uniform national regulations which eroded the special position of many southern firms. The national government also prohibited racial wage differentials, attacked the regional differential, and strengthened the position of organized labor in the region.

The National War Labor Board (NWLB) implemented these policies. Established by Executive Order to settle all wartime disputes between labor and industry, the NWLB retained the tripartite structure of the Industry Committees. By October 1942 the Board was authorized to determine compensation even where no formal disputes existed. "Wage stabilization," as the agency termed this function, so swelled the caseload that the national board was forced to establish twelve regional boards in the autumn of 1942. The regional boards were expected to clear away the backlog and to attune decisions to the peculiar economic conditions of the different sections.[36]

Within the regional offices, southern industry representatives led the opposition to NWLB policy. Southern business associations drew on the organizational resources they had mobilized to petition the Industry Committees. They posted lobbying groups at the regional offices to an extent unmatched outside the South. At the Region IV headquarters in Atlanta, business interests established an Industry Advisory Council "as an adjunct to the Industry group on the Board, to service companies having voluntary or dispute cases before the Board." With an annual budget of $50,000, contributed by trade associations, the Industry Advisory Council maintained offices in the same building as the regional board. No such formal apparatus developed in the Southwest, but the Texas State Manufacturers Association raised $30,000 a year to support its watchdogs at the Dallas board. It also joined on several occasions with the Louisiana State Manufacturers Association and the Associated Industries of Oklahoma to protest NWLB policy.[37]

The militance of industry groups contrasted sharply with the moderation of the labor members of the southern boards. Addressing a convention of southern AFL representatives in 1943, one Roosevelt administration official thanked the labor leaders for their "patriotism and splendid cooperation." Organized labor's moderation in the South, however, reflected more than unselfish patriotism. While union officials in Washington peppered the national office with protests, their colleagues on the Atlanta and Dallas panels sought to reduce conflict. They realized that the South remained, in the words of AFL representative George Googe, "a frontier in the matter of Labor Relations and Labor Organization." When asked to explain labor's preference for negotiation over dispute cases, Googe replied, "Some day the war will be over. Your Board will pass out of the picture and we will still be here. We want to show employers on every possible occasion that labor can make agreements with them and does actually make them." Southern labor leaders cooperated with the boards because they believed that the tide of national policy ran their way.[38]

Southern industry, on the other hand, found palpable threats in NWLB orders. Most menacing was the substandards policy, the mainspring of NWLB activity in the South. The same principle that had animated the *Report on the Economic Conditions of the South* and the Fair Labor Standards Act motivated the NWLB's injunction to correct substandards of living: the South was impoverished and low wages sustained regional underdevelopment. "Substandard wages are a handicap to all of us in the South," Frank Graham reiterated during a visit to Dallas. "Low wages in a region make that region an exploited colony of a high wage region." In one of its earliest decisions, the NWLB declared that the South must pay the "American wage of health and decency," despite its peculiar economic conditions. "A standard of living is a matter of human decency and not a matter of local or regional capacity to reach that standard," Graham instructed the Atlanta board. "We should not let the incapacity of a region lower our evaluation of human decency."[39]

The regional boards applied the substandards policy under a continuously evolving set of guidelines. The "Little Steel formula" laid the foundation for wartime labor policy. It forebade employers from raising straight time hourly earnings in their plants by more than 15 percent above January 1941 levels. Initially, the NWLB recognized three exceptions to the strict requirement of the Little Steel formula: those necessary for the efficient prosecution of the war, to correct inequalities and gross inequities, and to eliminate substandards of living. On April 8, 1943, FDR tightened the restrictions. He issued Executive Order 9328, the "Hold the Line Order," and narrowed the rules regarding inequalities and inequities. Thereafter, only the substandards principle could justify increases beyond the Little Steel formula or the going rates in a region.[40]

To implement the substandards policy, the NWLB adopted General Order 30 in February 1943. The order granted automatic approval to wage increases up to 40 cents an hour. Over the course of the war, the NWLB amended General Order 30 several times, raising the substandards level to 55 cents by V-E Day. The National Wage Stabilization Board, the successor agency to the NWLB, soon boosted the figure to 65 cents. According to the NWLB's *Termination Report,* the upshot of these measures in the South "was to alleviate the harsh effects of the Little Steel Formula and the hold the line order." Region IV chairman M. T. Van Hecke concluded that "the successive waves of Substandards and General Order 30 adjustments raised the common labor rate in this Region from less than 40 to 55 cents an hour."[41]

The effects of those directives were concentrated in the South. The substandards policy, little noted above the Mason-Dixon line, dominated WLB activity in the South. The policy not only drove up the basic wage for unskilled labor it increased wages generally across the region. All NWLB directives demanded that employers maintain occupational pay differentials when they eliminated substandard wages. In a few cases, notably in the textile, furniture, fertilizer, and lumber industries, the southern panels ordered across-the-board increases for all employees. Most decisions, however, "tapered" raises above the minimum, so that the higher a worker's starting pay rate, the smaller the mandatory wage increase.[42]

These substandards awards infuriated the representatives of southern industry, especially in the Southeast. Industry Advisory Council officials complained that the southern boards' decisions "approved increases without regard for local conditions or destabilizing effects." They noted that northern regional panels rarely considered substandards, deciding most cases on the basis of "sound and tested going rates" in order to control inflation. In 1944, two industry members of the Atlanta Regional War Labor Board threatened to resign because southern wages had climbed so much.[43]

Other NWLB policies, besides substandards, also sought to establish a rudimentary American standard of living in the South. Under its mandate to determine "sound and tested going rates," the NWLB extended national standards into southern industry. Washington authorities instructed the regional boards that "historical differentials . . . should be ignored." All employers must pay a "sound stabilized rate" and the going rates of many southern firms could not qualify as sound and tested on an up-to-date, national basis.[44]

Such directives from Washington proved to be the bane of southern employer representatives in the regional offices. Even when they could restrict reform—for instance, by tying wage adjustments to the region's "going rates"—representatives of southern industry found themselves frustrated by such "outside interference." Southern industry's disputes with Washington focused on two issues. First, the regional boards, especially the Dallas panel, demanded the discretion to set labor standards without review by the NWLB. The Dallas board objected to the NWLB's revision of some of its decisions. Second, the regional board opposed the creation of Industry Commissions that set prevailing rates on a national basis. The Dallas board rejected Washington's arguments for "a national approach to wage problems" in the iron and steel, petroleum, transit, lumber, and rig-building industries. In the rig-building case, the Washington board's assumption of jurisdiction had resulted in an immediate 25 percent wage increase in the southern region.[45]

David B. Harris, an employer representative on the Dallas board, most vociferously opposed the formation of national panels. He labeled them "a serious threat to industry of every kind and character." In Harris's mind, the national commissions were no more than a "maneuver" of organized labor whose "main objective is to have wage rates and working conditions nationalized." The movement, Harris concluded, was "best expressed in the activities of the National Steel Panel. It is an exceedingly dangerous movement."[46]

Harris's appeal fell on deaf ears. The enforcement of national labor standards did not threaten "industry of every kind and character." It menaced only southern industry. Northern manufacturers had no objection to removing the South's competitive advantage. Labor groups freely admitted to the "maneuver" which Harris condemned. The federal government, eager to ameliorate southern poverty, spark industrial modernization, and forestall future depressions, viewed the recomposition of the southern workforce favorably.

National policy further isolated southern industry by condemning racial wage differentials. Speaking for a unanimous NWLB, Frank P. Graham's opinion in the Southport Petroleum Company case abolished the classifications "colored laborer" and "white laborer" and reclassified "both simply as 'laborers' with the same rates of pay." The board ordered immediate wage increases for black workers to bring them parity with whites in the same occupations. Two weeks after the Southport judgment, NWLB chairman William H. Davis assured the Mexican Ambassador to the United States that the ruling applied equally to workers of Mexican ancestry.[47]

The NWLB amplified the ban on racial wage structures in the Miami Copper case. In that dispute, it not only condemned the payment of lower wages to blacks in the same occupations as whites, it also prohibited the discriminatory classification of non-whites in lower paying jobs when they performed the same tasks as whites in higher paying grades. In this 1944 case, the Non-Ferrous Metals Commission determined that less than one percent of the Miami Company's 2,319 "Anglo-American" male employees received less than $6.36 per shift, while more than a third of the firm's 902 "other employees" (Latin Americans, Negroes, Filipinos, Indians, and Anglo-American females) earned less than that rate. At the same time, only six of the 134 workers designated as "laborers" were Anglo-American. The Commission discovered no relation between wages and education, skill, length of service, or citizenship status. It concluded that all wages below $6.36 represented racial discrimination. The board in Washington upheld the Commission's decision. "We are convinced that the multitude of job titles and job rates below the $6.36 level lends itself to discrimination," a majority of the NWLB held, "and certainly to the suspicion of discrimination."[48]

The Miami and Southport cases marked a watershed in federal labor policy. True, the National Recovery Administration had discouraged racial wage differentials and the Roosevelt administration had established the Fair Employment Practices Committee to root out employment discrimination in war industries. But while these measures partially placated black leaders, they offered little to the black worker on the shop floor. The NWLB rulings, on the other hand, actually ordered pay parity for blacks in defense industries. From his office in the White House, Jonathan Daniels, FDR's wartime adviser on southern and racial issues, congratulated his fellow Tarheel Frank Graham on the Southport case. "It seems to me that we can never hope for any general prosperity in the South as long as we insist on poverty wages for Negro workers," Daniels concluded, "and therefore, I feel that your decision is important for the future of the South."[49]

For Daniels, the Southport and Miami decisions benefited the South not so much because they struck blows against racial discrimination as because they eliminated "poverty wages." Daniels and other southern liberals had long viewed racial discrimination as a byproduct of southern poverty. Thus, Daniels welcomed these decisions as an attack not on racism and segregation, but on low paid employment—a way to reduce the regional wage differential.

Daniels and his fellows could only be cheered by wartime progress on those lines. War policy had eroded the regional differential as a matter of course. Only substandard wages increased more than the 15 percent allowed by the Little Steel formula and substandards cases, of course, mainly involved the South. Beyond that, NWLB also issued explicit directives against geographic differentials. In the 1942 Aluminum Company of America case, the NWLB ordered special wage increases at Alcoa's southern plants in addition to a company-wide adjustment. The Board made this ruling because it was "impressed with the fact that the Aluminum Company of America stands in a very good position . . . to set an example and take the lead in the narrowing of the north-south wage differential." In a series of decisions, the NWLB gradually intensified its assault on the wage position of southern industry and rejected geographic differences in basic pay rates.[50]

The erosion of the regional wage differential cheered organized labor. At a 1943 convention of southern AFL leaders, one speaker concluded that "The New South is on the march." Unions were translating the war emergency into political and economic strength in the South. But in this frontier of unionism, labor owed its increased effectiveness entirely to the NWLB. The "filial-dependent relationship" between organized labor and the federal government might have compromised the independence and tempered the militance of unions in the nation's industrial heartland. But, in the hostile environment of the South, such nurture was essential.[51]

The basic policy of the NWLB encouraged labor and management to enter into collective bargaining agreements rather than appeal to the Board. "The struggle for freedom to organize," according to Frank Graham, defined the nation's objective in the war. Hitler had demolished democracy by destroying the parliament, the corporation, and the labor union. In Dallas, Graham defended the federal government's support of unionization in terms familiar to his southern audience: "As the devoted son of a Confederate soldier, I am reminded that we Southerners, by a more forceful means than a directive order of the War Labor Board, were compelled in the great struggle over sovereignty of the states to maintain our membership in the Union as a condition of existence. We were not given any fifteen day escape clause by which we could get out and stay out."[52]

Graham alluded to the basic union security guarantee of the NWLB. "Maintenance of membership," the standard NWLB formula, bound management to require union membership as a condition of employment for all workers who were union members at the end of a fifteen-day "escape period" at the commencement of a contract. In return, the union pledged not to coerce other employees to join during the life of the contract, a symbolic concession since such intimidation was impossible, if not illegal, in most parts of the South. The policy's practical effect was to enhance the power of organized labor by forcing employers to accept unions as the bargaining representatives of their workforce. Such legitimation was particularly crucial in the South. Organized labor remained weak below the Mason-Dixon line, and antipathy to labor organization was strong. Despite the opposition of southern and national business groups, maintenance of membership became a common feature of WLB directive orders. The Atlanta regional board, for example, granted the guarantee in 67 of 70 cases during its first year on the job. Eventually, many southern employers voluntarily entered into collective bargaining agreements rather than wait for the board to define the terms. "In this region," Atlanta board chairman Van Hecke concluded, "the Board's operations have made constructive contributions to the technique and content of the collective bargaining process."[53]

The improvements in the "collective bargaining environment," however, could not guarantee compliance with national labor policy. While the most celebrated cases involved northern and western companies, noncompliance was especially prevalent in the South and grew more frequent as the war drew to a close. Two large firms in Houston—the Hughes Tool Company and the Mosher Steel Company—refused to honor maintenance of membership orders in their contracts with the United Steel Workers. In response to them and other intransigent employers, a rash of strikes plagued the Southwest in the summer of 1944. The Dallas board attributed this labor militance to "the growing impatience of labor unions, their members and the workers in general" with the "refusal of the companies involved to comply with the directive order, even in the face of the rejection of their appeal by the National Board." Enforcement proved no easier in the Southeast. In February 1945, the Atlanta board found non-compliance so epidemic that it ceased to order striking employees back to their jobs. At that time 36 employers were disregarding Board decisions. "If there were 36 uncontrollable strikes," an Atlanta board press release noted, "we would see a tremendous clamor of public opinion."[54]

Southern state governments also resisted the incursions of organized labor. During the course of the war, five southern states enacted anti-labor laws. Florida and Arkansas took even more drastic action, embedding antiunion policy into their state constitutions. This anti-labor legislation established the precedent for the state "right-to-work" laws that would follow the enactment of the Taft-Hartley Act in 1947. But as long as war raged overseas, the federal authorities countermanded those statutes. The NWLB denied Alabama's claim that its Bradford Act forbade maintenance of membership rules and ignored Florida and Arkansas' "right-to-work" amendments.[55]

Federal protection proved decisive in organizing the hostile bastions of southern industry. During the war, unions enlisted the workers at Alcoa, Du Pont, and nearly all the mass transportation systems in the region. Unions that had failed to win contracts before Pearl Harbor found the war emergency an entrée into the region. So weak was the Oil Workers International Union in 1940, for example, that it could not afford even a single full-time organizer in the southern oil fields. With WLB support, however, the Oil Workers Union concluded its first major contracts in south Texas in 1942. By V-J Day, the union represented the employees of most oil companies in Texas and Louisiana. And Humble Oil, one of the few unorganized exceptions, closed down its Ingleside, Texas, refinery rather than obey a maintenance of membership order. The United Steel Workers and

United Rubber Workers had similar experiences. During the 1930s, the steel union achieved little success in the Birmingham District, while the rubber workers' early efforts in the South met violent opposition. After the NWLB assumed jurisdiction over those industries, the unions won collective bargaining agreements and established locals throughout the South.[56]

Wartime organizing drives fared worse in the textile industry. While membership doubled in the war years, unions represented only 20 percent of the southern workforce in 1946. At that date, organized labor counted more than 70 percent of northern millhands among its ranks. Inter-union rivalry between the AFL's United Textile Workers and the stronger CIO organization bore some responsibility for the failure, as did the lingering legacy of the failed 1934 General Textile strike. That strike, undertaken before the Wagner Act and the growth of the CIO, had been a fiasco for labor. All union supporters in the industry lost their jobs and other workers became forever suspicious of unions. Employer resistance—many southern mills refused to comply with NWLB union security directives—and high labor turnover also blocked unionization.[57]

Despite their difficulties in the traditional, low wage southern industries, union leaders believed that their wartime alliance with the federal government had laid a foundation for postwar union growth in the South. Ambitious labor PACs expanded their activities in the region; CIO-backed candidates won statewide office in Florida, Alabama, and South Carolina. Meanwhile, both the AFL and CIO also launched major southern organizing drives. The CIO's "Operation Dixie" enrolled about 150,000 new members in the two years after V-J Day. The AFL's program, a self-proclaimed "operation for Southerners by Southerners," claimed 500,000 new members.[58]

Southern unions certainly emerged from World War II much stronger than they had entered it. They nearly doubled the proportion of the southern workforce carrying union cards. Those successes, however, proved neither long-lasting nor widespread. In the South, unions depended on the federal government and that support waned after V-J Day. Beset by financial problems and unable to crack many of the region's major industries, the AFL discontinued its organizing drive in 1947. The CIO maintained its Southern Organizing Committee until 1953, but, after its initial successes, that body fought merely to retain existing membership.[59]

Anti-union sentiment ran deep in the South and that hostility manifested itself in the frustration of Operation Dixie and in subsequent declines in union membership. Without federal supervision, enrollments in CIO affiliates dropped by about 50,000 between 1945 and 1953. The more established American Federation of Labor lost nearly 700,000 members. Southern workers increasingly rejected organized labor; the "no union" alternative won in 35 percent of certification elections in the South in 1950. In 1946 it had prevailed in only 20 percent of those contests.[60]

After V-J Day, a national political backlash against unions weakened the alliance between organized labor and the federal government and, in turn, hindered organization in the South. Wartime gains, the intransigence of John L. Lewis, and a wave of strikes during the reconversion period inspired much anti-labor sentiment in the nation. This national hostility toward unions manifested itself in the election of the Eightieth Congress in 1946, the first national legislature with a Republican majority since 1930. Its principal accomplishment was the enactment, over President Harry S Truman's veto, of the Taft-Hartley Labor-Management Relations Act in 1947. This measure won the support of most southern representatives in Congress. Only three southern senators voted with their northern Democratic colleagues against Taft-Hartley. In the House, more than three-quarters of the region's congressmen approved the bill.[61]

With few exceptions, the South's political and economic leaders opposed organized labor. Taft-Hartley enabled them to confine unions to their northern strongholds. The law prohibited secondary boycotts—actions by allied unions in support of their striking brethren. It thus prevented unions from exploiting their strength outside the South in order to aid southern locals. Taft-Hartley also empowered individual states to outlaw the union shop, which crippled organizing campaigns in chemicals, textiles, and furniture. And the measure eventually led to a series of state "right to work" laws. By the end of the war decade, southern business leaders and their political allies had reasserted their dominance over labor-management relations. Some liberals likened that development to the return of the "Redeemers" after Reconstruction. In 1950, Florida senator Claude Pepper, the champion of southern labor whose election in 1938 had paved the way for the Fair Labor Standards Act, met defeat at the polls.[62]

ON THE MOVE: WAR AND SOUTHERN INDUSTRIAL WORKERS

Nevertheless, the war introduced changes in the southern economy that buoyed the hopes of union leaders. The defense boom lifted the South out of the Depression; as *Fortune* magazine noted in 1943, "for the first time since the War Between the States, almost any native of the Deep South who wants a job can get one." And federal policy assured those workers who remained in the region not merely a job, but the "useful and remunerative job" which President Roosevelt had demanded. But for those southerners who had filled the thousands of low wage, unskilled positions no provision was made. Some of those workers, mainly white men, moved into semi-skilled and skilled positions. Others left the labor force.[63]

Many such laborers left their homes or departed the region altogether. So many southerners took to the roads that local officials could not keep track of the migrants passing through their towns. To find these workers on the move, observers had to search in the juke joints of the roadside South. "There is a big population with no telephone number," one journalist explained. "The townspeople do not see them. You can go into Jacksonville and they are not aware of how many hundreds and thousands of people are camped around there."[64]

The recomposition of the labor force determined the migration paths of these "juke joint people." Between 1940 and 1945, three times as many people flooded out of the South each year as had during the previous decade. From Pearl Harbor to V-E Day, 1.6 million southern civilians left the land of Dixie, while another 4.8 million migrated within the region. The unskilled and unpropertied constituted the principal emigrants from southern fields and mill villages; farm owners and cash tenants stayed on the land. Sharecroppers and wage laborers, responding to the pressures of the AAA program and the call of war labor, departed for the cities and the North.[65]

The migration of the South's most downtrodden revealed a peculiar feature of the wartime exodus—its educational and racial selectivity. Most migrants came from the bottom of the educational ladder. Among southern whites, those with one to four years of schooling emigrated at seven times the rate of the better educated. The same was not true of blacks, among whom the high school educated migrated at the highest rate, but blacks at all points on the educational spectrum departed the region—and most black southerners were poorly educated. Many well-educated southern whites also left the region. They were replaced, however, by a stream of incoming white college graduates filling the region's new skilled positions.[66]

In fact, net out-migration from the South became an increasingly black phenomenon during the 1940s. The war accelerated the longstanding black exodus toward the economic opportunities of the North. But black out-migration represented more than an effort to enhance income. In 1940 and 1941 the defense boom soaked up white unemployment, leaving many black southerners without jobs and without relief, because public works expenditures were reduced. "When the period of defense preparation began in 1940," the U.S. Fair Employment Practices Committee (FEPC) concluded, "local white labor was absorbed and outside white workers were imported into centers of expanding activity, but the local Negro labor supply was not utilized to any appreciable degree." Government training programs in the South, designed to teach skills necessary for work in aircraft, shipbuilding, machine shops, electronics, and other war industries, enrolled far fewer blacks than their population share would justify, far fewer even than federal officials had expected. Southern offices of the U.S. Employment Service remained under state control and often refused to refer blacks to skilled positions. In one case, a southern bureau of the U.S. Employment Service placed black welders, welders recently trained at government expense, with employers outside the region, despite an acute local shortage of welders. Meanwhile, "during the construction of a camp near Petersburg, Va., hundreds of available Virginia Negro carpenters were barred from employment on this project while thousands of white carpenters from all parts of the country were imported to the site for employment." Discrimination was not restricted to skilled positions. The U.S. Bureau of Employment Security reported that even unskilled jobs in defense industries were closed to black applicants.[67]

This discrimination angered black rights organizations and they pressured President Roosevelt into issuing Executive Order 8802 in June 1941. With that order, FDR declared that "there shall be no discrimination in the employment of workers in defense industries or government because of race, creed, color, or national origin" and established the FEPC to investigate complaints and redress grievances. The FEPC, however, lacked enforcement powers of its own; it had to rely on the armed services, the Maritime Commission, and the War Manpower Commission to implement its recommendations. And the FEPC's jurisdiction was restricted to industries under government contract and government agencies concerned with job training. Already hamstrung by the Roosevelt administration, the FEPC met strong opposition when it ventured into the South. Before the opening of the FEPC office in Atlanta, the Atlanta City Council passed a resolution calling for its removal, and southern members of Congress pressured the Roosevelt administration to further curtail FEPC powers. While it scored an occasional success, its efforts to meliorate racial discrimination in southern factories proved a dismal failure.[68]

Not surprisingly, then, as war mobilization proceeded, whites claimed most new jobs in southern industry—positions at higher minimum wages and skill classifications. The wartime boom, and the simultaneous elimination of many low wage jobs, continued the South's heritage of *Herrenvolk* development. Whatever the intentions of federal policymakers and whatever the opportunities for blacks outside the region, the wartime growth of southern industry mainly benefited whites. This pattern persisted through the reconversion. In 1950, blacks received only 21 percent of new nonagricultural jobs in the Deep South, although they accounted for 43 percent of the population.[69]

Wartime migration, economic expansion, and federal policy left behind a southern labor force less black, less tied to the land, more skilled, and disproportionately male. From 1940 to 1950, agriculture's share of regional employment fell from more than a third to less than a quarter (Table 3-1). Women's participation in the labor force increased in the South as it did throughout the nation, but more slowly so that the southern rate became the lowest in the nation. The relative decline of women's work was all the more dramatic in that is occurred during a period of extensive urban growth. Southern urbanization had long been associated with higher female labor force participation rates (Table 3-2).[70]

Such changes bore the imprimatur of the FLSA and the NWLB. In place of many low-paying, unskilled positions, the region developed high quality jobs at the federal minimum, jobs that met the Administration's

TABLE 3-1
The Southern Labor Force, 1940–1950
Percentage of Southern Labor Force (%)

	1940	*1950*
Agricultural	34.9	23.8
Non-Agricultural	65.1	76.2
Male	76.5	73.3
Female	23.5	26.6
White	72.6	79.0
Non-white	27.4	21.0
White Male	58.0	59.3
White Female	14.5	19.7
Non-white Male	18.5	14.1
Non-white Female	8.9	6.9

Sources: U.S. Bureau of Labor Statistics, *Bulletin No. 898,* (Washington, 1947), 22. U.S. Bureau of the Census, *Statistical Abstract of the United States, 1952,* pp. 185–87. Calvin B. Hoover and B. U. Ratchford, *Economic Resources and Policies of the South,* (N.Y.: Macmillan, 1951).

TABLE 3-2
Women's Labor Force Participation Rates*
Percentage of Women in the Paid Labor Force, 1940–1950 (%)

	South	*North*	*West*
1940	24.2	26.2	23.8
1950	27.2	29.2	29.6

*Urban women were more likely to find employment in the paid labor force than rural farm women. In 1940 the labor force participation rates for urban women were 35.1 percent (South), 30.5 percent (North), and 28.4 percent (West). For rural farm women, 13.2 percent (South), 10.7 percent (North), and 12.0 percent (West). See Bureau of Labor Statistics, *Bulletin No. 898,* p. 26.

Sources: Bureau of Labor Statistics, *Bulletin No. 898,* p. 26. Bureau of the Census, *Statistical Abstract of the United States, 1952,* p. 185.

idea of remunerative employment at the minimal American standard. Employed white males could earn a living wage under the new industrial regime. But laborers on the margin—blacks, women, the unskilled—faced dim prospects in the postwar South. The drop in black employment in the South and the emergence of black teenage unemployment across the nation testified to the harsh side-effects of reform.[71]

"Full employment means not only jobs," President Roosevelt explained in his last State of the Union message, "but productive jobs. . . . Americans do not regard jobs that pay substandard wages as productive jobs." As early as 1936, F.D.R. had recognized two paths to southern industrialization, protecting the region's favored low wage position or promoting modernized facilities and replacing marginal workers with "productive jobs." Roosevelt and other federal policymakers chose the latter course. They determined to remake the South in the image of the nation.[72]

POSTWAR WAGE POLICY

Substandard wages denoted those below the "American standard of health and decency." That standard eluded most southerners. Still, the war had closed the economic gap between regions. Southern per capita income improved from a pitiful 59 percent of the national average in 1940 to 69 percent at the close of hostilities. Meanwhile, under NWLB strictures, southern wage rates rose more precipitously than those outside the region. Federal policy all but eradicated the geographic differentials in the traditional southern industries that had previously relied on abundant cheap labor.[73]

The erosion of southern wage differentials renewed the controversy that had raged during the NRA and FLSA debates. John V. Van Sickle of Vanderbilt University articulated the orthodox southern position. The defect in federal labor policy, Van Sickle asserted, "was not the principle of the minimum but the failure to provide for reasonable regional differentials." Van Sickle attributed that failure to the zeal of "Capital 'R' Reformers," in whose hands "regional wage differentials were simply declared to be undesirable and to be eliminated as rapidly as possible." Led by the lumber industry, low-wage southern manufacturers renewed their fight for regional wage differences and against increases in the national minimum wage. Southern pine producers also sought exemptions for the FLSA for small sawmills.[74]

As V-J Day approached, however, the Truman administration recommitted the federal executive to the elimination of southern low wage employment. Chester Bowles, head of the Office of Price Administration, linked higher minimum wages to productivity growth and economic expansion. NWLB chairman George Taylor pleaded with Congress and the White House that the disbanding of the NWLB not result in the abandonment of the campaign against substandard wages. To the delight of Taylor and like-minded officials, Lewis B. Schwellenbach, Truman's appointee as Secretary of Labor, resolved to continue the fight.[75]

Early in reconversion planning, the Office of War Mobilization and Reconversion named "maintenance of purchasing power" and the "establishment and preservation of equity among the different groups of workers" as its objectives. To achieve that equity, the National Wage Stabilization Board (NWSB) and the southern regional Wage Stablilization Boards, peacetime successors to the regional War Labor Boards, processed a rash of substandards cases in the South. These rulings, especially in the textile and garment industries, elevated the national minimum and the regional common labor rate to the 65-cent level sought by liberal and labor groups.[76]

The point man for this Administration policy in the Congress was Florida senator Claude Pepper. He sponsored a 1945 resolution demanding a nationwide basic hourly wage of 65 cents. "A minimum standard must apply to the Nation," Pepper's subcommittee asserted. "Congressional policy cannot perpetuate lower living standards for any part of the nation or any group of people in it."[77]

Although the Administration supported Pepper, most of his colleagues in the legislative branch opposed such plans. In 1946, Republicans won control of Capitol Hill and installed anti-labor senators like Robert Taft and Joseph Ball at the helm of the Labor committees. Southern Democrats joined the Republican opposition to labor and minimum-wage legislation. According to the Labor Department's Congressional liaison, all 32 anti-labor House Democrats and 41 of 59 undecided "Wobbly Congressmen" hailed from the South. With those votes in their column, the Republicans defeated initiatives to amend the Fair Labor Standards Act in 1946 and again in 1947.[78]

In 1948 the Democrats recaptured the Congress. They soon increased the minimum wage, with the South remaining the focus of debate on the Fair Labor Standards Act of 1949. Low-wage southern industry petitioned for regional differentials. Claude Pepper and northern Democrats fought them down. Similar regional concerns persisted when Congress raised the national hourly minimum to one dollar in 1955. In that debate, Tyre Taylor, president of the Southern States Industrial Council, complained "that this proposal to raise the minimum wage is aimed primarily at the South."[79]

But those shrill appeals, embedded in general attacks on government economic intervention, found little support in the southern business community. Large interregional corporations and mechanized plants formed a larger segment of postwar southern industry than they had before Pearl Harbor. Those employers had little use for low minimum wage rates. And, by the early 1950s, even segments of the textile and apparel industries approved of higher minima. Representatives of the textile industry, a principal opponent of the original FLSA, endorsed the one dollar standard in 1954. The Southern Garment Manufacturers Association did likewise.[80]

Modernization produced that change of heart. Textiles remained a low wage industry in 1955, but it was no longer entirely composed of small independent mills. Large concerns, many with diversified holdings, dominated the industry. Unskilled labor, especially women workers, still formed a large part of the workforce, but southern mills employed more skilled workers and machinery than they had before 1940.[81]

Skill upgrading proceeded throughout the region. True, it moved slowly in traditional southern industries. But during the war decade, many new capital intensive industries with skilled labor requirements arrived in Dixie. This new skill mix partially accounted for the region's acquiescence to the minimum wage. Skilled workers earned well above the proposed minima and skilled southerners received wages much closer to the national average than their unskilled counterparts. Employers of skilled labor had little reason to oppose the legislation.[82]

When the war mobilization agencies closed shop at the end of 1946, NWSB chairman William Wirtz congratulated the architects of the labor program in the South. In a letter to the Atlanta board, Wirtz averred that "You folks had the hardest job of any, facing as you did, a prevailing low wage situation which of necessity had to be improved as part of a 'stabilization' program." Conflict over federal policy in the South stemmed not only from this combination of two functions but also from the explosion of economic activity in the region. In a letter to a colleague in Washington, Alto Cervin, a member of the Dallas WLB, focused on the difficulty besetting the authorities dealing with the wages of Dixie: "The question of establishing job rates for new establishments has become an acute problem in this region."[83]

Cervin underlined the foundation of southern economic development in the World War II era. In the wake of the FLSA and NWLB, low wage sectors of the South declined despite an overall leap in manufacturing employment. This feat was accomplished by a forced upgrading of the workforce in pay and skill along with shifts in labor intensity, mechanization, and industrial specialization. By 1950, the federal government's efforts to lift the South from its colonial status forced retrenchment on much of the traditional southern economy. But contrary to the predictions of southern doomsayers, federal policy did not leave the South in ruins. A new, mechanized industrial economy was displacing the old labor-intensive South, mitigating the dislocations of the attack on low wage employment by introducing much more manufacturing to the region. "A bird's eye view of large scale Southern industry," War Production Board chairman and former textiles committee chief Donald M. Nelson remarked at the war's end, "makes you feel that the South has rubbed Aladdin's lamp." But if the South rubbed Aladdin's lamp between 1938 and 1950, the genie that emerged was the federal government. The growth of southern industry was not the product of basic economic factors alone. Under the direction of Nelson and other government officials committed to developing the region, those years would witness the agricultural South in retreat and the industrial South emergent.[84]

Chapter 4 ◎ "Bulldozers on the Old Plantation"

Thomas Wallner was distressed. Surveying the South's tumultuous mobilization for war in early 1942, the new president of the Southern States Industrial Council warned that "This war cannot possibly be won . . . for one fundamental reason and only one." The national government, Wallner complained, "meaning primarily the President of the United States—still stubbornly persists in the attempt simultaneously to fight a foreign war and to wage an internal economic revolution." Whatever the federal government's responsibility for the turmoil, the signs of such internal economic revolution were omnipresent in the South of the 1940s. John Dos Passos visited the region a few months after Wallner's outburst and discovered "Bulldozers on the Old Plantation," the long awaited realization of southern efforts at industrialization. "In two weeks a back country settlement with its shacks and barns and outhouses and horsetroughs and fences," Dos Passos observed, "all the frail machinery of production built up over the years by the plans and hopes and failures of generations of country people will have vanished utterly." And instead "among the freshcut pine stakes of surveyors, you will see in the making the long runways of an airfield . . . or the white concrete and glass tile oblongs of a war factory."[1]

For decades, many southerners had longed for the sight of bulldozers on the old plantation; the first proponents of a New South had promoted industrialization as far back as the 1880s. But the antipathy of the region's elites to capital investment that might threaten their hold on labor and resources had frustrated those efforts. The manufacturers who had found homes in the South before 1938 operated in precisely the low wage, semifinished goods lines which federal policy sought to eliminate in the decade after the enactment of the Fair Labor Standards Act in 1938. In the minds of many policymakers in the 1930s, such businesses only blocked the industrial path to regional economic progress.[2]

The industrial spurt observed by Don Passos and Wallner represented a departure from the traditional pattern of "New South" development.

This was a partial break from the "colonial" pattern of earlier regional industrial growth which had limited the South to the production and rudimentary processing of raw materials. It embodied instead the type of regional development championed by the southern liberals and their allies in Washington. Howard Odum had himself assailed the "nostalgic yearnings" of those who desired "to turn back to the agrarian ideal." And as war settled over Europe, the Southern Conference for Human Welfare, the leading voice of southern liberals, pressed the oncoming defense emergency as an imperative for southern industrialization along the northern model. Familiar with the observations of contemporary economists associating economic progress with the shift from agricultural labor to employment in manufacturing, federal policymakers believed that the establishment of durable goods industries like automobiles and steel would lay the cornerstone of regional development. President Roosevelt himself concurred that advanced industry would deliver economic progress to the nation's poorest section. "No small part of our problem today," the President explained to an Arkansas audience, "is to bring the fruits of this mechanization and mass production to the people as a whole."[3]

The South required such industrial development because it remained predominantly agricultural. Agriculture, forestry, and fisheries engaged more than a third of the southern labor force in 1940, a share more than double that prevailing in the other sections of the nation. Moreover, the region's agrarian character exerted more influence than the size of its rural workforce alone suggested. The agrarian South, as Odum described it, constituted "in area almost the entire landscape; in people, three-fourths and more of the population, either in actuality or in interest and experience." The paucity of industry complemented the dominance of agriculture. Even though southern industry gained ground on its northern competitors during the Depression, the South remained the country's least industrialized region as the 1930s drew to a close. With 27 percent of the nation's population in 1939, the region contained only 17.3 percent of its wage earners in manufacturing and contributed a meager 12.7 percent of the nation's value added by manufacturing. That limited capacity was restricted to a few industries operating in a few southeastern states.[4]

Furthermore, low wage, labor-intensive industries continued to dominate the relatively small southern manufacturing sector. From Henry Grady in the nineteenth century to the 1939 "Made in the South" campaign of the Southern Governors' Conference, regional development efforts had attracted employers primarily from the low wage, non-durable goods lines. Textiles alone provided 30 percent of the region's manufacturing employment; along with the food and lumber industries, the cotton mills were responsible for half of the South's industrial employment and a third of its value added in 1939. At the same time, the region housed little of the metalworking lines so crucial to the attraction of high-grade industries like electrical machinery and transport equipment (Table 4-1). The pre-defense economy of the South, the National Resources Planning

Table 4-1
Southern Industry in 1939

	*% of Value Added in South**	*% of Mfg Employment in South**	*% of Mfg Employment in U.S.**	*Mfg Employment in South, as % of U.S. Total*
Iron and Steel prods.	5.4	5.1	11.6	7.2
Electrical Machinery	0.4	0.1	3.5	0.6
Machinery, except Electrical	2.7	1.8	6.8	4.2
Transport Equip. except Autos	1.2	1.1	1.9	9.3
Automobiles and Parts	1.1	0.5	4.6	1.8
Non-Ferrous Metals	1.4	1.0	2.8	5.7
Lumber and Timber prods.	7.1	13.2	4.6	47.2
Furniture	4.2	5.9	3.8	25.2
Stone, Clay and Glass prods.	3.6	3.1	3.5	14.8
Textile Mill prods.	20.3	30.7	12.3	41.2
Apparel and related prods.	3.4	5.5	8.9	10.1
Food and Kindred prods.	15.3	12.3	11.8	17.1
Tobacco Manufactures	7.9	2.8	1.0	44.7
Pulp, Paper and prods.	3.6	2.4	3.2	12.4
Chemicals	9.5	6.1	4.2	23.9
Petroleum and Coal prods.	5.7	2.4	1.5	26.9
Rubber prods.	0.5	0.3	1.5	3.5

*Totals do not add up to 100 percent because of rounding.

Sources: Bureau of Labor Statistics, *Bulletin No. 898* (Washington, 1947), 29–31. And computed from Frederick Deming and Weldon Stein, *Disposal of Southern War Plants,* National Planning Association, Committee of the South, Report No. 2 (Washington: National Planning Association, 1949), 8.

Board concluded in 1942, featured firms which paid low wages, manufactured semi-finished goods, and produced relatively little value added per dollar of wages or material.[5]

Ironically, it was not the federal government of Franklin D. Roosevelt which first sought to unleash the predicted cycle of industrial growth, increased income, and further economic progress. Rather, the Sears, Roebuck Company of General Robert E. Wood ventured the pioneer effort. Like the President, General Wood maintained a personal attachment to rural Georgia; his wife hailed from Georgia and they operated a farm near Augusta. And, like the southern liberals, Wood believed that the South's rapidly expanding population offered a potentially rich sales base. Beginning in the late 1930s, Wood established more Sears stores in the South than the region's national low per capita retail sales appeared to justify. More than most other businesses, Sears, Roebuck's emphasis on the South promised economic benefits for the region because the chain's close, long-term connections with sources of supply influenced suppliers to locate new plants in the South. For example, Armstrong Rubber built a tire factory in Natchez, Mississippi, because Sears guaranteed Armstrong's profits: Sears promised to purchase the plant's entire capacity on a cost-plus basis for ten years. In March 1941, Charles A. Walter, Sear's vice president for Southern Industrial Development, went so far as to pledge a guaranteed market to all southern manufacturers who would build to Sears's specifications. Testifying before the Southern Governors Conference in New Orleans, Walters announced that Sears "wants to be a southern industrialist." In its effort to "increase the wealth and prosperity of the South," Sears was "bringing a message of expanding opportunity for every manufacturer in the South." That message so impressed southern officials that Sears successfully avoided the anti-chain store measures and high taxes debated in many southern capitals during the late 1930s and early 1940s.[6]

The federal government, however, was not content to forfeit the management of southern industrial development to Sears, Roebuck. Without federal intervention, Lowell Mellett, Frank Graham, and other national policymakers agreed, economic growth would only strengthen the South's existing order. Only the national government could end the region's reliance on low wage industry and plantation agriculture. Only Washington could break the hold of those economic interests and their governmental allies and establish a more progressive southern politics. The experience of private foundations seemed to underline the necessity of federal action. Private philanthropy had long supported liberal causes in the South. But the generosity of the Julius Rosenwald Fund and of the various Rockefeller boards never dented the established socioeconomic structure. Dr. Will Alexander, chairman of the Commission on Interracial Cooperation, served as the conduit for millions of dollars of foundation funds into the South. Alexander commanded a veritable empire of regional improvement projects in the 1920s and 1930s. Nonetheless, "Dr. Will" readily abandoned that work for service in the federal government. Alexander and his fellow southern New Dealers believed that effective reform required federal action. That view informed the *Report on Economic Conditions of the South.*[7]

Soon after the publication of the *Report,* the TVA led the federal government's drive for southern industrialization. TVA's first chairman, Arthur E. Morgan, had been ousted in 1938. Morgan's rival and successor was David E. Lilienthal, a talented orator and ally of the southern liberals who desired greater federal involvement in the region. Lilienthal hoped, he wrote in his journal in April 1939, that "TVA could become the medium for working out the long job of economic rehabilitation for the South." As Lilienthal's influence grew, he slowly steered the TVA away from the strategy of decentralization that had governed the agency's economic development efforts before 1938. Under Lilienthal the TVA abandoned the so-called "phosphate philosophy" of economic growth—the idea that fertilizer production and small, rural-oriented industries could improve agricultural productivity and raise living standards while preserving the area's rural character. As the defense build-up proceeded, TVA officials launched a program to bring large manufacturing plants to the region. They worried that the character of industrial expansion in the Tennessee Valley and in the Southeast did not "include a substantial proportion of high-order manufacturing." The Authority lent its services to local efforts at recruitment of such industry and to the preparation of plant sites. The TVA also accepted the responsibility for stimulating demand in the region, both by using electrification to allow and create consumer needs and by "awakening a desire for more and better things, for those things which we regard as part of our American standard of living."[8]

As the nation mobilized, Lilienthal moved to turn the TVA into an adjunct of the defense program. In May 1940 he worked out a plan for defense-related activities with Harry Hopkins. The plan called for the Authority not only to supply additional power and magnesium but also to train workers for war work and to actually build airplane, ordnance, and gun factories. Lilienthal had two purposes in enlisting the TVA in the defense program. First, he wanted to insulate the Authority from political attack, both from Congressional opponents like Tennessee senator Kenneth McKellar and from Interior secretary Harold Ickes, who hoped to absorb the TVA into his own national power agency. Second, Lilienthal sought to uplift southern laborers, to train them and employ them in lucrative manufacturing jobs. TVA continued to hire predominately local labor for its projects throughout the war years.[9]

The advent of war completed the reversal of philosophy within the TVA. After Pearl Harbor, TVA general manager Gordon Clapp predicted confidently that the region's very paucity of manufacturing would serve it well during and after the conflict. Clapp believed that wartime construction would bring technologically advanced industries into the South, industries that would better endure the postwar "scramble to determine what should survive in the more permanent industrial economy." Meanwhile, John Ferris, director of TVA's Commerce Department, lobbied before Congress for the establishment of fabricating and processing industries in the Southeast. The TVA, Ferris claimed, had faced no trouble in training surplus farm labor for skilled work on its "great projects." The defense program, he maintained, could do likewise and thus transform the lives of thousands of poor southerners. Throughout the conflict, the TVA pressed for further industrialization of the Valley region, especially the development of large-scale, finished products industries. Ferris explained the shift in philosophy in an address on "Industrial Development in the Southeast" at the war's end. "There is an increasing awareness of the fact that satisfactory income and living standards in the Southeast can be achieved only if, in addition to agricultural improvement, there is also sound and accelerating industrialization."[10]

Among the TVA's many wartime achievements was the development of new processes for aluminum manufacturing, new processes especially suited to southern resources and plant conditions. The Authority assembled a team of research scientists for this and other projects, and its success dramatized the economic benefits of industrial research. This stimulated the South's postwar love affair with scientific research. It helped to establish a homegrown technological community, something the South had long lacked, much to the detriment of its efforts to industrialize. Lilienthal himself recognized that before the TVA began its war work, the scientists of the South were principally science teachers, not researchers. TVA started to change that.[11]

The TVA transformed itself into an agency of southern industrial growth. It did so easily because the coming of war persuaded most Americans that rural life no longer offered the prospect of an American standard of living. As Ferris explained this realization: "As long as we have a raw materials economy, just so long will income be low." Contemporary economic sector theory buttressed the belief that conversion from agricultural to industrial employment would spark regional growth. Advanced economies, such theories held, devoted larger shares of their resources to secondary than to primary production. Meanwhile, the defense build-up restored the faith in industrial capitalism which the Depression had temporarily undermined. Analysts realized that agriculture could neither fuel a modern war machine nor sustain a high standard of living. Nearly two decades of farm depression underscored that observation. Poverty, poor health standards, and underemployment appeared to be chronic maladies of farm life, even when commodity prices held steady.[12]

Furthermore, the Tennessee Valley's record of economic progress in the decade after 1938 seemed to confirm the confidence in manufacturing-based growth. The TVA could not claim full credit for the Valley area's absolute gains in nearly every economic indicator, because, in fact, the increases were part of a national trend. The Authority could, however, point with pride to the Valley's relative gains over the nation and the rest of the South, both in indices of manufacturing growth and those of general economic progress. The early vision of a Valley inhabited by small, decentralized rural industries had faded into the reality of an industrializing region, in which, by 1946, most wage earners toiled in cities of 10,000 or more people. The expansion of manufacturing in the Valley, TVA chairman Lilienthal claimed a year after V-J Day, demonstrated that the TVA was "a success story of the old fashioned kind."[13]

Many southerners shared Lilienthal's enthusiasm. When former Nashville Agrarian H. C. Nixon wrote "The South After the War" in 1945, he credited the TVA with a "solid contribution to the life of the Valley" because "it has been the prime factor in the setting up of high-class industrial plants in the South." The Decatur (Alabama) *Daily* attributed to the TVA not merely great dams, new industry, and cheap power but also an advance in the thinking of a people. "They are no longer afraid," the paper claimed. "They can stand now and talk out in a meeting and say that if industry doesn't come into the Valley from other sections, then we'll build our own industry."[14]

Despite the TVA's headlong jump into industrial promotion, its efforts, and those of private groups like the Sears, Roebuck Company, paled before the gigantic size of federal investment in, and stimulation of, southern industry as part of the war effort. Indeed, rather than effecting industrialization itself, the TVA's principal role during the 1940s was facilitating the location of defense industry in the Valley and in the South as a whole. The courage of the southern people notwithstanding, it was their government in Washington that moved industry to the South from other sections and built them their own industry during the war decade.

The South, however, got off to a late start on the defense boom. In the year and a half before Pearl Harbor, it shared in relatively little of the government largesse. In 1940, for example, the South obtained only 10.5 percent of industrial allocations for defense. The South's slow start reflected primarily the lack of suitable facilities in place at the beginning of the war because businesses with existing facilities to convert or expand received the first government contracts. Consequently, WPA rolls held steady across much of the South, while the number of relief workers dropped precipitously in the country as a whole. One month before Pearl Harbor, the governor of Tennessee reported that his state had the dubious distinction of having more privately than publicly financed defense expansion.[15]

The administrative configuration of the War Production Board reinforced this initial bias against the South. First, the military won the upper hand in its long feud with civilian authorities over the control of procurement. That bureaucratic victory tended to slow the flow of war expenditures to the South because the armed services preferred working through established, reliable, and large contractors to hiring untested manufacturers in underdeveloped regions. The shortages of managers and skilled laborers in the South further exacerbated the suspicions of military planners. Second, alone among the war agencies, the WPB operated almost entirely out of a headquarters in Washington rather than through regional offices. Unlike the NWLB, the procurement authority had no branch especially responsive to southern aspirations.[16]

But as American involvement in the conflict intensified, the federal government catalyzed a tremendous expansion of southern industrial capacity. Federal investment in the region accelerated with each year of the war.

The headline "South's expansion breaks all records" became a monthly fixture of southern trade magazines like *Manufacturers' Record.* Total industrial construction in the South soared 25 percent higher in the first nine months of 1945, for instance, than it had for the same period in 1944. Accounting for just 11.8 percent of the nation's capital expenditures in manufacturing before the war, the region acquired 17.6 percent of the value of wartime manufacturing facilities expansions (Table 4-2). And in cases where the region's prewar manufacturing deficiency could not count against it, southern industry received an even greater share of federal dollars. The region housed 23 percent of the new plant built for the war effort. But only 10.5 percent of the investment for facilities expansions and 14.0 percent of those for factory conversions traveled below the Mason-Dixon line (Table 4-3).[17]

Even more familiar than the roar of war factories in Dixie during World War II was the sound of bulldozers ploughing military bases out of the "Old Plantation." The region received 36.5 percent of the total on-continent military facilities awards, playing home at one time or another to almost two-thirds of the domestic Army and Navy bases. So many servicemen took basic training in southern camps, swelling the base towns and the profit margins of local businesses, that one northern conscript suggested that "this whole draft business is just a Southern trick." The ruse, he explained, was "something put over by Southern merchants to hold the big trade they get from the training camps." Indeed, the wartime South resembled more a "military-payroll complex" than a miltary-industrial complex, an ominous pattern that would long outlast V-J Day.[18]

The clearing of new military bases caused a sort of agricultural enclosure movement across the South. Nearly half of the Army's major land acquisitions for bases and ordnance plants, and the vast majority of dis-

TABLE 4-2
Manufacturing Facilities Expansions, July 1940–June 1945

South as a Percentage of U.S. Total Value Put in Place by Industry and Source of Financing (%)

	Total	*FederallyFinanced*	*Non-Fed. Financed*
TOTAL	17.6	21.6	10.4
Aircraft	9.9	10.4	5.5
Ships	15.8	16.3	11.0
Iron and Steel	7.0	7.2	6.9
Non-Ferrous Metals	27.0	27.7	25.0
Chemicals, Coal, and Petrol. prods.	36.0	47.7	22.8
Combat and Other Motorized Vehicles; Guns and Ammunition; Explosives	22.9	25.1	5.3

Source: Computed from Deming and Stein, *Disposal of Southern War Plants,* 10–11.

TABLE 4-3
Manufacturing Facilities Expansions Authorized July 1940–May 1944*
Value of Facilities Authorized in South as a Percentage of U.S. Total by Type of Expansion and Type of Financing (%)

	New Plant	*Expansions*	*Conversions*
TOTAL	23.1	10.1	14.2
Publically Financed	23.0	9.8	11.5
Privately Financed	23.9	10.7	16.5

*Data by type of expansion are unavailable for the entire 1940–45 period.
Source: Computed from War Production Board, Program and Statistics Bureau, Facilities Branch, Report, 1 June 1945, Records of the War Production Board, RG 179, No. 221.3R, National Archives, Washington, p. 17.

placed farmers, were in the South. In Hinesville, Georgia, a small town forty miles south of Savannah, the Army purchased 360,000 acres for an anti-aircraft firing range. The range uprooted 713 families, half of them black farm families. Displacement particularly traumatized the 140 black Hinesville families who owned their own farms. A farm of one's own was always the triumph of a long and bitter struggle for black southerners, so once they owned a piece of land they fought tenaciously to keep it. As one of their white neighbors put it, "When a Negro buys a piece of land, it's out of circulation." Not until "the Army searchlights at Camp Stewart began to pierce the evening sky," a government official reported, "did many of these families realize the whole thing was real, that they would really have to get off the land and leave their homes to the mercy of the Nation's soldiers." Renters and tenants also encountered difficulties, particularly problems of "secondary displacement—tenants moving from the defense areas who owned their own tools and workstock displacing other tenants less able to bargain." Opportunities for defense work sometimes cushioned the blow, although blacks did not always find such employment. And some of the displaced preferred farming. One such man, a black sharecropper named Percy Bellman, moved off the land when the Army bought his landlord's holding for the new Huntsville Arsenal in Alabama. Bellman, after saving some money, tried to return to the land. "And they told me," he informed a Congressional committee, "there was so many people buying and selling land they wouldn't talk to a man about renting none. You just can't rent a farm."[19]

The Farm Security Administration supervised the relocation of displaced farmers. To relocate farmers and tenants from the Hinesville area, for example, the FSA bought an 18,000-acre tract in a neighboring county and two other small plots of land, enough to resettle about one-fourth of the displaced farmers. The agency had similar success across the region, fighting valiantly, but in vain, to slow the flight from the land. Army bases and defense plants sprang up throughout Dixie. Farms and forests became factories and arsenals.[20]

By 1945 the Southern Conference for Human Welfare could proudly claim that "there is not a material of war that does not come from the Southern states." And Walter Matherly, a liberal Florida educator who had served under Frank Graham at Chapel Hill, joined the chorus of appreciation for the new large manufacturing enterprises on the southern scene. "These industries are of many kinds and are widely scattered or located in many places," Matherly explained in 1943. "They range all the way from the building of ships and the fabrication of steel to the assembly of aircraft and the making of munitions." Aircraft factories and shipyards generated the most excitement. McDonnell, Vultee, Bell Aircraft, Bechtel-McCone, General Motors, and Higgins Ship-Building all opened large aircraft plants in the region. The South played an even larger role in shipbuilding. During World War II, the U.S. Maritime Commission built five of its eleven new merchant shipyards in the South, even though before the war the region contained only six of the eighteen yards capable of building Class I merchant vessels.[21]

The most dramatic example of wartime southern industrial growth was Higgins Industries—brainchild of the colorful Louisiana entrepreneur Andrew Jackson Higgins. Higgins transformed a small shallow-draught boat company of 400 workers and less than a million dollars in annual sales in 1939 into a diversified transport equipment corporation of eight plants, 20,000 employees, and annual sales upward of $100 million by 1944. He promised and met incredible production schedules. In the process, he won the devotion of Louisiana blacks, whom he hired and paid on an equal basis with whites. Uncharacteristically for a southern industrialist, he also secured the allegiance of labor unions, with whom he cooperated enthusiastically. So congenial was Higgins's relationship with the workers—whom he often cajoled through a bullhorn on tours of his yards—that a union leader immortalized him in verse. "Fire the forge now, light the torch/ Uncle Sam is on the march," rhymed Holt Ross. "Papa Higgins has the plans/ the Federation has the men/ With Higgins ships we'll fill the ocean/ Through hard work and true devotion."[22]

Higgins also won the grudging respect and the reluctant cooperation of officials in the national capital, which he called the "District of Confusion." The NWLB hounded him for keeping his wages too high. And in the spring of 1942, the Maritime Commission finally approved his $200 million plan to produce Liberty ships on an assembly line basis at a swampy site just outside of New Orleans. When a shortage of steel forced the cancellation of the contract in July 1942, Higgins protested. His "bellow from the bayous" was heard in the White House. After agreeing that Higgins was difficult to work with, President Roosevelt reminded WPB chief Donald Nelson "that the fact remains that he has production facilities and can do a good job." "I have no idea what he should make," FDR told Nelson, "airplanes or corvettes or something else, but I do want his plant utilized."

Not only did Higgins possess a peculiarly persuasive personality, but his mechanized, unionized, high wage operations embodied the very sort of economic organization the Roosevelt administration sought for the South. A few days later the government awarded Higgins a contract for 1200 cargo planes.[23]

The South's domination of a new industry—synthetic rubber—and the section's new areas of concentration in the steel and non-ferrous metals fields proved to be less spectacular, but more significant. Combat had cut off most supplies of natural rubber. Synthetic rubber could be made from either alcohol (an agricultural commodity) or petroleum, both of which were widely available in the South. It became a crucial component of the war effort. So pressing was the shortage of rubber supplies that the national government established the new industry wholly with federal funds and almost entirely within the South; 41 percent of the synthetic rubber industry's capacity stood in Texas alone. The metal lines followed the rubber factories into the South. Although the heaviest concentrations of the steel industry remained in the Pittsburgh-Youngstown and Chicago-Detroit corridors, much of that capacity shifted south and west, creating a new steelmaking center along the Gulf Coast in the Southwest. The steel factories of southeast Texas sprang up so fast that they confounded the attempts of the Dallas regional board to establish wage rates for the industry. Similarly, the region won more than 25 percent of the facilities contracts for the production of non-ferrous metals.[24]

The rapid development of industrial centers like the Texas steel belt illustrated another aspect of the procurement program—the preferential treatment of the Southwest. The Southwest, of course, possessed many economic advantages—raw materials, cheap, local energy sources, a secure location in the nation's interior, and a dry climate (of particular importance to the aircraft industry). But, despite those advantages and considerable economic progress over the four decades since the discovery of Gulf oil, the Southwest had not yet reached its take-off point until wartime investment provided the lift. The four southwestern states—which in 1939 contributed 19 percent of the South's wage earners in manufacturing and 26 percent of its value added—obtained 55 percent of the awards for facility expansions in the South. Texas alone acquired more than one-third of the region's total, joining the "Big 9" of states with over a billion dollars in war facilities. So richly did the Lone Star state benefit from its expanded steel, petroleum, and aircraft industries, that in 1945 it temporarily replaced North Carolina as the South's leading industrial state. A few years after the war's end, Robert Wood of Sears, Roebuck prophesied that "within fifty years, Texas will lead all other states in the union in population and wealth, that it will have the most economic and political power of any state, and that Houston will be the fourth or fifth largest city in the United States." Wood proved perspicacious. Within a half-century, Houston became the nation's fourth largest city and Texas rivaled the nation's leaders in economic and population growth. Economic and political power emerged even sooner. Within a decade, Texas politicians controlled many positions of influence in the national government. From those bulwarks, they expanded the alliance with the national defense establishment, forged during World War II, that transformed the Texas plains into a burgeoning national armory.[25]

North Carolina, on the other hand, suffered the fourth lowest ratio of war facilities to prewar manufacturing of all the states in the union. It was one of only nine states wherein private investment accounted for more than 30 percent of war facilities. Unlike the businessmen and politicians of Texas, North Carolina's leaders made only lackluster efforts to woo war industry. Long the South's leading industrial state, North Carolina owed that position to its leadership in traditional southern industries like tobacco and textiles. Representatives of those industries dominated state government. During the war years, those industries kept up the battle against federal wage policy. They fought to forestall mechanization, to preserve regional wage differentials, and to safeguard their supply of cheap labor. Competition from high-paying defense contractors would have undermined their efforts. The distribution of war industry within the South was not solely, or even principally attributable to the behavior of local elites. But Texas leaders exploited their state's advantages in ways that North Carolina's did not. The wartime failure to recruit defense industry later haunted North Carolina boosters. North Carolina never received a large volume of defense work, even after the state government lauched a vigorous industrial recruitment campaign in the 1950s.[26]

The flow of capital expenditures into the Southwest highlighted another facet of wartime federal spending—the development of new manufacturing areas. The assignment of two large aircraft contracts to Detroit might create more jobs and involve larger expenditures than the construction of one small factory in Marietta, Georgia, but the marginal effect of those dollars was less dramatic in established manufacturing areas than in towns where few plants had stood before the war. The factory really changed Marietta. *Time* did not exaggerate when it announced that "From Georgia's Tobacco Roads through the Mississippi Delta out to the oil and cattle country in West Texas . . . new towns had sprung up where none had been before." Of twenty-six manufacturing areas wherein 100 percent of war facilities represented new plant, fifteen were in the South. These were mainly explosives factories, but they included the shipyards of Panama City, Florida, and the aircraft town of Marietta, Georgia. Eldorado, Arkansas, an oil town and another city receiving all of its wartime expansion in the form of new facilities, became a regional haven for the new synthetic rubber industry. The South also harbored fifteen of the thirty-six manufacturing areas with 90 to 99 percent of their war facilities in new plant. When the WPB calculated the ratio of war expansion to prewar employment, the scale of southern industrial growth proved breathtaking.[27]

Despite the defense program's obvious stimulus to southern manufacturing, many contemporaries believed that the region owed little of its progress to the intentions of federal policymakers. They were wrong. Although the exigencies of war remained the national government's top priority and the institutional cards were stacked against the relatively undeveloped South, the Roosevelt administration strove to overcome the considerable obstacles to locating industry in the region. And these efforts helped to account for the preferential treatment of the South. Indeed, the South's record was amazing considering the military's well-known preferences for established firms operating in areas with requisite skilled labor supplies. Many procurement officials feared that the South's labor supply was too unskilled for war work; for example, an official of the War Plant Site Board told David Lilienthal that southern labor was not readily teachable for aircraft production. Despite the TVA's success with local labor, and the nation's success with training "farm boys" for factory work, this suspicion of southern capabilities persisted. Officials commonly reasoned that midwestern farmers, while lacking industrial savvy, had plenty of transferable experience with agricultural machinery. But not so southern farm boys. As one congressman characterized this view, "they can fix a mule but not a tractor."[28]

Despite these constraints, the Roosevelt administration used the war emergency to develop the South. After all, it was the effort to modernize the southern economy that brought officials such as Donald Nelson of the WPB and George Taylor of the NWLB into the national government in the first place. FDR's affection for his "other home" and his horror at southern underdevelopment led him to look kindly on efforts to build the South during the war. So did the conviction of war agency administrators that industrialization of the South would stave off future economic depressions and secure national prosperity. Prominent southerners recognized the President's bias. Virginius Dabney, editor of the Richmond *Times-Dispatch,* declared that "no other President of modern times devoted an even remotely comparable amount of thought and effort to southern problems, and he alone among them had the vision and the courage to initiate a broad program addressed to the economic and social ills of the region." And in April 1945, Florida senator Claude Pepper mourned that the South "in the death of President Franklin D. Roosevelt has lost its best friend." Eleanor Roosevelt also sought preferential treatment for the South because she believed that war plants might provide opportunities for black southerners.[29]

The Liberty ship program embodied FDR's hopes for the South. President Roosevelt planned to use ship contracts to develop the southern seacoast, despite the opposition of Admiral Emory S. Land, chairman of the Maritime Commission. Admiral Land resisted such decentralization on principle because it required the diffusion of scarce resources and managerial talent. Nonetheless, the lobbying of the President's uncle, Frederic Delano of the National Resources Planning Board, assuaged the admiral's fears of shortages of skilled labor and possible racial strife in southern ports. The Maritime Commission built nearly half of its new yards in the South.[30]

Such decisions ratified the intentions voiced early in the war by the President, by defense counselor Chester C. Davis, and by the Plant Site Committee of the Office of Production Management. These policymakers wanted to move industry south and west in order to exploit reservoirs of "ineffectively used labor." After Pearl Harbor, the WPB more explicitly affirmed this region-building dimension of its site selection policy: "On the sites which are selected depends not only the strategic security of our defense industries and much of their efficiency for defense production, but also important and permanent consequences for the economic development of different parts of the nation." Echoing the proposals of Howard Odum and Rupert B. Vance, war administrators asked "that every possible preference be given to locations where large reserves of unemployed or poorly employed people are available and where industrialization during the defense period will contribute to a better long-run balance between industry and agriculture." The WPB reiterated this policy near the war's end; reconversion plans, the agency demanded, must consider the "relative need for industrialization of certain regions."[31]

Nonetheless, many contemporaries (and historians) feared that the federal government had failed to ensure long-term southern economic development. While these critics of federal policy did not deny the impact of war expenditures, they worried about the problem of convertibility—whether war plants would lead to sustained growth or merely yield, in *Time*'s words, a "hangover." Fears of such a "hangover" were nourished by the South's large share of explosives and ammunition production, enterprises which would almost certainly disappear after the hostilities ended. The WPB Planning Division believed that the war experience had laid a "goundwork for the continued industrialization of the South," but feared that "while the South has obtained many large new facilities during the war it will face very difficult problems in retaining a lasting increase in industrialization from the new plants themselves." The immediate aftermath of victory exacerbated these concerns; in the year after V-J Day, manufacturing employment in the South fell off faster than in the rest of the country, and the region experienced a short-lived migration back to the farm.[32]

The fears proved unjustified. The process of southern industrialization was not reversed. Industrial expansion, and indeed the defense spending which had energized it, continued after the war. Whatever the intent of federal policies, those policies had catalyzed dramatic economic change during the war years. The growls of bulldozers on the old plantation sounded the leveling of the region's red hills and cotton fields; the thunder of new industrial workers teeming into the once quiet towns of the South signaled the demolition of an old way of life. As one Alabamian explained to John Dos Passos, "Looks like the war has speeded up every kind of process, good and bad, in this country."[33]

Channeled by federal policy, the war sparked a thoroughgoing transformation of the southern economy. The mobilization program accelerated the growth of southern industry, so that in 1944, manufacturing surpassed agriculture as a source of income payments in the South. It fostered the diversification of southern manufacturing, stimulating the more robust growth of high wage, high productivity industries. At the same time, the region's industrial and economic hub shifted from North Carolina, and the cotton and textile states of the Southeast, to Texas and the Southwest. And, as labor and capital poured into these new industrial enterprises, the region witnessed the reorganization of southern agriculture.[34]

On the farm, long the cultural and economic capital of southern life, the war consolidated the "reforms" of the New Deal era. Rural southerners flooded into the cities, in Dos Passos's words, "to be doing something toward winning the war, to make some money, to learn a trade." Together with the call to military service (and rural southerners rarely accepted farm deferments), this "Gold Rush Down South" drained away 22 percent of the southern farm population. Opportunities for war work ameliorated the plight of many tenants and croppers who had been forced off the land by the incentives of AAA policies, but they also encouraged the further departure of farm population, despite the sudden agricultural labor shortfall. The migrants were drawn overwhelmingly from the ranks of sharecroppers and wage laborers, those who had benefited least from federal agricultural programs; farm owners and the higher classes of tenants remained on the land. Furthermore, the rural population plummeted at a much faster rate than did agricultural employment, indicating the departure of several underemployed family members with each migrating farm worker. The sharecropper exodus also furthered the breakdown of paternalism, or rather it exposed how "paternalism" merely reflected the landlords' needs to protect their labor supplies. One Alabama planter expressed worries about his croppers who "had never been further than the county seat Saturday afternoons in their lives." Most of his tenants had left the fields for lucrative construction jobs in nearby towns. "But what in the world are they going to do when they get turned off?" the landloard wondered. "I'm not going to be in a position to look after them."[35]

Such yearnings for southern paternalism demonstrated the severity of the rural labor shortage. Despite enormous increases in farm wages, workers streamed out of the countryside. The resultant labor shortages were so acute that prisoners-of-war were sent to harvest southern fields. At the insistence of Louisiana senator Allen Ellender, the Army held German prisoners in the South during the spring of 1946, conscripting them for farm work a year after the surrender of the Third Reich. The labor shortage fostered a revolution in the way the Cotton Belt was farmed, intensifying the organizational changes initiated in the 1930s. As the tenant displacement of the Depression gave way to voluntary departures during the war, mechanization of the Cotton Belt accelerated. Not only did the proportion of southern farms operating tractors skyrocket, but the sudden shortage of peak-season help also inspired concentrated attention on the harvest bottleneck. Without abundant cheap labor, the potential market for mechanical cotton-pickers was large enough to galvanize manufacturers like International Harvester into developing and marketing mechanical harvesters. The new machinery diffused rapidly through the Cotton Belt, so that by the end of the war decade, the traditional southern mule farmer had all but disappeared from the regional landscape.[36]

And to a large extent, the cotton culture itself followed the mule farmer into oblivion. Before Pearl Harbor, the AAA had already engineered a decisive decline in cotton acreage, as cotton dropped from 39.6 percent to to 22.4 percent of the region's harvested acreage between 1929 and 1939. Thereafter, the share of arable land devoted to cotton slipped only slightly, but during the war years, for the first time, southern farmers planted fewer acres of cotton than the maximum approved by the federal allotment. And when the Congress lifted all controls on cotton planting in 1944, many cotton farmers, especially in the Southeast, converted to livestock, hay, and other crops. The war also marked the beginning of a gradual substitution of soybeans for cotton; while cotton production declined in eight southern cotton growing states from 1940 to 1945, the output of soybeans more than doubled.[37]

Despite the movement toward mechanization and away from reliance on cotton, reorganizations long sought by federal officials, the rural South remained the nation's poorest region. Even with increasing productivity, government analysts concluded, southern agriculture could never remedy regional poverty. The Federal Reserve Board's 1945 study of American agriculture concluded: "Although rural poverty can be found in almost every section of the country, the degree to which it is concentrated in the South makes this region a national problem." That study and many others like it noted that freedom from King Cotton had largely been supplanted by dependence on federal programs. "The rural South," confessed the Federal Reserve report, "will probably have to be supported to a considerable extent by Federal Subsidies of one kind or another."[38]

The authors of the report, however, like so many analysts of the South, foresaw another antidote to southern rural poverty—the industrialization of the region. And the liberal application of that medicine was evident throughout the region at war's end. The state of Tennessee, for example, advertised with great fanfare in January 1945 that manufacturing had overtaken agriculture as the leading source of employment in the Volunteer State. Regionwide, all indices of industrial growth—number of establishments, manufacturing employment, wages, and value added—showed large wartime gains. Significantly, the increases in payrolls and value added outstripped the rise in employment, so that the predicted rise in income did accompany industrial growth. "The nation is money-prosperous," one southerner explained to Jonathan Daniels early in 1946, "but the South and Southwest are the relative gainers."[39]

This rapid industrialization appeared all the more promising to southern liberals like Daniels because the pattern of wartime development appeared to depart from earlier spurts of regional economic growth. Between 1940 and 1946 the composition of income payments in the region shifted substantially. Income from manufacturing, from the federal civilian payroll, and from trades and services grew faster in the South than in the rest of the country, while agricultural income rose more slowly. This at once continued a longstanding pattern, and accelerated it. And within the booming manufacturing sector, the restructuring of the regional economy was most pronounced. The higher wage industries expanded faster in the South between 1939 and 1947 than did the traditional lower paying southern lines. In the decade and a half after the onset of defense production, the grip of textiles and sawmills over southern manufacturing steadily eased. At the same time the durable goods lines accounted for a bigger share of the region's employment and output. By 1954 the industrial economy of the South more closely resembled that of the nation as a whole than it had in 1939 (Table 4-4).[40]

Even where the war effort exerted little direct stimulus on a particular industry, the wartime planting of new lines of manufacturing in the South promoted later economic diversification. "War industries," one Texan observed, "have demanded new skills and technical knowhow that were not here in adequate supply." They also provided physical plant for reconversion, trained managers from outside the region, and for many large national corporations, the first of many substantial investments in the South. The government financed the facilities for many of the postwar South's biggest firms: shipyards for Litton Industries and Tenneco, aircraft plants for Lockheed, General Dynamics, and LTV, and aluminum factories for Alcoa and Reynolds Metals. Along with the regulations of the NWLB, the introduction of these new industries ensured that the region's industrial workforce would be better paid and more highly skilled than before the conflict. This recomposition of the southern labor force validated John Dos Passos's claim that "New Industries Make New Men."[41]

New industries also make new industrial regions, and they did so in the Gulf Southwest during World War II. Fueled by tremendous federal investment, the four southwestern states of Texas, Louisiana, Arkansas, and Oklahoma experienced the greatest relative gain in manufacturing capital expansions of all the sections of the country. And the Southwest's share of southern industrial capacity jumped accordingly (Table 4-5). "The center of gravity of American economic life," *Manufacturers' Record* boasted in 1944, "has shifted markedly South and West during the war." In particular, the growth of Texas was so phenomenal that in 1946, one friend of Jonathan Daniels declared that he could no longer recognize the Lone Star State as part of the South.[42]

These economic transformations owed a great deal to the efforts of federal policymakers. Industrial capacity shifted South and West, along the trail blazed by federal investments. Southern manufacturing diversified, as

TABLE 4-4
Distribution of Manufacturing Employment, South and United States, by Industry

Percentage of Southern Manufacturing Employment in Selected Industries, 1939–1954 (%)

	1939	*1945*	*1954*
Food and kindred prods.	12.3	11.3	12.2
Tobacco Manufactures	2.8	2.0	2.4
Textile Mill prods.	30.7	22.0	21.7
Apparel and related prods.	5.5	5.3	8.3
Lumber and Timber prods.	13.2	11.4	9.5
Furniture	5.9	4.6	3.7
Pulp, Paper, and prods.	2.4	2.4	4.0
Chemicals	6.1	7.7	7.8
Petroleum and Coal prods.	2.4	2.8	2.6
Rubber prods.	0.3	0.6	0.7
Leather and Leather prods.	1.1	0.9	1.1
Stone, Clay, and Glass prods.	3.1	2.0	2.9
Machinery, except Electrical	1.8	3.1	3.6
Electrical Machinery	0.1	0.6	1.5
Transport Equipment	1.6	12.0	5.7
Iron and Steel prods.	5.1	6.7	---
Non-Ferrous Metals	1.0	1.5	---
Primary Metal prods.	---	---	3.6
Fabricated Metal prods.	---	---	3.6

Percentage of U.S. Manufacturing Employment in Selected Industries, 1939–1954 (%)

	1939	*1945*	*1954*
Food and kindred prods.	11.8	9.6	10.5
Tobacco Manufactures	1.0	0.6	0.1
Textile Mill prods.	12.3	7.8	6.6
Apparel and related prods.	8.9	6.9	7.6
Lumber and Timber prods.	4.6	3.7	4.1
Furniture	3.8	2.6	2.2
Pulp, Paper, and prods.	3.2	2.6	3.4
Chemicals	4.2	5.1	4.7
Petroleum and Coal prods.	1.5	1.3	1.4
Rubber prods.	1.5	1.6	1.6
Leather and Leather prods.	3.8	2.3	2.3
Stone, Clay, and Glass prods.	3.5	2.5	3.1
Machinery, except Electrical	6.8	9.2	9.9
Electrical Machinery	3.5	5.4	6.1
Transport Equipment	6.5	16.7	10.9
Iron and Steel prods.	11.6	11.9	---
Non-Ferrous Metals	2.8	3.0	---
Primary Metal prods.	---	---	7.1
Fabricated Metal prods.	---	---	6.5

Sources: Computed from U.S. Bureau of the Census, *Census of Manufactures 1954, Vol. III, Area Statistics* (Washington, 1957), 4–5, 71–76, **107**-4–**107**-5, **119**-4–**119**-6, **108**-5, **147**-6–**147**-7. Bureau of Labor Statistics, *Bulletin No. 898*, pp. 29–31.

TABLE 4-5
Geographic Distribution of Southern Manufacturing, 1939–1945
Percentage of Wage Earners in Manufacturing and of Value Added by Manufactures (%)

	1939		*1945*	
	Wage Earners	*Value Added*	*Wage Earners*	*Value Added*
Texas	9.3	14.4	13.2	17.3
Louisiana	5.2	6.3	5.8	6.8
Arkansas	2.7	2.1	2.4	2.1
Oklahoma	2.1	3.3	3.4	4.3
Alabama	8.6	7.8	9.0	8.5
Florida	2.4	3.8	4.2	4.1
Georgia	11.6	9.0	11.5	9.2
Kentucky	4.6	6.0	4.7	5.6
Mississippi	3.4	2.3	3.4	2.6
N. Carolina	19.8	17.3	15.5	13.6
S. Carolina	9.3	5.4	7.6	5.3
Tennessee	9.7	10.2	10.7	10.7
Virginia	9.8	12.1	8.5	9.9
4 Southwest	19.3	26.2	24.8	30.5
9 Southeast	79.2	73.8	75.1	69.5

Source: Computed from Deming and Stein, *Disposal of Southern War Plants,* 7, 21. Totals do not add up to 100 percent due to rounding.

the federal government simultaneously introduced new industries and restrained the growth of traditional, low wage employment. And the changing landscape of the rural South betokened the continued influence of agricultural policy in the face of severe labor constraints. "The South, by whatever comparative test," H. C. Nixon observed as the war drew toward a close, "will emerge from this war with more social change and more unfinished business than any other section of the country." Cataloguing the transformed landscape, Nixon prophesied that the South "will have fewer share-croppers, but more welders and pipefitters." And the former champion of the agrarian way conceded: "It will have more industry and industrial capital, less rural isolation and more urban sophistication."[43]

The nature of that unfinished business spawned considerable concern in the South and in Washington after V-J Day. The WPB wondered whether the South would "revert to its position as the nation's number one problem area." The Southern Conference for Human Welfare voiced concern for those who had left the fields to work in war plants. "Does anyone believe," the Conference demanded, "that these workers whose wages have been so good are going to be willing to return to an average income of only $550 per farm family." Others worried about the convertibility of southern war industry, how much of the plant capacity, managerial knowhow, and worker training could be translated into peacetime production. But whatever the concern, there was widespread agreement that the region's postwar fate hinged on further industrialization. The National Policy Association's Committee of the South spoke for southern liberals and government policymakers when it asserted that the "region's economic growth and stability rest in very important part upon the growth of manufacturing industry."[44]

Sustaining wartime momentum, those analysts believed, would require the emergence of robust southern markets for consumer goods. And industries geared to the market indeed followed the returning veterans into the South. Sears, Roebuck stepped up its efforts in the region by opening a Southern Merchandise Office in Atlanta during the war to guarantee that southern manufacturers "jump out of the firing line into the frying pan business." Atlanta, which boasted many new consumers of its own and could offer access to Florida's burgeoning consumer market, became the nation's newest major manufacturing center. In 1946, George Healy, Jr., of the New Orleans *Times-Picayune* observed that demand for new manufacturing sites in Atlanta was overwhelming the "'city that time forgot.'" One study reported that the thirteen southern states received 35 percent of the nation's manufacturing buildings contracts in the three years after V-J Day.[45]

An influential economic study of the postwar years confirmed the optimistic prognoses of Healy and Sears, Roebuck. Commissioned by the National Planning Association, two government economists, Stefan Robock and Glenn McLaughlin, conducted an investigation of industrial location in the South based on a sample of eighty-eight plants established in the region between 1945 and 1949. Robock and McLaughlin discovered that the South's burgeoning market formed the single most important factor in management decisions to locate in the South, with access to materials second, and labor considerations third. Moreover, labor-oriented plants tended to be small businesses. They involved smaller capital investments and created fewer jobs than either materials or market-oriented firms. This finding, of course, reflected the southern liberals' bias against low wage plants, a bias which the report's authors shared. It also reaffirmed the belief that regional wage differentials hindered rather than helped southern industrial development by curtailing purchasing power.[46]

But whatever Robock and McLaughlin's conclusions on the wage issue, the data collected in the report revealed much about the character of southern industrialization after 1945. Most dramatically, the report uncovered a burst of market-oriented manufacturing growth in the postwar South. The expansion of the automobile industry led the way. Ford and General Motors both built new facilities in the Atlanta area after the war. Ford, with its uniform national wage policy, and GM, with its union contracts, paid the same wages in their northern and southern plants. Following the automakers into Dixie were four battery companies, three tire and tube plants, and several retail dealerships.[47]

The expanded market for agricultural machinery also spawned a new industry in the South. International Harvester opened a $52 million tractor factory in a converted Louisville war plant and a mechanical cotton-picker plant in Memphis. IH immediately implemented its nationwide fair employment policy in those plants, hiring and promoting black workers with little regard to local mores. Allis-Chalmers responded by converting a Gadsden, Alabama, ordnance plant to cotton picker production. The Memphis and Gadsden plants alone provided the region six times as many jobs as the entire agricultural implements industry had in 1939. These new market-oriented enterprises, and others like them, reduced the influence of labor-oriented industries for the first time in the history of southern manufacturing.[48]

The South, of course, continued to support many low wage firms. But, while the absolute size of the low wage, labor-oriented sector maintained the South's reputation as a haven for cheap labor, the relative dominance of those lines over the regional economy diminished in the decade after V-J Day. In the long run, this proved a boon to the region. The South, it turned out, was hitching its fortunes to the leading sectors in the American economy because low wage lines stagnated during the postwar decades, while high paying ones prospered. One study of the postwar economy discovered that, with only one exception, declining and slow growth industries paid lower than average wages, while the faster growing lines paid wages in excess of the national standard.[49]

The diversification of southern industry stemmed in part from the proliferation of branch plants in the South. Every one of the firms studied by Robock and McLaughlin maintained its home offices outside the South. These branch plants of large national concerns dominated the manufacturing output of the South. They also contributed to regional development by bringing managerial knowhow, research facilities, national wage scales, and insurance and retirement plans to the region.[50]

In so doing, however, they may have also diminished the imperative to develop managerial talent and financial services within the South. Before 1938, southern elites expended little to educate the region's population, train its workforce, or uplift its needy citizens. They were intent on preserving the reservoir of unskilled labor which had fueled the regional economy. But even after the federal government weaned the South from its economic dependence on unskilled labor, the South continued to avoid investing in education and social welfare. Integration of the South into the national economy drove up the wages of employed unskilled southerners. It also opened national markets for skilled labor to southern employers, drawing educated people southward and postponing development of the South's own human resources.[51]

Such reliance on outside capital was hardly novel to the South of the 1940s and 1950s, but the dependence on federal programs did represent a new force in southern life. "The industrialization of the South is a certainty in the next two decades," former Georgia governor Ellis Arnall intoned in 1946. "The form it will take depends on the national policies that are adopted in the next few years." The government of Franklin D. Roosevelt had whetted a regional appetite for national assistance—a hunger which only additional high-grade industry could sate. And the administration of FDR's successor did nothing to diminish such expectations. Just after V-J Day, the Truman administration financed the relocation of the United Aircraft Company's Chaunce-Voight Division from Bridgeport, Connecticut, to Dallas, Texas. Fifteen hundred workers, 2000 machines, and 50 million pounds of equipment made the move.[52]

As the Cold War intensified, public investment in southern industry resumed. Defense spending, especially in the wake of the Korean conflict, solved the nagging problem of unused war plant in the South. The government returned surplus factories to defense production. The synthetic rubber industry, for example, was maintained and expanded by the Truman administration for security reasons, even after the return of natural rubber supplies in 1945.[53]

Indeed, the United States followed a policy of promoting economic growth through military expenditures. And the adoption of this strategy—"military Keynesianism"—had broader implications than the reopening of surplus war plants. The strategy derived from the so-called "Growth Keynesianism" of Leon Keyserling, the chairman of the Council of Economic Advisors who championed an ever-expanding economy. When war broke out in Korea the Truman administration decided to use national security needs to enlarge the economy. Reluctant to reinstitute wartime rationing and price controls, Truman and Keyserling believed increased military spending would forestall inflation and promote future national economic strength. NSC-68, the defense policy adopted in 1950, committed the United States to a massive, long-term military buildup. It included an appendix on national economic policy drawn up by Keyserling and his colleague Hamilton Dearborn. Keyserling and Dearborn cited the necessity of exploiting the nation's full economic capacity to finance the defense program. They outlined federal policies "to promote a stable and equitable distribution of purchasing power" and "to promote the full utilization of the United States potential for growth."[54]

Such a policy had a particular resonance in the South. As the nation's least developed region, the South stood to benefit more than other sections from Keynesian demand-stimulating policies, military or otherwise, because it offered untrammeled ground for public investment. The policy also encouraged the growth of a new political leadership in the South. In 1946, Howard Odum noted that a new generation of southern politicians, a cohort bent on economic progress, had emerged from the war. That leadership would renew the wartime alliance with the defense establishment which had sent so many war plants South.[55]

By the end of the 1940s, a southern economy was taking shape, radically different from the one described by the *Report on Economic Conditions.* In 1951, *Manufacturers' Record* proudly announced that the South had kept pace with the nation during the Korean war boom, a departure from the region's history of responding less vigorously to boom conditions than the more industrialized areas of the country. That healthy performance signaled that the foundations of a new economy had been laid. One observer even chided Jonathan Daniels for bothering to inquire about the region's problems. "Aren't you starting your mind in the wrong groove when you talk about a problem?" she asked. "It seems sort of dated and old fashioned to me."[56]

The South, however, had not yet eradicated the conditions that earned it the sobriquet "Economic Problem No. 1." It had narrowed the differentials in income, value added, and wages that distinguished it from the rest of the nation, but large gaps persisted. In 1947, for example, per capita income in the South reached only 63 percent of the national standard. In 1949 a Joint Congressional Committee published another report, *The Impact of Federal Policies on the Economy of the South.* That report, coming eleven years after the *Report on Economic Conditions,* showed that "agriculture is much more important than in the remainder of the country," and that the region's labor system was still beset by a shortage of "industrial capital equipment and relatively simple processes." The Congressional study ended with the familiar call for still more highly mechanized, high productivity industry. To accomplish that purpose, the report concluded, more federal funds to stimulate industrial investment were required.[57]

The region's lingering poverty in 1950 proved that gleaming new auto plants and crowded military bases were not sufficient to ensure regional development. By itself, industrial growth could produce neither the economic progress nor the social and political upheaval which many federal officials had prophesied in the late 1930s. The South's rate of industrial growth, after all, had consistently outpaced the rest of the nation since the 1880s. Successful development required roads and airports to serve the factories, municipal services to support the labor force, and education to train workers and managers. And economic progress also depended on local political systems amenable to economic growth and capable of managing it. In short, the sort of economy envisioned by the New Dealers demanded a new balance of power in the South. The assistance and the prodding of the federal government—the assault on low wage employment and the promotion of industry—had eroded the power of traditional elites and encouraged the emergence of a new leadership.

While this new order was not yet ascendant in 1950, another component of federal intervention hastened its appearance—the proliferation of government entitlement programs after 1938. These programs steered disproportionately large shares of their funds into the South, but threatened the region with further dependence on and control by federal authorities. National programs built a regional constituency for economic development, but they also exacerbated southern fears of federal "interference." Nothing dramatized the emergence of the new regional consensus on the desirability of economic growth, or demonstrated the potential for conflict within it so vividly, as the decision of the state of Alabama to alter its Great Seal a few months before Pearl Harbor. Bowing to wishes of businessmen concerned about the South's reputation for laziness, the Alabama state legislature removed from the seal the motto "Here We Rest." But portending the struggles over the politics of entitlement which lay ahead, the legislators replaced it with "We Dare Defend Our Rights."[58]

Chapter 5 ❂ Persistent Whiggery: Federal Entitlements and Southern Politics

In a celebrated remark of December 1943, President Roosevelt replaced "Dr. New Deal" with "Dr. Win-the-War." Few observers, then or later, realized that the President had not retired the old medic, but had dispatched him below the Mason-Dixon line to reorganize the economy of the South. Wage policy shaped a new workforce. It raised labor standards and eradicated the sweatshop, but also eliminated the jobs that had furnished employment to countless workers. At the same time, the federal government constructed thousands of new factories in the South, catalyzing economic diversification and enhancing the region's allure to private industry. No effort, however, so pointedly revealed the continued vitality of reform after 1938 as the proliferation of federal entitlements in the South and the principle of equalization in the disbursement of federal funds. That principle dictated the distribution of grants on the basis of need. Along with the enormous increase in the extent of federal expenditures, equalization channeled a vast share of national resources toward the improvement of the South's deficient public services and economic infrastructure. But these benefits also acted like a cancer on the established political economy of the South. They fomented rapid growth, but simultaneously eroded the traditional bases of political and social leadership.[1]

These spending programs constituted the most visible component of the Southern New Deal. The federal government attempted nothing less than the redress of the economic unbalance between the regions, the readjustment which the *Report on Economic Conditions* had promised in 1938. The *Report* itself had recognized that the "solution" to the South's economic distress required special attention from the federal government, attention in the form of differential aid to the South. Prominent southern allies of the Roosevelt administration echoed that call for federal assistance. Before the Southern Governors Conference, Frank Graham maintained that the "federal government is the only agency which can redress this economic and educational unbalance." Former Georgia governor Ellis Arnall warned that "Discriminations" against the South "in the distribution of federal funds for highways, education, and public health" must end, or southerners would remain "hewers of wood and drawers of water to imperial masters in the East." Lyndon Johnson crusaded for federal intervention in the Southwest, demanding not just "single, isolated projects," but a comprehensive program "related to the whole needs of the region."[2]

But while such complaints of discrimination were hardly novel in the 1940s, the reliance of Graham, Arnall, and other southerners on the federal government to "eradicate the economic plague spots in America" demonstrated a new relationship between the authorities in Washington and growing segment of southern society. The advocates of regional development, in the nation's capital and below the Mason-Dixon line, were casting their lot with the resources of the federal government. Even more striking, they were implying their willingness to accept federal supervision of southern economic development.

During the late 1930s and early 1940s, nearly all the actors on the southern political scene desired increased federal spending in the region. As long as federal dollars could be administered by local units without federal control, the region's political and economic elites shared the enthusiasm of reformers like Graham and Arnall. Influential southern legislators like Texas senator Tom Connally and Mississippi's Pat Harrison, men estranged from the Roosevelt administration by the purge and FLSA fights, joined southern New Dealers Claude Pepper and Lister Hill in the Congressional battle for legislation that would apportion federal grants-in-aid on the basis of need. Public officials like Connally, and the business interests they loyally represented, perceived federal aid as an unconditional bonanza. Proceeding as if nothing had changed from the early New Deal days when southerners controlled federal legislation, Connally, Harrison, and other southern politicians failed to appreciate the potential connections between funds and control. Connally did not heed the warning of a prophetic constituent worried about federal encroachment on his state's politics, economy, and society. "With federal grants coming into the states, it is only proper that the Federal Government should have some right to say what that money is to be expended for," a Dallas realtor explained to his senator in Washington. "If the State of Texas is to preserve its state's rights to the Nth degree, then Texas should be prepared to pay the freight on its own functions instead of relying upon governmental aid and then denying that the government has a right to say anything about the conduct in the state. The old axiom that "the one who pays the fiddler calls the dance," B. F. Farrar cautioned, "applies most poignantly in the question of state's rights."[3]

THE POLITICS OF EQUALIZATION

The pervasive desire for federal aid, even among economic interests and political leaders who were fearful of regulation, advertised two facts about fiscal relations between the southern states and the federal government in the 1930s. First, the widespread support for assistance registered the overwhelming needs of the South. The region lacked both the economic capacity and the political willingness to provide adequate social services to the region's people. Transportation, the best developed of the South's public services, remained substandard; salesmen could still boast that their products had endured the "test of Arkansas mud." Where less effort was directed, the region fared even worse. In 1940 the southern states' per capita expenditures for public health facilities were the most meager in the nation, a bitter fact reflected directly in the region's desperate shortage of hospital beds. The region placed last as well in educational attainment, even among white males. Liberals appealed for improved services for humanitarian reasons, but they, along with business boosters of less noble mettle, recognized that without decent health care, adequate transportation systems, and educated workers, no amount of enticement would lead to industrialization. In fact, as national manufacturing concerns moved into the South in the two decades after Pearl Harbor, displacing some of the low wage businesses that wilted under the minimum wage regulations, the voices of southern industry became increasingly concerned with these accoutrements of economic development.[4]

Accentuating the South's plight was a second feature of fiscal federalism in the 1930s—the bias in favor of wealthier states in the distribution of intergovernmental grants. By distributing funds on the basis of population and requiring states to match the federal contribution on a one-to-one basis, federal grant programs discriminated in favor of the wealthy, urbanized states. So while the South required relatively more aid per capita to bring its services in line with the rest of the nation, it actually received smaller slices of the federal pie. To some extent, this imbalance reflected the region's own suspicion of federal programs, an impulse to safeguard regional politics from national concerns and national administrators. Sam Rayburn, for instance, warned against matching grants in the early 1930s because they made communities "come crawling up and agree with every condition" to receive their share of the money. But it also reflected the mechanics of the grant system. The upshot of the system was that, in 1940, a state's per capita income directly determined its share of federal grant funds. This arrangement had performed well enough until the onset of the Depression, but drew criticism after the scope of the payments and the South's desire for them mounted in the 1930s.[5]

Consequently, a movement to reform this system accompanied the drive for more federal aid to the South. "We would like for the Federal Government not to require the Southern States to put up dollar for dollar," one southern county official requested in 1942. Ellis Arnall assailed the practice of matching funds and implored that the distribution of federal grants reflect need rather than ability to match. President Roosevelt and other Administration leaders approved of such principles in broad outline, and gradually the practice of calibrating matching requirements to local per capita income found its way into many federal programs. Most commonly, the grants set aside a portion of their resources to aid poorer regions. Such provisions—equalization principles—became so prevalent that, by 1962, the relationship between state income and grants had reversed itself and slightly favored poorer regions like the South.[6]

Like the FLSA and other national measures aimed at the one-third of a nation at the bottom of the economic ladder, the introduction of the principle of equalization into the grant system particularly affected the impoverished South. And like those other efforts to restructure the southern economy, the exceptional impact of federal aid programs was not an incidental effect of the region's proverty, but rather one of design. Congressional debates, both the successful and the unsuccessful efforts to legislate preferential treatment for the South, revealed the deliberation undergirding the politics of equalization. Proponents of equalization plans had to face down the opposition of both fiscal conservatives and many representatives of urban areas. The latter believed that variable aid plans discriminated against their districts, which had high incomes, but great demand for public services.[7]

The first such controversy to reach a head involved a southern amendment to the Social Security bill in 1939. The fight centered on federal grants to the states for old age benefits. With the exception of Oklahoma, the southern states paid benefits below the national average. The South occupied the nine bottom spots on the 48-state scale. To rectify that regional deficiency, a group of southern senators initially proposed a variable aid formula, but it failed in committee. Unable to secure preferential assistance, the same bevy of southerners came forward with the so-called Connally Amendment, a plan which simply increased the federal share of the monthly benefit from one-half to two-thirds for the first $15, preserving the fifty-fifty matching requirement for any amount above $15. The proposal favored the South, where benefits stood well below the $15 threshold, but held out hope for Congressional approval since it offered relief to the wealthier states as well. Arguing eloquently in a vein that could have been used against them during the battle over the wages and hours bill, Connally, Harrison, James Byrnes, and Walter George insisted on the federal government's responsibility to maintain minimum nationwide standards. "If the Federal Government has any sense of obligation or responsibility," Connally lectured his fellow senators, "how is that met by paying a man in Arkansas $3.08, as against $15 to a man similarly situated residing in the State of California?" The amendment carried in the Senate by a vote of 43 to 35 with all but three of the 24 voting southern senators in favor of the amendment.[8]

In the Conference Committee, however, the House managers refused even to send the amendment back for a roll call vote. They reported the Social Security bill without the provision. This so enraged the southern leadership that it voted against the Conference report in protest. Connally railed that "the doctrine of the present system" is "to the rich, powerful, and opulent state which is able to pay $30 a month the Federal Government says, 'We will give you $15 per head.'" But, "to the poor State which is in penury and rags, it says, 'We will give you $3.08.'" Nonetheless, the Conference report carried the Senate. Not until after World War II did the Congress adopt a variable aid system.[9]

Successful efforts to expand the federal highway and airport systems further testified to the relentless southern pressure for equalization in the two decades after Pearl Harbor. Southern legislators, especially those from the Southwest, supported the 1941 and the 1944 Federal-Aid Highway laws. These measures helped to finance new highway systems across the South, but they retained the traditional one-to-one matching formula. States received money from the federal treasury only if they were willing and able to ante up significant funds themselves. In 1956, however, Congress departed from the fifty-fifty financing scheme for the first time. Championed by southerners like Texas senator Lyndon Baines Johnson, whose state stood to place second only to New York in its share of the interstate highway spoils, the law apportioned 90 percent of the cost of an extensive system of super-highways to the federal government. The injection of billions of dollars of federal highway funds would realize unprecedented benefits in the South, where the interstate system proved to be a crucial component of regional industrialization. The new highways connected the South with the industrial Midwest, forging the first economical trans-mountain links between the Southeast and areas like Ohio, Pennsylvania, and Illinois. The interstate network also offered prospective southern manufacturers the locations they most desired. It allowed businesses to locate away from cities with their regulations, taxes, crowded conditions, and unions, without sacrificing access to distribution centers like Dallas and Atlanta. After 1940, the South increasingly relied on federal dollars for highway construction (Table 5-1).[10]

TABLE 5-1
The Federal Highway Program in the South, 1940–1960

	South's Share of U.S. Total	*Federal Funds as % Total Funds, South*	*Federal Funds as % Total Funds, U.S.*
1940	27.6	9.6	8.9
1945	34.1	3.1	2.7
1950	27.9	7.9	8.4
1955	30.1	8.4	7.0
1960	31.1	23.1	20.5

Sources: Computed from Bureau of the Census, *Statistical Abstract of the United States, 1941*, p. 450. Census Bureau, *Statistical Abstract, 1947*, p. 493. Census Bureau, *Statistical Abstract, 1952*, p. 493. Census Bureau, *Statistical Abstract, 1957*, p. 549. Census Bureau, *Statistical Abstract, 1962*, p. 558.

A similar pattern prevailed in the evolution of the Federal Airport Program. Speaking for the majority of southern representatives in 1945, Louisiana senator John Overton inveighed against efforts to reduce federal support for local airports. "Why should we be parsimonious in this matter," Overton beseeched his Congressional colleagues, "Why should we undertake to hamper the development of this great modern method of transportation?" Overton and other southern politicians supported this "great modern method of transportation" because a generous national program promised the hitherto undeveloped South a transportation system on a par with any other in the nation. It also cemented southern legislators' alliance with the military establishment. The Pentagon favored airport development, and amity with the Defense Department promised southern politicians access to defense contracts and military bases. On the final Senate ballot on the Airport Act of 1946, a 49 to 32 tally in favor of the measure, 19 of the 24 southerners who recorded preferences approved the bill. The lonely opposing voices belonged to the region's one Republican, three die-hard opponents of all federal spending, and two South Carolina senators, whose state was overlooked by the measure.[11]

Supporting votes quickly bore fruit: in 1952 the region received 26 percent of the authorized projects (accounting for 28 percent of the federal expenditures). Texas received more dollars than any other state in the union. By 1962 the region's share of federal disbursements dropped off slightly, but a fivefold rise in the level of expenditure more than made up for the loss. Thanks to the federal program, southern boosters could entice businessmen seeking industrial locations with promises of modern air transport systems.[12]

As the airport and highway programs demonstrated, southern legislators recognized that equalization measures could better negotiate the Congress if they made some concessions to other regions. Occasionally, southerners like Tom Connally and Lyndon Johnson merely decided to swell the federal government's share of a program's costs. All states benefited from such a scheme, of course, but the poorer southern states gained the most. For without substantial federal assistance, the South could not have provided even a minimum level of services. Furthermore, the South provided much virgin ground for federal investment in transportation and municipal services.

In addition to enlarging Washington's financial participation, the South often secured more straightforward preferential treatment. Measures targeted entirely at the South fared poorly in the halls of Congress, but ones that designated part of their funds to the needy states and reserved a part for general distribution found their way into law. The Public Works Acceleration, Community Health Services Grants, and Hospital and Medical Facilities Construction Program, all directed substantial portions of their allocations to states with low per capita income. The Hospital Program, for instance, enacted in 1946, focused on the South's deficient health care facilities. By 1962, the South received more than 40 percent of the appropriations under that program.[13]

In the two decades after the *Report on Economic Conditions* recommended intensified federal assistance for the South, the government in Washington stepped up the size of its investments in the region and augmented the South's share of those payments. More telling, however, was the region's growing fiscal dependence on those disbursements. In 1940, federal grants accounted for 14 percent of total southern state government revenues, roughly equivalent to the national average. By 1955, the southern states relied on federal grants for 20 percent of their revenues, tripling the percentage gap between the region and the national average. By the end of the 1950s, in every southern state except Florida, the state government's reliance on federal dollars exceeded the national standard (Fig. 5-1).[14]

Those figures pointed to the fact that while the direct economic effects of federal grant programs remained relatively modest throughout the 1940s and 1950s, federal programs played a crucial strategic role in the South, functioning as catalysts for state action and regional growth. States undertook new programs. In Texas, for example, a constitutional stricture had long prohibited statewide relief programs. The state amended its constitution in order to participate in the Social Security programs. While matching grants gently encouraged southern localities to provide new services, projects with full federal financing just went ahead and provided them. Working with Lyndon Johnson, Undersecretary of the Interior Abe Fortas pressed for federal development of the Texas-Oklahoma river valleys, not just aid to states. While the states proposed single-purpose dams for water usage, the feds had the funds and the inclination for multi-purpose projects—ones that would provide for cheap power, irrigation, flood control, and general regional development.[15]

Federal investment affected the economic as well as the political environment. Where the logic of private investment dictated that no transportation system be constructed without a reasonable level of demand for the service, federal programs reversed this logic. In some cases, the government simply built facilities ahead of demand; in others, matching funds altered relative prices so that an airport became feasible despite little local demand. The airport in turn permitted the recruitment of private businesses and government contracts that never would have considered a location without air transportation. Similarly, federal dollars often provided the critical mass for public services like health care. Equalization formulas steered funds to poorer localities where they had the greatest marginal impact, eradicating the region's worst public health problems. If those improvements did not directly advance economic development by enhancing worker productivity, they certainly did so by removing the stigma of rampant disease and inadequate health care that frightened many potential employers from a region. This dependence on the national government

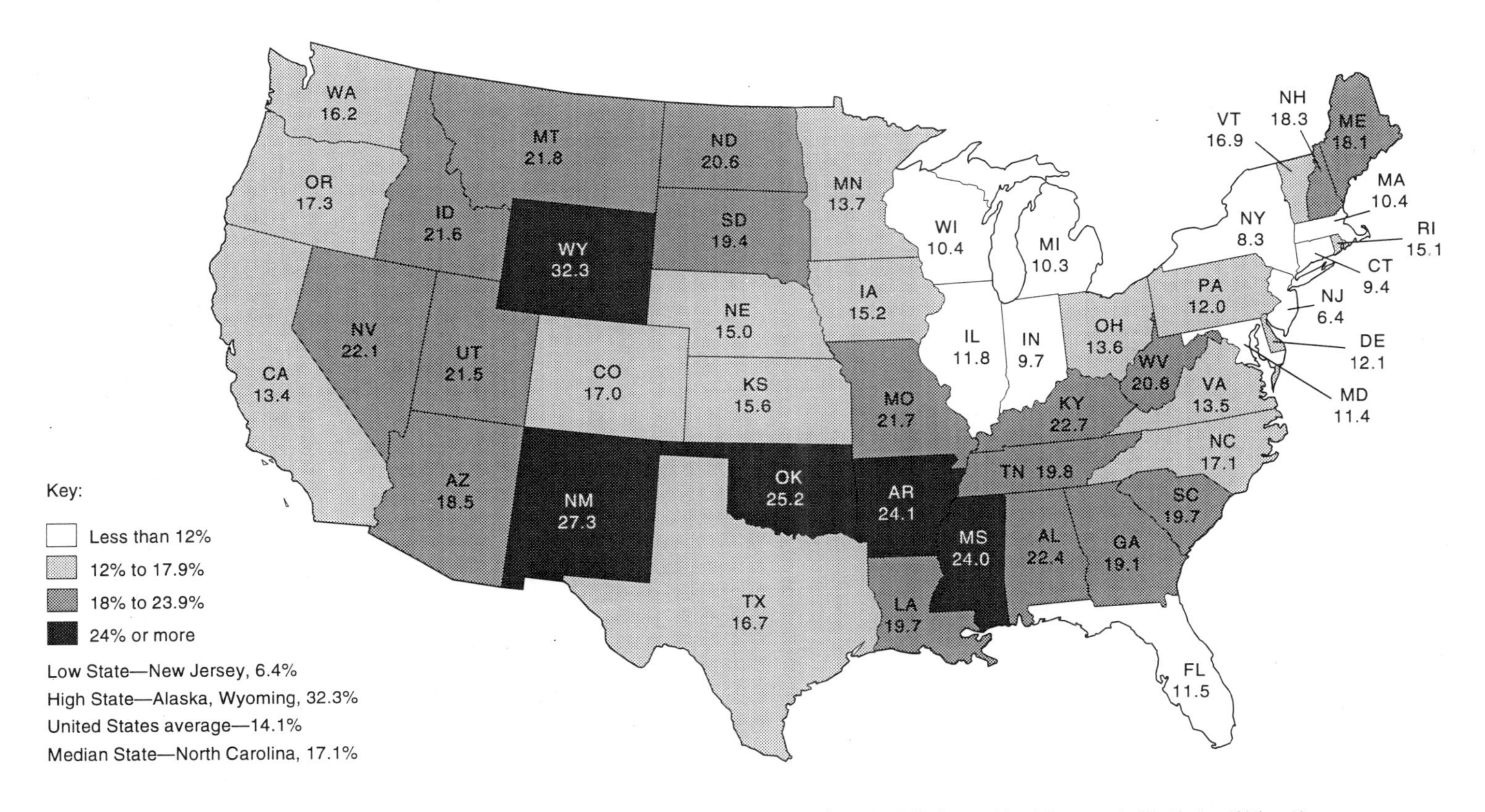

Source: U.S. Advisory Commission on Intergovernmental Relations, *Periodic Congressional Reassessment of Federal Grants-In-Aid To State and Local Governments*, (Washington: 1961), p. 16

FIGURE 5–1 Federal Payments to State and Local Governments As a Percent of All State and Local General Revenue, 1959.

affected not only the consumers of federally provided services, but also the growing army of government employees who administered the programs in the South, and the proponents of development who advertised them to investors outside the region.[16]

Furthermore, the economics of grants conveyed particular benefits to the South. Intergovernmental grants exert both income and relative price effects. Income effects enhance the income of a given locality. Since receiving grants (without reducing taxes) is a more palatable political alternative than raising taxes when private income increases, grants tend to increase the funds available for public spending more than an equivalent rise in the community's private income. This "flypaper effect," or "money sticks where it hits," had particular resonance in the postwar South, where a powerful consensus against higher taxes dominated the political landscape. Both the traditional elites, who preferred minimal government expenditures, and the advocates of development, who sought improved public services but feared the effects of high taxes on industrial recruitment, shared the proclivity to spend federal money with as little local contribution as possible. The sliding scales of many matching requirements reinforced that political impulse, furnishing southern states the means to purchase some minimum level of services at lower relative prices than those paid by governments in other regions.[17]

"THY POVERTY AND NOT THY WILL"

Most important, the federal presence introduced a new force into southern political economy, one which greatly outstripped the merely economic effects of the grants. While national labor standards and the defense boom stripped the South of its economic isolation, the grants broke down the political isolation of the region. They pried open the closed one-party systems that had prevented the projection of national issues into state arenas and had protected the entrenched elites from any viable challenge. Despite the considerable economic advancement of the war era, the region experienced a sort of cultural lag, a period during which political institutions and social attitudes straggled behind economic change. Most southerners desired economic progress in the form of federal financial assistance, but resisted national intervention in and control of that progress. William Nicholls captured this tension between "southern tradition and regional progress" with a quotation from Shakespeare's *Romeo and Juliet.* "My Poverty, but not my will, consents," the desperate apothecary warns Romeo, before doing the stronger man's bidding. To which Romeo, representing for Nicholls and the South the powerful federal establishment, replies, "I pay thy poverty and not thy will."[18]

After 1938, the federal government involved itself in a southern political landscape which presented a complicated mélange of powerful elites, bitter personal factionalism, and legislative malapportionment. Assembled under the umbrella of the Democratic party, southern politics at the end of the Depression decade confirmed in great measure W. J. Cash's doleful description of the South's "barren" political landscape, "made up on the one hand of honest but complacent Tories and too few men of liberal sympathy to have much practical effect, and on the other hand by a horde of outright demagogues." Although state alignments varied, ranging from the unchallenged supremacy of the Byrd machine in Virginia to the free-for-alls of Florida elections, southern politics maintained certain consistent regional characteristics. Southern states all shared a low level of voter participation. Turnouts for decisive contests (the Democratic primary in most southern states; the general election outside the South) ranged from one-half to one-fifth of the proportion of eligible voters casting ballots in Ohio or New York. Active disfranchisement of black voters accounted for part of this low turnout. The white primary remained the regional practice until it was invalidated by the Supreme Court in 1944, while many states retained poll taxes, understanding requirements, and other barriers to voter registration through the 1950s.[19]

Second, a bias in favor of rural areas distorted the apportionment of southern state legislatures. In Alabama and Florida, fewer than one-third of the people could select majorities in the state legislature. In Georgia, the infamous county unit system gave the state's most populous counties, home to more than 30 percent of the state's population, just 12 percent of the unit votes. The system ensured control of the unit votes, and thereby of all state elections, for the smaller rural counties. A travesty of democracy, the system awarded rural Chattahoochie County a unit vote for just 132 popular votes, while Fulton County (Atlanta) received a single unit vote for every 14,092 votes cast.[20]

This malapportionment demonstrated a third feature of the regional political order—the continued domination of Black Belt elites. The Black Belt—a crescent of land running across the mid-section of the Deep South named for its dark, rich soil—was an area of sprawling plantations, with a large population of vote-less black southerners. Black Belt counties gained the most from disfranchisement and rural over-representation, and through those devices maintained a secure grasp over most of the region's state capitals. They had consolidated that power by protecting it from the only potential challenge to their domination—interference from outside the closed state systems by the national government. The regional elites' strength in the U.S. Senate had long guaranteed that immunity. The New Deal, with initial reluctance and eventual force, assailed that barrier. It was hardly accidental, therefore, that when Southern Democrats bolted the national party in 1948 to protest federal intervention in the region and reassert their regional power, these "Dixiecrat" votes were almost completely confined to the Black Belts.[21]

In fact, the Dixiecrat revolt was only the most celebrated incident in a long, slow response to sustained national pressure on the region's political system. In the twenty years after FDR first undertook his 1938 "purge," the South wrestled with the stresses of federal intervention that it had so long managed to avoid. The ensuing contradictions, conflicts, and confusion led the political scientist V. O. Key to declare that "The South may not be the nation's number one political problem . . . but politics is the South's number one problem."[22]

The most apparent of these stresses was the rift between the southern leadership and the national Democratic party. As the South became increasingly dependent on the federal government, the region's control over national affairs deteriorated. Discontent over specific policies highlighted the South's principal political grievance—the Democratic party's courting of the labor, black, and urban votes that began during the 1936 election, accelerated with the attempted purge of 1938, and became the mainstay of national Democratic campaigns during the 1940s. After the purge, Southern Democrats slowly edged away from their northern colleagues, building a Congressional alliance with Senate Republicans and threatening secession from the party during FDR's third term. The warnings came from high places, from North Carolina's Josiah Bailey on the Senate floor and from Governor Sam Jones in the Louisiana State House, and burst into outright rebellion during the 1944 election. These defections from the Democratic party met with little success. The electoral victories of New Deal champions Claude Pepper and Lister Hill dampened the insurgency in the Southeast, while in Texas, the regular Democratic ticket defeated a slate of unpledged anti-Roosevelt electors called the "Texas Regulars."[23]

Even had that rebellion succeeded, it could not have curbed the ongoing evolution within the national Democratic party. By 1944 the arithmetic of national elections discounted the South; the votes of organized labor and northern blacks were an essential component of Democratic victory. Old Guard southerners briefly held out hope for a return to the older Democratic coalition with the ascension of Harry S Truman. Southern members of Congress regarded Truman as a loyal party man less interested in ideological alliances than FDR, especially after the rout of liberals in the 1946 mid-term elections. But those hopes soon evaporated after Truman vetoed the Taft-Hartley Act and announced his civil rights program.[24]

Indeed, Truman alienated the South more than Roosevelt ever had, adopting the electoral strategy outlined by his adviser Clark Clifford in a memorandum prepared a year before the 1948 elections. Clifford recommended bold and concerted efforts to protect the progressive, labor, and black votes, calling for a dramatic program "tailored for the voter," rather than for Congressional compromise, since the Republican Congress would never consider any such program. Accurately predicting Thomas E. Dewey as the Republican standardbearer and the third party challenge of Henry A. Wallace, Clifford's memo rejected any conciliation of the South, which Clifford believed "could be considered safely Democratic." The prescient White House aide miscalculated on that last point, losing four southern states to the Dixiecrats, but proved wise on Election Day when Truman scored a shocking upset. Black votes proffered the narrow margin of victory in the pivotal states of California, Ohio, and Illinois. Truman's surprise victory also completed the de-southernization of the National Democrats. "I was happy and pleased to be elected to the presidency by a Democratic party that did not depend upon either the extreme left-wing or the southern bloc," Truman maintained. "The fundamental purpose of our campaign in 1948," Truman reflected in his memoirs, "was to put the Democratic party on its own feet and leave it intact. This was achieved."[25]

Perhaps Truman achieved his fundamental purpose in 1948, but an "intact" Democratic party had come to mean one without the support of southern political leaders. Truman's sweeping endorsement of a liberal "Fair Deal" in his 1949 State of the Union address exacerbated the growing disaffection of southern Democrats in general and the Black Belt leadership in particular. Many such southerners, while remaining Democrats for local elections, cast their first Republican votes in the 1952 presidential election. In 1952, Dwight D. Eisenhower carried four southern states, becoming the first Republican to dent the "Solid South" since Herbert Hoover in 1928. Ike not only swept the traditional Republican bastions in the hill country, but also triumphed in the citadels of traditional Democratic loyalty, the Black Belts of South Carolina, Louisiana, and Mississippi. In the five presidential contests before 1952, no Republican had won more than 6 percent of the vote in Mississippi or South Carolina; Eisenhower garnered more than 40 percent. And this "Presidential Republicanism" persisted after 1952. Eisenhower, who had won Texas, Tennessee, Florida, and Virginia in his first election, added Louisiana to his column in 1956. With Lyndon B. Johnson on the ticket, Texas and Louisiana returned to the Democratic fold in 1960, but the Republican ticket showed surprising strength even in LBJ's home state.[26]

Outside of the hill country, which had remained Republican since Reconstruction, and the dissident Black Belt, the South's growing cities and suburbs provided the most fertile ground for Republican gains. Eisenhower's support from the cities of Texas, for example, had accounted for his victories in the Lone Star State. Republican gains in the region's cities were symptomatic of the impact of economic change on political competition. Eisenhower had, after all, counted the South's most urbanized and economically vital states—Texas and Florida—among his electoral prizes.[27]

In fact, the growth of manufacturing directly fueled the political opposition to the traditional southern ruling class. Industrial workers formed decisive elements of the victorious coalitions for Lister Hill and John Sparkman in Alabama, and Claude Pepper in Florida. They also helped Gordon Browning defeat the candidate of powerful Memphis boss E. H. Crump for the Tennessee governorship. Nowhere was the link between industrialization and opposition politics tighter than in Georgia. There, in 1946, James V. Carmichael, the candidate endorsed by Ellis Arnall to succeed the reformist governor, won a plurality of the popular vote but lost the election to Eugene Talmadge on the county unit vote. In that election, the map of Georgia counties with substantial manufacturing employment corresponded almost exactly to the list of counties casting their ballots for Carmichael. Indeed, in several signal Georgia elections during the 1950s, urban areas posted the strongest opposition to the ruling Talmadge faction, with booming Atlanta furnishing the staunchest dissent. The voting trends seemed to validate the Southern Conference for Human Welfare's prediction that the postwar South was "ripe for a political turnover. It will happen whenever the new army of industrial workers realizes the power within its reach."[28]

Outweighing the votes of the "new army of manufacturing workers," however, was the disposition of their new employers. Before World War II, southern manufacturers formed a major segment of the southern elite. They joined with planters in opposition to federal intervention, municipal services, and investments in education. Those industrialists approved rural areas' domination of state political systems. They were content to reap, in low wage rates, the benefits of the region's economic isolation. Disheartened by the durability of this alliance, Jonathan Daniels lamented his failure to enlist southern businessmen in the service of reform. "I was a little disturbed to see the list of signers restricted to such do-gooders as editors, preachers, labor leaders and women," Daniels wrote to Virginius Dabney about one southern liberal communiqué in 1943. "Do you think there will ever be a time," he grumbled, "when we can get the businessmen and politicians in the South interested." This mutuality of interest between manufacturers and the rural-based political elites weakened reform efforts. In Virginia, for example, the Byrd machine owed much of its success to its cooptation of the state's business organizations.[29]

The development of new industry in the postwar South, however, divided the business community from the entrenched leadership. The larger durable goods manufacturers were less dependent on cheap labor. Often the branches of northern corporations, they sought thorough regional economic development. Unlike their predecessors, these employers required access to distant markets, as well as to skilled workers and experienced managers. Training such laborers and drawing such managerial talent required adequate health care, competent schools, clean cities, and reliable transportation systems. These demands placed a premium on efficient, growth-oriented state governments that could accommodate the federal government in the era of growing national expenditures and entitlements. As Rupert Vance put it in 1955: ". . . the dominant psychology of the South is no longer agrarian; it is Chamber of Commerce." A new "Whiggery," a development party not hesitant to use government to accomplish its end, emerged in the South along with the process of industrialization.[30]

The alliance between business and reform that Daniels desired arrived soon after V-J Day. The first manifestations of this new Whiggery were the so-called "G.I. Revolts" which captured several southern cities. In New

Orleans, a group of reformers bent on promoting the city's economic development defeated the Maestri machine, a conservative oligarchy of Black Belt planters, sawmill owners, and oil barons. Comparing the rubble of Europe to the "the rubble of New Orleans with its deplorably cracked and dirty streets," the reformers' candidate, an ambitious returning veteran named DeLesseps S. Morrison, wrested the mayoralty from Maestri in 1946. Once in power, Morrison pursued a campaign of civic improvements and aggressive public relations, designed to lure industry to New Orleans. Across the border in Arkansas, another veteran, Marine Lieut.-Col. Sid McMath, attacked the corrupt leaderhip of Hot Springs in 1946 and turned that local victory into a successful quest for the governorship two years later. McMath devoted his administration to vigorous highway improvement and industrial development campaigns. A similar movement for efficient modernized government by the veteran-led Independent party ousted the leadership in Augusta, Georgia. Sensing this widespread political insurgency, one respondent to Jonathan Daniels's query about the state of the postwar South saw "a fight to the death between the old guard poll-taxers and the young Democrats."[31]

Alabama also turned to a G.I. reformer, Big Jim Folsom, even if Folsom owed more to the populist than to the Whig or business progressive tradition. Just thirty-seven years old when he won the governorship in 1946, Folsom was as much a veteran of the New Deal as the war. He had built his political career as a CWA and WPA administrator during the 1930s and only enlisted in the military to sweeten his political résumé. But even that feat took some doing; the six-foot eight-inch Folsom was initially rejected as too tall. After a brief stint in uniform, Folsom returned to the political trenches, recruiting many young war veterans into his campaign, men who shared Folsom's dissatisfaction with the established order. Personal scandal eventually sullied Folsom's career—he was too fond of wine, women, and song—but, as governor, he pressed for old age pensions, canal and road development, and legislative reapportionment to weaken conservative control of state government. Unlike many of his contemporaries, flashy Deep South politicians of a similar populist bent, Folsom never exploited the race issue. He pressed instead for Alabama's economic renewal.[32]

These "G.I. rebellions," significant as they were, did not yet represent the ultimate victory of the new drive for growth. Indeed, the Old Guard continued to hold sway through most of the postwar South, largely because economic change had not then proceeded all that far. The G.I. revolts signaled, however, the emergence of a new breed of southern politician. Extensive federal intervention had shattered the foundations of the regional political economy. Labor legislation closed low wage plants; agricultural policy encouraged mechanized farming and migration off the land; defense spending and grant programs injected new industry and new government restrictions into the region; and the Supreme Court threatened racial segregation. At no time since the Civil War had southern political leaders more acutely felt the need to shield themselves from the interference of the national government, but at no other time had they been so powerless to do so. Undercut by these national policies, which integrated the South into the national economy, and by shifts in the political balance of power, which vitiated the region's effective veto over national policy, the Old Guard found itself driven toward an untenable position.

The crumbling foundations of traditional southern state's rights conservatism were exposed in 1949 when Richard Russell (D-Ga) rose from his seat in the Senate chamber to introduce Senate Bill 708, a massively subsidized plan for the relocation of black southerners. Leader of the southern conservatives in the 1940s, the austere Russell was hardly the sort of man to bother his beloved Senate with what he knew to be a ridiculous plan. For Russell, the Senate was his monastic order and the South—the New South of economic oligarchy and white supremacy—was his church. Moreover, Russell was no friend of big-spending social policy. The proposal of S. 708 betrayed deep frustration, a sense that the rules of the game were changing.[33]

"If the rest of the Nation is determined to force their views upon the southern people and use the Federal power to revolutionize the social and political relations in the South," Russell declared, alluding to Truman's civil rights program, "common fairness would demand that they assist in equalizing our racial problem with that of the rest of the Nation." The dour Georgian proposed the establishment of a "Voluntary Racial Relocation Commission," armed with an enormous $4.5 billion budget, to move blacks out of the South and whites into it. This commission would guarantee migrants superior housing and better job opportunities than they had enjoyed in their native region. It would continue operations until all the states of the Union, on a percentage of population basis, had the same racial make-up.[34]

Russell's bill quickly died in committee. It was protest, not policy, even if the national government eventually encouraged just such a migration by other means. But the form of Russell's protest, a massive federal program, demonstrated that the order of battle in the South had changed. Competition between southern liberals favoring federal intervention and an Old Guard devoted to state's rights and small government no longer defined the field. Southern conservatives did not demand freedom from all federal interference; Russell, who as chairman of the Senate Armed Services Committee lobbied tirelessly for defense contracts for his home state, recognized the danger in that course. No, Russell and his allies wanted federal intervention, as long as it did not menace their political power, their economic domination, or the system of white supremacy. Still, that was a strange potion to brew, even if Russell, as we shall see in Chapter 6, would soon concoct it with some success. In the late 1940s, with the national Democratic party continuing to alienate its once stalwart southern wing, that brew seemed beyond the Old Guard's devices.[35]

THE ECLIPSE OF THE SOUTHERN LIBERALS AND THE RISE OF THE WHIGS

Ironically, southern New Dealers found their situation equally vulnerable. The women and men who had entered the federal government to reform the South during the 1930s left government service soon after FDR's death. They had advocated far-reaching socioeconomic change—unionization, industrial development, public power, resettlement of tenants, federal relief programs, minimum wages. They also believed that such reforms would ameliorate the economic distress at the root of racial tensions. Still, most southern liberals, whatever their personal feelings about white supremacy, thought segregation impregnable. They generally opposed integration and feared, in Jonathan Daniels's words, that the nation seemed "to be back to the extreme abolitionists and the extreme slaveholders in the lines of discussion." One southern liberal captured this sentiment in 1943. "No white southerner," he said, "can challenge the statement that the Negro is entitled . . . to full civil rights and economic opportunity," but "there is no power in the world—not even in all the mechanized armies of the earth, Allied and Axis—which could now force the southern white people to the abandonment of the principle of segregation."[36]

Such a halting, indirect program for improving race relations became obsolete under the onslaught of the civil rights movement. Northern liberals, black rights organizations, and the national administration rejected the southern liberals' conservative course on race relations. Many southern New Dealers continued to emphasize economic underdevelopment as the source of black discontent, but this view became increasingly outmoded. It was soon associated with a conservative, even a recalcitrant reactionary position.[37]

At the same time, however, the advancing Cold War shifted the economic program of American liberalism to the right. Southern liberals found themselves behind national opinion on matters of race, but far in front of it on economic issues. The emergence of militant pro-civil rights sentiments coincided with a declining constituency for New Deal-style economic reform. Claude Pepper found little support for his health care initiatives, Lister Hill for his rural telephone program. Like many of his fellow southern liberals, Arkansas congressman Brooks Hays felt hemmed in on both side during the 1940s. "Northern liberals," Hays lamented, "were making more difficult the role of Southern progressives who wanted real progress."[38]

In the 1940s and 1950s, American liberalism stressed economic growth rather than redistribution, consensus rather than conflict between the "economic royalists" and the working man. Liberal policymakers no longer felt an emotional bond with the masses or distrusted the capitalist system. Keynesian economic management had become the tool of their trade; prosperity their principal economic object. This renewed faith in

American capitalism, like support for civil rights, was linked to the era's pervasive anti-communism. Liberals sought to demonstrate the superiority—economic, political, and spiritual—of democratic capitalism to totalitarian communism. They lacked the desire to rebuild institutions and to remold the economy that had animated liberals of the New Deal generation.[39]

Not surprisingly, then, the southern liberals who seemed poised to take on a pivotal role in national policymaking in the late 1930s found themselves virtually without influence a decade later. Seeing the war as a main chance for the economic rehabilitation of the South, some southern liberals knit their objectives too closely into the defense program. Aubrey Williams attempted to revitalize the National Youth Administration by making it an adjunct of the defense build-up. So did the directors of the TVA and the FSA. War plants and military bases followed, as did public works justified by military necessity. But such boodle had little peacetime justification and came without liberal politics, unions, or public welfare services. It benefited the South as a region, not its ill-housed, ill-nourished, ill-clad people, and it drained some of the reformist energy from southern liberalism. After 1945, those southern liberals who resurrected the old economic agenda faced charges of truckling with communism. Claude Pepper's opponents branded him "Red Pepper." The Southern Conference for Human Welfare, and its educational wing, the Southern Conference Educational Fund, were similarly tainted. The leaders of those organizations, Clark Foreman and Aubrey Williams, escaped relatively unscathed, but their influence as policymakers had evaporated.[40]

Old Guard conservatives felt besieged by the national Democratic party and the civil rights movement, and so did the southern New Dealers. Into this chaotic landscape, then, stepped the new Whigs, southern politicians who numbered economic development as their first priority and who looked to the federal government to underwrite the effort. These new Whigs diverged from the established leadership in their recognition that social change would accompany economic development. They differed also from such New Dealers as Claude Pepper and Frank Graham in that they sought assistance for the downtrodden only as a secondary means to promote regional industrialization and to remove the stigma of southern backwardness. Accordingly, they favored development-related expenditures like airports and education more intensely than welfare programs, a preference revealed on the ledgers of the state governments they ran during the 1950s and 60s.[41]

These "Whigs" eventually became the pivotal figures in southern politics. The "Other South" of the twentieth century—white voices that dissented from the savage ideals of unbending segregation, massive resistance, and a return to complete political and economic isolation from the rest of the nation—was composed not so much of the small coterie of white southerners who actively opposed the racial caste system, as by the much larger and growing faction of development advocates. Like the business progressives of the 1920s, these politicians emphasized industrialization as their top priority. But unlike their 1920s forebears, the Whigs recognized the South's interdependence with the rest of the nation—its dependence on federal largesse and on the private investment that invariably followed. They realized that industrialization combined with economic and political isolation—the recipe of the business progressives—would fall flat in the post-World War II South.

Still, these development-oriented politicians almost never overtly challenged white supremacy. They simply opposed self-defeating resistance to desegregation, especially visible resistance, arguing that economic progress either required or was certain to accomplish some changes in race relations. The increasingly familiar sight of southern governors personally recruiting new industrial installations and campaigning for increased federal aid demonstrated not only their commitment to industrial growth, but also the increasing potency of economic development as a campaign issue. A candidate had to avoid being "out-segged," as many victims of the ever-powerful race issue learned, but they could no longer afford to be out-developed either. That new political imperative offered potential opponents of the established leadership an escape route, a tunnel out of the tyrannous grip of the race issue that had so long afflicted southern politics, thwarted reform efforts, and isolated the region from the rest of the nation.[42]

This tunnel opened slowly. Georgia's Ellis Arnall first ventured through. Born in the North Georgia foothills in 1907, Arnall grew up in Newnan, Georgia, a town he later described as "exceptional among Southern cities of its size only because there was a larger, more firmly established middle class than in most communities that were farm and textile towns." Determined to build an independent, prosperous middle class throughout the state, Arnall launched his political career at the tender age of twenty-five. He won election to the state legislature in 1932, soon after taking his law degree at the University of Georgia.[43]

Arnall owed much of his philosophy to the regionalism of Howard Odum. He believed that the "narcotic that has been keeping the South in her twilight step is poverty." For Arnall, economic deprivation formed the sole barrier between the South and the fruits of national prosperity and democracy. "There is no problem in the South," Arnall proclaimed, "that does not have its origin in the poverty and exploitation of this region. Injustices, instances of racial friction, inadequate opportunities for education and lack of opportunity for economic advancement are all attributable to the low income of the people of our section."[44]

In Arnall's mind, economic deprivation was at the root of racial conflict, and of the undemocratic reactionary politics which thrived on it. "Wipe out poverty, and the friction will become negligible," Arnall promised. "And the demagogue will be robbed of his chief stock in trade." Development would not "solve all the South's social and economic problems," the Georgian conceded, "but gradually, with individual incomes on a parity with the rest of the country, state government will decrease their tendency to be one-ring and sometimes one-man circuses."[45]

The one ring Arnall knew best was his native Georgia, and the one man he had in mind was Eugene Talmadge, the "Wild Man from Sugar Creek." Like Mussolini, Talmadge deliberately arrived late at political gatherings. His dramatic entrances stole the show from other candidates. Talmadge also held rallies in small venues to ensure "overflow crowds." Self-proclaimed Messiah of the dirt farmer, Talmadge was a notorious racial demagogue, foe of federal programs, and opponent of expanded municipal services. "The poor dirt farmer ain't got but three friends on this earth," Talmadge told his impatient, cramped campaign audiences. They were "God Almighty, Sears, Roebuck, and Gene Talmadge."[46]

This strategy twice won Talmadge the governorship. But in 1942, Arnall exploited Talmadge's conflict with the University of Georgia to dramatize the costs of obstinacy and the benefits of economic progress and defeated Talmadge for the governorship. Darling of northern liberals, Arnall cultivated close ties with FDR through Jonathan Daniels, and supported federal intervention in the region. He avoided the taint that might derive from inviting Yankee carpetbaggers by portraying increased federal aid not as generous or necessary assistance, but as an obligation to end the "colonial exploitation" of the region. Arnall vigorously criticized "discrimination" against the South in federal grant outlays.[47]

While in office, Arnall extended the suffrage to younger voters, secured the independence of Georgia schools, and abolished the poll tax. Despite his popularity, he failed to amend the state's constitutional prohibition against a governor succeeding himself. Arnall threw his support in the 1946 election to James V. Carmichael, whose strong candidacy accentuated the progress of the Whig coalition and whose defeat confirmed how long a road it had yet to travel.

Carmichael won the popular vote, polling his strongest pluralities in urban, industrial areas, but he lost in the decisive county unit vote. And although the Arnall program won widespread public support, it proved no match for the race baiting of Carmichael's opponent. After Eugene Talmadge's unexpected death, his son Herman won the chaotic battle to succeed his father by literally seizing the State House. The Talmadge machine thereafter consolidated its power and held the governorship until 1962.[48]

Despite the defeat of the Arnall forces in Georgia, other challenges to the region's reigning leadership soon followed. In Virginia, Francis Pickens Miller, one of the original Southern Policy Committee members in the 1930s, attacked the Old Guard, narrowly losing the 1949 contest for governor. Opposition to the Byrd machine reemerged in 1954, when the "young Turks," proponents of expanded public services in the state legislature, rebelled. In Alabama, Big Jim Folsom stressed economic issues to gain the governorship in 1946; forbidden by statute to succeed himself in 1950, Folsom won a second term in 1954, successfully ignoring his opponent's race baiting. In North Carolina, Agriculture Commissioner Kerr

Scott became governor in 1948. Scott embarked on a road expansion program and appointed Frank Graham, the South's most prominent liberal, to a vacant U.S. Senate seat. In Tennessee, Governor Gordon Browning and Senator Estes Kefauver overcame the ruthless opposition of the Crump machine to win their statewide offices.[49]

Even after the Supreme Court's decision in *Brown v. Board of Education* temporarily arrested the trend toward moderation, making political conflict over segregation unavoidable, this Whiggery survived, waking from its dormancy with enhanced potency. While *Brown* initially weakened candidates who were unwilling to advocate massive resistance, the ensuing strife, with its closed schools, boycotted businesses, and deployment of federal troops dramatized the benefits of moderation. Resistance also threatened the region's dependent relationship with the federal government. Cutoffs of federal funds to recalcitrant states loomed closer after desegregation became the law of the land. In that charged atmosphere, many pro-development racial moderates won election at the height of the desegregation conflict. Pledging to put prosperity for the Sunshine State ahead of all else, Leroy Collins defeated a committed segregationist for the Florida governorship in 1956. In Georgia, Carl Sanders, a segregationist, "but not a damn fool," defeated the Talmadge candidate, arch-segregationist S. Marvin Griffin, with the promise that "I won't cause your state to be spread across the headlines all over the nation and cause you embarrassment."[50]

In North Carolina, Luther H. Hodges entered office promising to resist integration, but he muted his opposition to *Brown* when it threatened his industrial recruitment efforts. Hodges possessed unquenchable energy; indeed, one associate withdrew from one of Hodges's industry hunting expeditions because his doctor warned him against the folly of trying to keep up with Governor Hodges. A self-proclaimed "Businessman in the Statehouse," Hodges claimed that his purpose in seeking public office was to implement his "pet theory" that "the sound principles of good business could and should apply to government." This ordinary businessman proved a deft politician during the school desegregation crisis. Without actively advancing integration or derailing economic progress, Hodges defused black protests. He preserved his own and the state's reputation for "moderation" and kept its aggressive economic development program on track. "Industrialization, then," Hodges claimed, repeating the creed of the southern Whigs, "with all its advantages to the people and the state, became the number one goal of my administration."[51]

A few years later, TVA director Frank E. Smith celebrated the emergence of this "new breed of Southern politician." No longer "a prisoner of race," the southern leadership was removing the political obstacles to regional economic development. From the vantage point of 1965, with most of the national civil rights legislation in place, Smith could safely tout the arrival of the new breed. But during the conflicts of the 1950s, despite the victories of Hodges, Collins, Browning, and others, that outcome remained uncertain, and many southern politicians of moderate sentiment remained prisoners of race. While business moderates fared well, the region's few genuine liberals suffered setbacks. Claude Pepper met defeat in 1950. So did Frank P. Graham in a bitter contest distinguished by his opponent's race baiting. V. O. Key did not exaggerate in 1949 when he ruefully concluded that "the race issue broadly defined thus must be considered as the number one problem on the southern agenda."[52]

Eventually, Key believed, the abatement of the race issue would uncover a liberal southern political landscape. In Key's view, racial conflict masked a "politics of economics." Without the race issue, the South, with its vast population of impoverished people, would rally behind New Deal-style liberalism. Southern liberals, like those associated with the Southern Conference for Human Welfare, shared Key's view. In 1944, Clark Foreman and James Dombrowski, the two principal leaders of the SCHW, asserted: "There is good ground for maintaining that the South can become, in a very short time, the most liberal region in the Nation." But already by 1950, a race-free southern politics did not promise a liberal future. The victories of Whig politicians testified to the emergence of a conservative, business-oriented politics. The first wave of Whigs—Browning in Tennessee, McMath in Arkansas, Governor Dan McCarty of Florida—had already entered office. They and successors like Collins and Hodges pursued industry, improved highways, and upgraded higher education. They did not concern themselves with the problems of the downtrodden or expand welfare services. And over the long run, regional economic development would only augment southern Whigs' political strength.[53]

While it had yet to reach ascendancy, in the 1950s a new impulse animated southern politics, one derived from the interplay between local politics, development priorities, and federal aid. In his prophetic essay "The South after the War," H. C. Nixon had predicted that the postwar South would "have and desire close ties with the national government as great public programs take shape in the Southern region." Nixon's prediction only hinted at the cycle of dependence that actually emerged. The success of southern politicians came to rest on their ability to advance industrialization. That task in turn hinged on the region's provision of reliable services, adequate schools, and other public goods. And the South's capacity to provide those facilities depended on both the funding and the approval of the federal government. Federal funds guaranteed a minimum level of social services so that localities could commit their entire resources to industrial recruitment. On the other side of the register, the refusal to award any given contract to the South because of regional intransigence on labor standards or race relations cost the region more than the relatively small value of the contract, for such rejections risked future access to federal grants and frightened prospective private investors. And indeed, the region's expanding sector of government employees formed a constituency with an extraordinary interest in maintaining peace with the authorities in Washington. As early as 1948, a group of federal employees led

Mississippi's "loyal Democrats," the faction which stood by the national Democratic ticket during the Dixiecrat insurgency.[54]

The federal government had paid the fiddler, and as B. F. Farrar had warned, it chose to call the dance. In so doing, it helped nurture the new "Whig" breed in southern politics. The quest for economic development need not have carried political moderation in its wake. Indeed for nearly a century the "New South" movement had lived in symbiosis with rigid White Supremacy. But under the pressure of black-rights advocates and federal desegregation orders, the combination of business boosterism and nominal acquiescence in the law proved an acceptable alternative to the politics of massive resistance.[55]

In 1960 the southern economy still lagged behind the rest of the nation. By that time, however, two decades of federal intervention had sown the seeds of the oncoming economic revolution. Federal research facilities and defense contracts were towing new industry into the region, eroding its dependence on low wage, low productivity enterprises and unskilled labor. Along the new highways and air corridors of the South flowed streams of people, black and white, educated and uneducated, into and away from the South. Most important, a new relationship with the federal government had been forged. While many white southerners remained watchful of the federal government, and resentful of its growing involvement in race relations, the southern people needed federal dollars, and their leaders went out to acquire them. When the national government eventually demanded that the South dance to the tune federal dollars had called, whether that melody involved labor standards or segregation, a growing consitituency within the southern polity agreed.

The South followed more willingly for many reasons, not the least of which was that national policy was changing. Federal intervention, stripped of the reform motives of southern liberals, was more in harmony with the new Whig ruling class of the South. The nagging problem of civil rights remained, of course, but as long as southern states muted their resistance to integration, they could usually count on continued federal largesse. No federal spending was more desirable, no tune was sweeter to southern ears, than defense-related programs. As early as 1949, *Business Week* recognized the strategic distinction between military spending and reformist social programs, between "military pump priming and "welfare pump priming." Defense spending, the magazine explained, "doesn't really alter the structure of the economy. It goes through the regular channels. . . ." But "welfare and public works spending . . . does alter the economy. It makes new channels of its own. It creates new institutions. It redistributes income." In short, military spending, understood broadly, offered development without political reform and social change. The leadership of the postwar South would soon appreciate this lesson.[56]

As the 1950s closed, observers of the South like William Nicholls could justifiably view southern traditions as a brake on regional progress, but they could not deny the extent of recent economic change. In 1962, the

Alabama Research Council declared that the "Cotton Economy, which came into Alabama at about the same time as statehood, has ended." The announcement was justified; in 1959, cotton farming accounted for a trifling 1.5 percent of personal income in Alabama. Agriculture in general accounted for but one-third of the share of state income it had in 1939. Manufacturing, finance, communications, and transportation all contributed larger proportions. The "Benighted South" of the 1930s, whose high annual precipitation seemed to combine tears with rain, was about to arise as the "Sunbelt."[57]

Chapter 6 ◎ Missiles and Magnolias

"Our economy is no longer agricultural," William Faulkner observed in 1956. "Our economy is the Federal Government." In characteristic fashion, Faulkner likened the South's surrender of its pastoral heritage to the Edenic Fall. The South, he contended, had "sold state's rights" during the Depression in return for the fruits of the New Deal. In Faulkner's conception, the historical experience of his native land after 1945 recapitulated the path of his characters, the denizens of fictional Yoknapatawpha County. Southern leaders, motivated by pride and avarice, reenacted the drama of the Original Sin. They exchanged the richness of the region's land and people for the foolish promise of progress.[1]

Although Faulkner may have exaggerated, many of his contemporaries shared the essence of his vision. His metaphor captured the region's metamorphosis from guardian against the federal government to suitor of it. Indeed, while some southerners would claim the credit for the economic change Faulkner attributed to the national government, no southerner would deny the extent of economic progress or the federal role in achieving it. By the mid-1950s, federal intervention had sparked an increase in income, an industrial surge, restrictions on the sweat shop, and an expansion of political participation. During the following three decades the national government continued to push in those directions, along a path familiar to southern reformers of the 1930s.[2]

The vehicle of those transformations, however, would be unfamiliar to them. After 1950, federal action in the South did not spring from the welfare agencies New Dealers championed, programs geared toward the bottom one-third of the nation like the Tennessee Valley Authority and the Farm Security Administration. Increasingly, whether it be planting industry, expanding universities and research facilities, or restructuring the regional labor market, the representative of the national state in the South was the military. When Faulkner surveyed the suffusion of national authority through his native region, he could only suspect the intimate ties between the South's future and the nation's masters of Cold War.

Social welfare activities waned as the defense intervention waxed. The South aggressively sought military spending, including those civilian programs that fell under the rubric of national security or promised development without direct aid to the impoverished. It paid homage to and reaped benefits from the defense establishment in the forms of prime contracts, the space program, highway and airport funds, and research grants. But at the same time the South ceased to be the prime target or the principal beneficiary of federal social programs. With the exception of the minimum wage, the effects of which were largely absorbed by 1960, the national welfare state had little impact on southern poverty. And, indeed, the region has been a bastion of hostility toward social programs, even when promulgated by southerners like Lyndon Johnson and Jimmy Carter. Only initiatives like the Appalachian Regional Commission—state-run, growth-oriented agencies—found a warm welcome in southern capitals.[3]

The South's relations with the national government—at first, so peculiar—more and more came to define national attitudes and experiences. United States domination of postwar international affairs inevitably expanded the activities of the national government in some realms. But those same political, military and economic objectives across the globe constrained it in others. In the 1950s, Cold War diplomacy governed national policymaking. This "Soviet effect," as one economist termed it, penetrated all aspects of economic decisionmaking, from tariffs and foreign aid, to educational subsidies and research outlays. Spiraling military activity did not so much displace welfare programs as submerge them below the surface of international tensions. The government and the private sector would go to any extreme to meet the Soviet threat, but refused to take a step in any other direction. As Admiral Hyman Rickover put it in 1959, "If the newspapers printed a dispatch that the Soviet Union planned sending the first man to Hell, our federal agencies would appear the next day, crying, 'we can't let them beat us.'"[4]

Some policymakers, notably in the Truman administration and the Democratic party, welcomed this political atmosphere. They saw in expanded military appropriations a means to ensure full employment, to smooth out the business cycle, and to relieve distressed areas. After the inauguration of Dwight D. Eisenhower, national officials generally rued these developments, but they found themselves unable to resist them. Despite his suspicions of the military-industrial complex and his belief that military spending strained the private economy, President Eisenhower acceded to the expansion of the defense establishment, not merely in size, but in scope. The Eisenhower administration may have rejected Keynesianism in principle, but practiced it through the Department of Defense. It even pioneered the Keynesian technique of manipulating the rate of disbursement of previously appropriated military contracts as an economic stabilizing device. When Eisenhower finally succeeded in restraining military spending, the nation's overall growth rate slumped, even though aggregate consumption remained stable.[5]

In its public pronouncements, the Eisenhower administration justified domestic economic policy in terms of the anti-communist struggle. Ike's Secretary of the Treasury, George M. Humphrey, repeatedly felt compelled to assert the "strategic importance" of the nation's economic health. Such remarks, when combined with its opposition to welfare expenditures, committed the Eisenhower administration to maintaining full employment and a depressionless capitalism, not on the basis of social security, but of international security.[6]

In 1962 a prominent midwesterner assailed the curious division of federal functions and its byproduct—the shift of regional economic vitality engendered by the defense establishment. From the unlikely pulpit of the Select Committee on Small Business, Minnesota senator Hubert H. Humphrey launched an attack on the national security apparatus that had built up the South at the expense of his native region, and the Pentagon at the expense of the New Deal.[7]

Humphrey, one of the leading liberal politicians of the postwar generation, recognized the barriers to the construction of a European-style welfare state in the United States. But despite those obstacles, and all the commentaries on the underdevelopment of the American state they have prompted, Humphrey appreciated the extensiveness and the potency of the defense establishment. The American government might be attenuated in its provision of social services, but the military and its surrogates acting in the name of national security deeply influenced national life. They nurtured leading industries, steered migration, financed physical infrastructure, expanded higher education, and promoted the births and deaths of regional economies. Humphrey neither intended nor attempted to alter those facts. His purposes in 1962 were dual: the narrow objective of claiming for Minnesota and the Middle West what southern senators procured for their home states, and the larger purpose of insinuating the activities of the welfare state within the protected haven of the security state. He hoped to insert aid to depressed areas, jobs for the unemployed, and federal support for public education under the defense umbrella.[8]

In the early 1960s, Humphrey had some cause for hope. The military had desegregated during the Korean conflict, conservative southerners quietly accepting the change as necessary for national security. Military bases soon became "scattered isles of integration" in the South. In June 1962, President Kennedy established the President's Committee on Equal Opportunity in the Armed Forces. The Gesell Committee, as the body was known, inspired liberals and angered southern conservatives, precisely because it threatened the distinction between the national security state and the welfare state. The Gesell Committee recommended that the federal government use its financial might to combat off-base racial discrimination in the South, that it deprive the South of military dollars unless the region desegregated. "The military purse will be used, not to increase our combat readiness and military striking power," complained the New Orleans *Times Picayune,* "but as a means of economic strangulation." In another editorial, the paper committed itself to maintaining the separation between military spending, which it approved, and social spending, which it reviled.[9]

The New Orleans newspaper and the Minnesota senator were not alone in this understanding of the national government. Marvin Hurley, vice president of the Houston Chamber of Commerce, realized the inconsistency of his organization's, and indeed his region's, attitudes about federal intervention. "Can it in good faith accept federal highway aid and oppose federal transit aid," Hurley asked of Houston's business community. "Can it in sound logic accept federal funds for airport construction and oppose airmail subsidies? Can it justify federal research and development contracts for colleges and universities and oppose federal aid to education?" At whatever price in sound logic and good faith, Houston and the South had no trouble answering those questions in the affirmative.[10]

Few Americans even recognized the contradiction, not because of hypocrisy, but because of a binary attitude toward the national government. Social programs, with the exception of Social Security, remained subject to constant political debate in the postwar era. The defense establishment, on the other hand, maintained considerable independence. The Budget Bureau, for instance, hardly supervised military spending. The Pentagon shared the administrative autonomy enjoyed by the European welfare states. Even on the occasions when defense outlays commanded public scrutiny, the demands of secrecy and the technical nature of modern weaponry kept most defense activity outside debate. As Georgia senator Richard Russell explained, ". . . there is something about buying arms with which to kill, to destroy, to wipe out cities and to obliterate great transportation systems which causes men not to reckon the dollar cost as closely as when they think about proper housing and the care of the health of human beings." That situation benefited certain regions, industries, and segments of the labor force. These advantaged sectors composed the political and economic phenomenon that has been called the "Sunbelt."[11]

The South's participation in the sunbelt boom altered its course of development. It transformed southern economic development from a program of catch-up to one of follow-the-leader. From the 1938 *Report on Economic Conditions* until the mid-1950s, federal intervention had been remolding the South into the shape of the rest of the nation. But the sunbelt South of the 1960s and 1970s, while never approaching the national norm in many indices, became an economic and political trendsetter. A shift in its relationship with the federal government steered the South into that second phase. Federal policy reinforced broader economic trends, such as the growth of tourism. And federal activity itself switched from social welfare efforts to reform the South in the nation's image to defense programs that helped to install the region as national leader in aerospace, electronics, and "business climate." The intentions of federal intervention in the South shifted as well. With some important exceptions, most notably the space program, deliberate federal efforts to reform the southern economy halted by 1960. But national policy, however unintentionally, particularly benefited the South. In the region's state capitals, this new brand of federal action delighted Whig politicians. It induced doubts and fears among the remaining southern liberals of the New Deal school.[12]

The continued neglect of the conditions that had earned the region the sobriquet "Economic Problem No. 1" in the first place was the unfortunate consequence of these developments. The South's progress from national economic disgrace to rapid growth region proved not so much a transformation, although considerable transformation occurred, as the importation of a new regional economy. Sitting atop the considerable remnants of "Economic Problem No. 1," the new prosperity did not affect all southerners nor did rapid economic growth eradicate Dixie's heritage of poverty and deprivation. Parts of the South, both geographic and demographic segments, remained nearly untouched by economic development and federal action. Southern development efforts in the 1960s and 1970s may not have brought "juleps for the few, pellagra for the crew" as the New South movement of the nineteenth century had, but the peculiar character of those efforts and their relation to defense spending left a large crew in want and squalor.[13]

DEFENSE ACTIVITY IN THE SOUTH

"We can quickly recognize the influence that defense spending exercises over the economy of the Nation," New Jersey senator Harrison Williams noted in 1964. Any shifts in that government foray into the "economic stream of the Nation," he argued, "whether it be in the number of procurements given out, in the geographic distribution of such procurements, or in the size of the yearly defense budget will have its repercussions throughout the Nation." The senator from New Jersey was not idly repeating the obvious. As representative of a state whose share of military contracts had dropped by almost a third between the Korean conflict and the mid-1960s, Williams desired to expose the military's role in the elevation of the South and the industrial decline of the Northeast. Williams's efforts, however, failed to interrupt the southward migration of defense dollars. As overall military expenditures surged upward, the South's share of prime contracts doubled during the 1950s from roughly 7 percent to approximately 15 percent of the total. After another steep climb in the 1960s, the region's share of defense spoils came to rest at about a quarter of the nation's total during the 1970s (Table 6-1). Texas was the biggest gainer, but the contracts, sub-contracts, bases, and space installations spread around the region. Even Mississippi, the nation's poorest state, jumped from 0.1 percent of prime contracts in fiscal year 1951 to 2.6 percent in

Table 6-1
Military Prime Contract Awards, Selected Years, Fiscal Years 1951–1980
South as a Percentage of the United States Total (%)

Fiscal Year	*South's Share (%)*
1951	7.6
1953	12.5
1955	12.5
1958	16.1
1960	15.3
1963	15.8
1967	23.5
1970	25.4
1975	22.0
1980	24.2

Sources: U.S. Dept. of Defense, Directorate for Information Operations and Control, *Prime Contract Awards, by State, Fiscal Years 1951–1977* (Washington, n.d.) and U.S. Dept. of Defense, Directorate for Information Operations and Reports, *Prime Contract Awards, by Region and State, Fiscal Years 1978, 1979, 1980* (Washington, n.d.).

fiscal 1978. With the NASA test facility in Bay St. Louis, Mississippi, "the closed society" became one of the gateways to the moon.[14]

The instincts of northern legislators like Williams and Humphrey proved accurate. The military had anchored southern growth in the decade between 1952 and 1962. According to estimates by the Brookings Institution and the Department of Labor, defense income contributed more than 20 percent of income growth in Mississippi, and 10 to 20 percent of income growth in Texas, Alabama, Georgia, Florida, North Carolina, and Virginia. Only the Far West felt a deeper influence. Humphrey's Minnesota was among the states where losses of defense work substantially depressed income growth. The various indices of defense dependency, for both employment and income, placed the South above the national average throughout the postwar era. The region finished second only to California and the Mountain states. Even in parts of the South with low dependency ratings, states like Tennessee and Louisiana, the defense industry became those states' largest employer by 1976.[15]

The South so boldly proclaimed its association with defense that *Southern Exposure,* a liberal southern magazine, devoted its entire 1973 inaugural issue to an exposé and critique of southern militarism. "Time was," the first number of *Southern Exposure* recalled, "when the only way for a poor southerner—black or white—to escape the region's poverty was to join the Army. Now, if he's lucky, he can get a job in a defense plant in one of the small, rural towns that punctuate the southern landscape." Or, an impoverished southerner could migrate to "a New South metropolis and look for a factory or warehouse job; but even here, he would likely find himself working in an industry that got (or gets) its stimulus from the demand of nearby military bases or far-off wars." In 1973 more southerners worked in defense-related industries than in textiles, synthetics, and apparel combined. Defense dollars permeated nearly every town in the region.[16]

Defense dependency perpetuated itself over the postwar years. For practical reasons, such as the existence of already constructed facilities, defense dollars tended to stick where they first hit. The dormant Michoud Aircraft plant, a holdover from Andrew Jackson Higgins's wartime exploits, drew several military ventures to New Orleans, including the Apollo rocket assembly factory. The government-owned Bell Bomber plant in Marietta, Georgia, then the largest factory in the South, was reopened by Lockheed in 1951.[17]

Political considerations reinforced the geographic concentration of defense industries. Local constituencies and favored contractors fought to keep defense facilities operating. Moreover, regional specialization in defense resembled other forms of economic specialization, gearing the local political economies of many southern communities to the war plant in the same way they had previously centered around the textile mill or the cottom plantation. Marietta, Georgia, for instance, the home of Lockheed-Georgia, the largest single industrial organization in the Southeast in the 1960s and 1970s, employed workers in fifty-five of Georgia's 159 counties. With more than 90 percent of its business in defense contracts, Lockheed-Georgia not only paid several hundred million dollars a year in wages and salaries, but also served as the state's largest customer and supporter of technical universities. The aircraft giant purchased everything from soft drinks to metal parts from local suppliers.[18]

In addition to subcontracts and expanded demand for industrial, consumer, and public goods, federally subsidized industries occasionally introduced innovations which altered what economists term the "production functions" of extant regional industry. That is, they changed the relative proportions of capital and labor used by a firm. The aircraft industry, for example, pursued and perfected numerically controlled computer automation of production in the 1950s and 1960s. The new processes offered advantages in performance and in management control of the workplace, but were not cost effective. Without cost-plus military contracts, the industry might not have pursued numerical control. Whether or not the new technology represented the most efficient course is debatable; considerable evidence suggests that it did not. But developments in aircraft manufacture pointed up the peculiar nature of the defense market, with its barriers to exit and entry, guaranteed profits, and subsidies for innovations that could not proceed in the open market.[19]

Defense production was also vulnerable to cyclic fluctuations. Partially shielded from the business cycle, defense orders ebbed and flowed as political factors influenced the military budget. For instance, in 1964, Defense Secretary Robert McNamara closed 95 military bases and embarked on an aggressive cost cutting plan. A wartime build-up followed in the late 1960s, succeeded by another round of cuts in connection with procurement scandals in 1970.[20]

Southern politicians fought to insulate the region from these fluctuations and maintain a steady flow of contracts. This strategy, however unwittingly, preserved the region's ancient role as commissary to the nation and its armed forces. The South provided the Pentagon with textiles, tobacco, coal, and food. The region did so happily because the military's demand for such items remained stable; it was far more secure and predictable than the lucrative "weapons roller coaster," even if less stimulating economically. In no other region did weapons account for so small a share of total defense contracts.[21]

Ultimately, these peculiarities of the defense economy might derail southern and national economic progress. Certainly, the heady proclamations that space and defense research would spin off new consumer goods and industrial technologies proved false. With the exception of the aircraft industry, efforts to translate military technology to commercial use failed. Defense contractors lacked marketing skills and the ability to produce high volume at low unit costs. By the late 1960s, commentators acknowledged the "locked-in nature of defense resources." They argued that space and defense industry drained talent from the private sector and railroaded the American economy down a dead-end track. Defense spending did not set off the chain reaction of interactive innovation and economic growth that propelled the American economy in the nineteenth century. Nonetheless, the South had no choice but to hitch its economy to the defense train. No other source offered the technological community and the skilled employment opportunities the region desired.[22]

The effects of that intervention were so pronounced as to lend credence to the warning of John Crowe Ransom, the unregenerate Nashville Agrarian of the 1930s. With Cassandra-like prescience, Ransom had beseeched the South to avoid participation in the nascent military-industrial complex. "Our vast industrial machine," Ransom had warned in 1930, "with its laboratories of experimentation, and its far-flung organs of mass production, is like a Prussianized State which is organized strictly for war and can never consent to peace."[23]

Whatever its economic dependence on the national security state, Ransom's region, the agrarian South, became fortress Dixie. During the 1970s, the *New York Times* reported that defense was the single largest employer in four southern states, outpacing agriculture, textiles, lumber, and all the others. And of America's top ten defense contractors, seven, including Lockheed, McDonnell-Douglas, General Dynamics, and Rockwell, operated large installations in the South.[24]

The changing geography of defense production reflected the changing character of defense demand during the postwar decades. In fiscal year 1953, tanks, vehicles, weapons, and ammunition constituted more than half of military hard goods delivered; they made up only 12 percent of prime contract awards in FY 1961. Meanwhile, expenditures on missiles soared from less than one percent to more than one-third of the military budget. The shift from conventional ordnance to missiles, electronics, and aircraft coincided with the regional migration of defense work and the enlistment of new firms as major contractors. In the decade after the conclusion of the Korean War, contracts did not leave the North for the South and West. Rather, the Pentagon terminated standing contracts for unnecessary tanks and vehicles, while offering new contracts for missiles and electronics. The conspiracy of international tensions, weapons developments, and increasing military budgets launched a competition for entirely new defense-related industries.[25]

This competition accelerated broader developments in the national economy. Changing military demand reinforced the shift from heavy manufacturing to electronics and services. The economic stream was flowing away from industries such as steel, which required proximity to raw materials and to rail or water transport. Combined with the expansion of the highway system, this shift allowed the migration of economic activity. "Footloose" businesses could take advantage of the South's cheap land, surplus labor, and warm climate. These "dynamics of deconcentration," as a federal government report termed the phenomenon, were intensified by the long period of postwar economic growth. This sustained prosperity, driven in part and regulated by federal macroeconomic policy, kept up continual pressure for expansion of existing plant and the construction of new facilities. Over time, the obsolescence of old plant in the Frost Belt and the high costs of doing business there pushed new capacity toward the Sunbelt. Defense procurement was one part of this process. For instance, when the Kennedy administration heated up the economy by stepping up spending for military equipment without raising taxes, electronics and aerospace firms built new capacity in central Florida to exploit the area's cheap land and amenities.[26]

The South and West won the decade-long battle for new contracts. New England held its own and conducted a major comeback after 1970. The mid-Atlantic and Midwest regions suffered the greatest losses. Texas's success in securing the new military aircraft industry was so great that visitors, like Secretary of Commerce Luther H. Hodges in 1962, could hardly distinguish between what was "new in Texas and what's new in air power." Since the aircraft and missile businesses were rapid-growth, research-based, automated industries with a history of low unionization, the South's victory in the battle for new military capacity proved a harbinger of its future.[27]

Federal regulations contained provisions to influence this "victory," to steer defense purchases toward lagging regions. Authority for such preferential treatment stemmed from the Armed Services Procurement Act of 1947 and the Defense Production Act of 1950. The former allowed the Pentagon to forgo bidding and conclude negotiated, non-competitive contracts to maintain the "mobilization base" of depressed areas; the latter recommended the geographic dispersal of defense facilities to maintain a sound economy. The Truman administration, bent on using the military budget to relieve unemployment, directed procurement officers to increase procurement in depessed areas. This policy, formalized as Defense Manpower Policy No. 4 (DMP#4) in March 1952, allowed contractors in "labor-surplus areas" to match the low bid determined by competitive bidding. In 1953 the Administration went further, establishing a set-aside program that reserved procurement quotas for contractors in depressed areas. These contracts were to be awarded to depressed areas even if firms in other regions submitted lower bids.[28]

American taxpayers and their elected representatives, however, balked at such a program. They applauded benefits for depressed areas as long as they cost no additional money. Beginning in 1954, the Congress attached a rider to every annual appropriation bill barring the payment of a price differential under DMP#4. In 1960, DMP#4 was revised to incorporate that restriction, and to grant preference to areas of "chronic and persistent labor-surplus" over other depressed areas. The Kennedy administration then attempted to reverse that decision. It extended preferential treatment to all low income areas, ushering in a twenty-five-year-long series of amendments to DMP#4. The revisions established an elaborate ranking scheme for depressed areas, including preference for small business.[29]

In 1959 a group of senators introduced a bill requiring regional quotas for all defense expenditures. Dissatisfied with DMP#4, the proposed legislation guaranteed a share of contracts for depressed areas and required the Department of Defense to place contracts in regions where "relatively small proportions of procurement have been purchased." This proposed "Armed Services Competitive Procurement Act of 1959" was introduced by Senators Jacob Javits and Kenneth Keating of New York, and Clifford Case of New Jersey. The entire New York State delegation sponsored a companion measure in the House. The bill required procurement agencies to limit advertising for contracts to designated regions. It deliberately avoided the question of whether the armed services could award contracts in excess of low bids. During the hearings, the Defense Department adamantly opposed the bill, labeling it a threat to both national security and interregional harmony. Senator Keating mocked the Pentagon's objections; he also taunted the southerners on the panel, declaring, "Our country can't exist half rich and half broke any more than it could exist half slave and half free." Still, the bill never emerged from the southern-dominated Armed Services Committee.[30]

The failure of the Javits-Keating-Case bill ensured that DMP#4 would exert little influence on procurement. The Defense Department simply refused to apply it. As early as 1952, one procurement official told the Joint Congressional Committee on Defense that he had no intention of paying a higher price for the same goods under DMP#4, even if he was authorized to do so. And neither the Congress nor the American people approved of paying an extra dime for defense, even if the jobs such a contract provided might lead to net savings on social welfare and unemployment payments. The money may have emanated from one treasury, as Senator Humphrey insisted, but it came from two different parts of the government, with different objectives and different popular mandates. And never the twain did meet. From 1952 to 1970 the Pentagon disbursed approximately 4 percent of its annual budgets under DMP#4, a paltry sum that in no way reflected the proportion of population or manufacturing output located in depressed areas.[31]

Southerners actually applauded the Pentagon's reluctance to enforce the preferences for depressed areas. Responsible government finances and support of the military, of course, had long defined the agenda of southern legislators. But they had other reasons besides these principles for opposing DMP#4 and similar regulations seemingly directed at the South. First, the allotments for "labor surplus areas" benefited only a small segment of the South—the old industrial towns of the Appalachian and Piedmont regions—and conferred most of the advantages on the industrial North. The South was poor and underdeveloped, but it was not "de-industrializing." Second, the South required neither an executive order nor a Congressional directive to steer military investments below the Mason-Dixon line. The region already received a sizable chunk of those disbursements, an especially large one considering the relatively sparse resources and industrial experience the region had to offer. As Humphrey's subcommittee reported, the Midwest, with its extensive university resources and heavy annual production of professionals, failed to garner prime contracts, Research and Development grants, or non-profit research funds proportional to its share of the nation's technical and scientific skills. The undernourished South got more than its share.[32]

Indeed, from the moment of its promulgation in 1952, senators complained about the non-enforcement of DMP#4 and the geographic concentration of defense purchases. But southern senators were never among the protestors. The southern politicians who petitioned for war plants in 1939 had traded places with midwesterners by 1960. Philip Hart of Michigan led the unsuccessful fight for a Congressional watchdog committee on the regional impact of defense activities in 1959. Humphrey's investigating committee, a special panel of the Committee of Small Business, had no southern members.[33]

Given the region's power in the Senate, the absence of southerners on those committees showed that the South had no desire to reform military procurement. For the South had developed an alternate system, a political alliance with the Pentagon that brought economic bounty to the region in return for Congressional support of the Defense Department. In part, that alliance merely continued the heritage of World War II. Federal officials had made a concerted effort to deploy war industries in the service of regional development. During the war, they built an infrastructure of industrial plant, testing facilities, and military bases that remained after

1945. The Defense Department could not be faulted for developing that foundation.

The arrangement also betrayed the region's lingering suspicion of federal intervention. It reflected the South's traditional desire to control federal activity, to steer the benefits of federal aid toward the region's leadership and away from its dispossessed people. The South's Congressional delegation resisted the tide of whiggery longer than its statehouses. And even the new Whigs preferred federal aid without cumbersome regulations and Washington administrators. Only the Pentagon offered relatively unrestricted federal dollars. The South's "dependence on the federal military dole," proclaimed the liberal journal *Southern Exposure* in 1973, "is deepening the push from this region, not for humane domestic programs, but for a continued policy of growth that is both stimulated and secured through massive military spending."[34]

The cooperation between southern politicians and the Pentagon was also a natural alliance. It expressed the region's traditional patriotism and zest for a strong defense. Since World War II, American foreign policy has borne the stamp of southern leaders. Southerners like James F. Byrnes, Tom Connally, Lyndon Johnson, and J. William Fulbright fashioned much of postwar American diplomacy. National foreign policies, including military actions, consistently received their strongest popular and electoral support in the South. In the 1960s and 1970s, protestors demonstrated outside the facilities of defense contractors in other regions, but such actions were rare in the South. Lockheed cited the region's hospitality toward military programs as one reason for moving a missile research facility from California to Texas in the 1970s.[35]

Moreover, the Congressional seniority system granted the South the political power to deliver large defense budgets. In so doing, southern hawks saw to it that their roosts were well-furnished. The postwar growth of the Pentagon and its demand for new goods and services like missiles offered potential bonanzas to southern legislators. They had little to trade but their influence in the Congress and they dealt that influence shrewdly. Such efforts made South Carolina congressman L. Mendel Rivers legendary. As one journalist put it, Rivers, as the second-ranking member and eventually as chairman of the House Armed Services Committee, "transmogrified Charleston into a microcosm of military-industrial civilization." Into his district, Rivers poured an Air Force base, a naval base, a Polaris Missile maintenance center, a naval shipyard and ballistic submarine training station, a naval hospital, a Coast Guard station, a mine warfare center, and the Sixth Naval District Headquarters. McDonnell-Douglas, Avco, GE, and Lockheed all established factories in the district during Rivers's chairmanship. By some estimates, as much as one-third of the income in the area and half the employment was defense-related. Representative Carl Vinson (D-Ga), Rivers's mentor and predecessor as Armed Forces chairman once joked that "you put anything else down there in your district, Mendel, it's gonna sink."[36]

Rivers gave as well as he received; he was an influential and steadfast supporter of the military. He repeatedly squelched debate on the defense budget and occasionally even appropriated more funds than the military brass had requested. Before military audiences, Rivers routinely mocked liberal civilians: "The only powder those people have ever smelled is talcum powder. The only war they have been in is bood-war."[37]

Rivers's only rivals for Pentagon spoils hailed from neighboring Georgia—the home of both Senate Armed Forces Committee chairman Richard Russell and his opposite number in the House, Representative Carl Vinson. They made defense the largest employer in Georgia. By 1960 the two men had secured fifteen military installations for their home state. After Vinson's retirement, Russell ensured a steady supply of prime contracts for the Lockheed plant in Marietta, the largest private employer in the state.[38]

While some southern politicians, like Rivers, viewed this alliance as just another game of pork barrel, many others saw it as a main chance for their region. Military spending not only promised jobs, but permanent high paying jobs, which demanded and fashioned a skilled workforce. Defense plants also pioneered new production techniques and brought industrial prestige and additional investment to the region.[39]

The space program was a case in point. Before *Sputnik,* southern politicians displayed little enthusiasm for rocketry. In fact, Albert Thomas (D-Tex), the Houston area congressman and the chairman of the subcommittee that handled space appropriations, waged a one-man war against the program. Thomas was the nemesis of the National Advisory Committee for Aeronautics (NACA), the principal space agency before NASA. Thomas believed in a balanced budget and found rocketry unimportant. NACA's failure to place a facility in his district reinforced that animosity. But "No sooner had *Sputnik*'s first beep-beep been heard," one Defense Department official recalled, "than the nation's legislators leaped forward like heavy drinkers hearing a cork pop." Texas naturally hosted the biggest celebration for the biggest "drinkers." Charles Brewton, a former assistant to Alabama senator Lister Hill, journeyed to Austin and passed a memorandum to Lyndon Johnson's closest aide. That memo suggested that *Sputnik* could cripple the Republicans, save the Democrats from a divisive battle over race relations, and land LBJ in the White House. Seeking such political advantages, Texans led the fight for an expanded space program. Johnson, the Senate Majority leader, held public hearings and roasted witnesses like NACA chief Jimmy Doolittle. In his inimitable style, LBJ taunted the old flying hero that "he'd rather be dead than broke" and remain "a solvent if moribund bloke." Across Capitol Hill, Johnson found allies in Speaker Sam Rayburn, Representative Thomas, a sudden convert to the importance of space, and another Texan, House Space Committee chairman Olin Teague. They steered a space bill through the Congress in 1958. In addition to its southern constituency, the principal feature that distinguished the new space agency—the National Aeronautics and Space

Administration—from its predecessor was its enormous budget. As one pundit put it, NA¢A became NA$A.[40]

But the space program represented more than a political coup for southern Democrats, and more than a chance at palm-greasing. Even though NASA would invest only a fraction of the resources that the Pentagon invested in the South, southern leaders saw the space program as a conduit to a new "New South" of science and technology-based enterprise. The glamor of the program in the early 1960s, especially of the desperate race to reach the moon before the Soviets, convinced many Americans that space research held the key to the nation's economic and political future. Southern politicians as diverse as Mississippi senator John Stennis, Lyndon Johnson, and Florida congressman Donald Fuqua envisioned NASA as the Moses who would lead the South to a high-tech promised land. If NASA administrator James Webb, an Oklahoman who owed his appointment to LBJ, did not share that view, he did nothing to disabuse southern legislators of it. NASA even pledged to include blacks in the regional reconstruction. The space agency promised to employ blacks in technical capacities at all its southern installations.[41]

The hopes for space-based economic development were largely borne out in several southern communities. First, NASA almost single-handedly supported scientific research in universities across the South. In 1962, despite the shortage of universities and doctorates in the South, NASA placed fifty-seven of its 130 grants and research contracts in the southern states. In terms of dollar awards, NASA placed the largest chunk of its research dollars with the nation's "Big 20" colleges and universities, but the public universities of Alabama, Georgia, and Florida stood among the top recipients.[42]

No southern university benefited more than the University of Alabama-Huntsville. The space program practically established the university in the first place and did indeed sustain the booming metropolis that housed it. A sagging textile town of 16,000 when Werner Von Braun and the German rocket team arrived in 1950, Huntsville claimed 72,000 residents, but little else, when the Marshall Space Flight Center opened in 1960. The Huntsville Research Institute followed in 1961. The Cummings Research Park opened a year later. Two industrial parks, housing IBM and Rockwell, opened in 1964 and 1965. The University of Alabama-Huntsville, which NASA officials began lobbying for in 1964, held its first classes in 1969. In 1966, six years after NASA arrived, per capita personal income in Huntsville outdistanced the rest of Alabama by 20 percent. The city's population nearly doubled between 1960 and 1970. The influx converted an overwhelmingly agricultural workforce into a diversified one, with many government and service workers. Almost a third of the city's 1966 labor force had come there since 1960. Huntsville, according to the Information director at the Redstone arsenal, "was high tech when high tech wasn't cool." The city represented the economic treasure chest that southern leaders sought with space.[43]

Houston was a more luminous jewel. In September 1961, NASA announced its decision to construct the Manned Spacecraft Center (now the Johnson Space Center), a $60 million installation on land donated by Rice University for the purpose. Politics certainly played into the location decision; the space agency did not underestimate the influence of Johnson, Rayburn, and Thomas. But whatever the political calculation, NASA policy dictated a dispersal of facilities away from the East and West coast, and the necessity of water transport probably ensured a Gulf site. Twelve of the 18 locations NASA considered were in the South. A year after the decision, before construction of the Center had been completed, 29 space-related companies had opened up shop in the Houston area. By 1965, Houston ranked first in the Southwest in population, manufacturing payroll, value added, and as an industrial and consumer market. Moreover, it began a dozen-year spurt of growth in science and market-oriented industries that promised to end its dependence on natural resources. Unlike Huntsville, Houston did not depend exclusively on the space center for its growth, but business leaders agreed that the center both accelerated and reshaped that development. The city's boosters never failed to remind visitors that the first phrase spoken from the surface of the moon was not "the Eagle has landed," but "Houston?"[44]

The Apollo program, the moon expedition, was the grand prize in the aerospace bonanza. Largest of the space programs, Project Apollo spread its operations in a crescent from Florida to Texas. And after completing its lunar expeditions, NASA maintained its southern base. In 1985, five of the space agency's nine facilities were in the South, including the Marshall and Johnson Centers, the Kennedy Space Center at Cape Canaveral, the Michoud Assembly facility in New Orleans, and the Mississippi Test facility at Bay St. Louis. According to NASA administrator James Webb, space programs, whether on the steppes of the Soviet Union or the South's Gulf Coast, dragged "the whole economy up behind this penetrating wedge of advanced technology."[45]

If any symbol of the national defense state's involvement in the South shone brighter than those space facilities, it was the region's surfeit of military bases. As it did during World War II, the region remained the nation's boot camp throughout the postwar era. In fiscal 1975, for example, the South drew one and a half times the national average in defense salaries. Forts, naval stations, and airfields translated into numerous service jobs in the base towns; the Defense Department estimated that every soldier stimulated three service sector jobs. Accordingly, base towns reported high ratios of service to nonservice employment. Military bases also carried increased federal aid to the affected communities. Beginning in 1950, the federal government picked up the costs of educating the children of servicemen. Washington also provided subsidies to compensate for loss of tax revenues that federally owned lands represented.[46]

But military bases have exerted an uncertain influence on regional economic growth. When a military facility closed, the guidelines for federal aid ensured that the community lost its financial support just as its largest employer closed shop. In Mobile, Alabama, the closing of the Mobile Matériel Area at Brookley Air Force Base shocked the local economy. The number of employed dropped almost overnight by more than 5 percent, the unemployment rate soared, and many residents departed the area. The catastrophe highlighted the fact that no major employer had moved to Mobile in the decade before the base closure. The shutdown of Fort Donaldson in Greenville, South Carolina, on the other hand, sparked an aggressive and successful redevelopment campaign so that Greenville escaped Mobile's fate. The experience led the Greenville *News* to editorialize, "From where we look at it now, we wish we had another military installation like Donaldson for the Pentagon to close, and for the city to redevelop for the good of the local and regional economy." As different as they were, in both cases, the presence of military bases had arrested industrial growth. Without pulling dynamic industries in their wake, military installations contributed little to regional development. Georgia politicians came to that conclusion in the late 1970s. State leaders realized that Georgia's five major bases provided few economic dividends for Georgia. They successfully lobbied the Pentagon to shift base operations toward army electronics and to work more closely with the research institute at Georgia Tech in Atlanta.[47]

Despite their limited economic influence, military bases constituted a large share of defense spending in the South from 1955 to 1980. That fact underlines the limits of defense spending in fostering regional economic growth. The South, while faring quite well in the distribution of prime contracts, never rivaled California. The resurgence of New England in recent years only reinforced the South's leadership in bases and salaries rather than contracts. Moreover, the South continued to receive a large portion of its contracts for food, textiles, cigarettes, and coal. These contracts neither developed new skills in the area nor challenged the dominance of the traditional low wage southern industries. So, until the 1970s, defense nourished high-technology industry less in the South than the region's leaders had supposed. Los Angeles, not Atlanta or Houston, represented the nation's biggest defense city. Not surprisingly, then, journalists, mesmerized by the highly visible bases, tended to exaggerate the direct economic effects of defense spending in the South, while economists, in search of quantifiable measures of economic stimuli, downplayed the role of military spending in Dixie.[48]

Neither the journalists nor the economists, however, emphasized the marginal impact of defense activities in the region. The net level of spending obscured the pivotal role of Pentagon involvement. Unlike the industrial North, the South received far more in federal spending than it paid in federal taxes. Defense accounted for almost all the regional disparities in that "balance of payments." Moreover, while the South did not win the most defense dollars between 1955 and 1980, the funds it received mattered more than in other regions. They energized the South's take off, transformed its political institutions, and shaped the direction of its development.

The crucial point is not that defense dollars developed the South at the expense of other regions, as Frostbelt leaders charged during the 1970s. Nor is it that the South received the lion's share of defense largesse because of its political influence. It is not even the marginal impact that matters most—the fact that dollar for dollar, a defense base meant more for Alabama than a lucrative prime contract meant for California, even if the base (unlike the contract) did not contribute to an economic take off. It was not defense spending per se that so reshaped the postwar South, but the peculiar operation of a series of programs, many only tangentially related to defense but all under the aegis of the defense establishment.

This constellation of defense-related programs provided manuevering room for Whig politicians. It provided access to funds and jobs in a region of slender resources, a region deeply suspicious of non-defense programs. It allowed for close cooperation between defense agencies and influential southern Congressional committee chairmen. And it disbursed federal dollars with a minimum of interference—either in race relations or in economic arrangements. Southern business progressives could develop their states without providing economic opportunities for blacks or the poor.

Federal action, then, forged a new sort of southern political economy. It permitted the South to pursue development through federal investments, as southern New Dealers had envisioned, but without liberal politics or redistributionist economic policy—without support for welfare, labor, blacks. The alliance between the South and the defense state brought with it highways and airports—old favorites of southern business progressives—but also funds for research facilities and higher education. It provided little in the way of welfare, job training, or primary education.

Furthermore, the activities of the national military establishment directly affected the various components of growth. Southern politicians recognized that relation, even if other analysts neglected it. In 1977, Florida governor Reubin Askew noted that tax levels, wage rates, degrees of unionization, transportation facilities—all of the seemingly independent determinants of economic development—were linked in some way to government policy. And no part of the government exerted a stronger influence on these components of growth than the national security establishment.[49]

DEFENSE AND THE COMPONENTS OF GROWTH

Askew directed his remarks to a southern forum, but in 1977, they no longer applied exclusively to the states below the Mason-Dixon line. Indeed, their general applicability revealed a South in more or less equal competition with other regions. That parity undoubtedly would have pleased the architects of the assault on the colonial economy of the 1930s.

Their primary goal was to enlist the region in the American mainstream—to raise deficient southern incomes and public health to the national standards and to end the region's dependence on agriculture and "colonial" industry. Those New Dealers even recognized the role of the national defense program in remolding the regional economy. In 1939 and 1940, the TVA, NYA, and FSA painstakingly argued the case for regional economic development through national defense.[50]

Much of that economic convergence had occurred by the mid-1950s. At the time of his death in 1954, Howard Odum was at work on a sequel to his monumental *Southern Regions of the United States.* The pace of change in the two decades after the publication of *Southern Regions* so impressed Odum that he planned an updated testament to that southern development. The original volume, published in 1936, had charted the region's deficiencies and aimed the South toward more complete integration into American economic life. More than any speech or report, its unrelenting collection of maps had documented the region's deviation from national norms. Staring out from every even-numbered page of Odum's book, the South of those maps, measured always in solid black or all white, stood in stark contrast to the dots and stripes and plaids that characterized the remainder of the country. Presumably, the maps of *Mid-Century South: The New Southern Regions of the United States* would have melded the South with the rest of the nation.[51]

That convergence continued after 1960. Between 1959 and 1980, the South led all American regions in economic growth. The region registered impressive gains in its share of the nation's value added and manufacturing employment (Table 6-2). Total non-agricultural employment expanded rapidly, at a rate second only to the Far West. Employment in manufacturing firms expanded faster in the South than in any other part of the nation. This changing geography of American industry stoked the engine

Table 6-2
The South in the National Economy, 1958–1982
South as a Percentage of United States Total, Selected Indices, Selected Years(%)

Index	*1958*	*1967*	*1977*	*1982*
Non-Agricultural Employment	22.5	24.0	27.7	29.5
Manufacturing Employment	18.8	20.7	22.2	28.3
Value Added by Manufactures	17.2	20.1	25.2	27.4

	1960	1970	1980
Population	27.2	27.5	30.0

Sources: U.S. Bureau of the Census, *Statistical Abstract of the United States, 1960,* p. 212. *Statistical Abstract, 1963,* pp. 782–83. *Statistical Abstract, 1968,* p. 220. *Statistical Abstract, 1969,* pp. 722–23. *Statistical Abstract, 1979,* p. 409. *Statistical Abstract, 1980,* p. 811. *Statistical Abstract, 1986,* pp. 10–11, 411, 750–51 (Washington: 1960–87).

of growth in the Sunbelt. The manufacturing boom revved up major industrial centers in Texas, Florida, the Carolinas, Georgia, and Oklahoma.[52]

The spectacular growth restructured the regional economy. In 1940 the South devoted a far larger proportion of its labor force to agriculture and the extractive industries than did any other section. Plantation, mill, and lumberyard defined the workplace of the South. At the same time, services, manufacturing, and trade occupied an unusually small share of the region's workforce. By 1975 most of those differences had evaporated (Table 6-3). During the 1950s and 1960s the southward migration of industry closed the gap in manufacturing. The 1970s witnessed rapid growth in the service, finance-insurance-real estate, trade, and government sectors. Even the fall off in manufacturing growth between 1970 and 1975 served to reshape the South in the nation's image. The decline was concentrated in lumber, food, textiles, and primary metals. Machinery, instruments, electrical equipment, and fabricated metals continued to expand. No longer a satellite in the national economic universe, the South of the 1970s set nationwide industrial trends.[53]

The decline of agriculture dramatized the process of convergence. In the decades after 1950, the region threw off its centuries-long dependence on crop production. "We no longer farm in Mississippi cotton-fields," William Faulkner explained. "We farm now in Washington corridors and Congressional committee-rooms." Faulkner hardly exaggerated when he linked the surrender of the rural heritage to New Deal agricultural reforms. And the forces set in motion by the New Deal continued to hold sway in the postwar era. Modifications in farm policy hastened the retreat from cotton farming. The 1949 Agricultural Adjustment Act introduced a modernized parity formula. While it retained the old 1910–14 base as the index for all farm products, it geared individual commodity "parity" levels to the average price of the most recent ten years. This revision corrected support levels to reflect contemporary intercommodity price ratios. It also created new incentives to abandon cotton, since the old formula had overvalued cotton and undervalued livestock and livestock products.[54]

Much of the rural-dominated economy waned along with the cotton culture. Mechanical harvesters shortened the cotton harvest season, so that cotton gins demanded larger capacities and larger capital investments. In much the same fashion that small tenant plots gave way to large machine and wage-labor units on the farm, the ginning industry contracted to fewer, larger firms. In 1900 the South supported 32,000 cotton gins. In 1973 only 2,880 remained. Much of the rural culture—the tradition of the furnish, the tenant family's wood and hunting rights—followed the path of the vacant gins. New crops, especially soybeans and livestock, covered the fields that had once been devoted almost exclusively to cotton. The giant Delta and Pine Land Company of Scott, Mississippi, demonstrated the shifts in the rural economy. Before World War II, under the direction of AAA factotum Oscar Johnston, the plantation employed 1200

TABLE 6-3
Industrial Structure, U.S. and South, 1958 and 1982 Employees and Value Added (VA), Selected Industries
Percentage of the U.S. and Southern Totals (%)*

	1958 South		*1958 U.S.*	
Industry	*Employees*	*VA*	*Employees*	*VA*
Food and Kindred Prod.	12.50	13.66	10.70	12.50
Tobacco Manufactures	1.17	2.38	0.50	1.00
Textile Mill Prods.	18.59	11.40	5.60	3.40
Apparel and related Prods.	9.04	4.32	7.40	4.20
Lumber and Timber Prods.	7.99	3.67	3.60	2.20
Furniture	3.90	2.68	2.20	1.70
Pulp, Paper and Prods.	3.90	5.62	3.50	4.00
Printing and Publishing	3.74	3.62	5.40	5.20
Chemicals	6.61	14.67	4.40	8.70
Petroleum and Coal Prods.	2.24	4.06	1.10	1.80
Rubber Prods.	0.92	1.43	2.20	2.30
Leather and Leather Prods.	1.13	0.73	2.20	1.30
Stone, Clay and Glass Prods.	3.63	4.20	3.50	3.90
Primary Metal Prods.	3.45	5.11	6.80	8.20
Fabricated Metal Prods.	4.17	4.12	6.60	6.70
Machinery, except Electrical	3.54	3.91	8.50	8.80
Electrical Machinery	2.67	3.59	7.00	7.30
Transport Equipment	5.85	5.59	9.70	10.80
Instruments and Related Prods.	0.36	0.40	1.90	2.00
Miscellaneous Manufactures	0.03	0.23	3.60	3.40
Other	0.19	0.27	3.80	---

*Columns do not add to exactly 100 percent because of rounding.

Sources: U.S. Bureau of the Census, *Census of Manufactures,* 1963, Volume III *Area Statistics* (Washington, 1966). *Census of Manufactures,* 1982, *Geographic Area Series* (Washington, 1984). *Statistical Abstract, 1963* (Washington, 1963), 745–49, and *Statistical Abstract, 1986* (Washington, 1986), 776–81.

1982 South		*1982 U.S.*	
Employees	*VA*	*Employees*	*VA*
8.22	10.24	7.80	10.70
0.96	2.90	0.30	1.10
9.92	5.94	3.80	2.30
9.91	4.44	6.20	3.20
4.41	2.54	3.00	1.90
3.48	2.03	2.30	4.60
3.39	4.92	3.20	4.00
5.12	4.58	6.80	6.60
6.36	12.77	4.50	9.40
1.19	3.93	0.80	2.70
3.44	3.87	3.60	3.40
0.78	0.40	1.00	0.60
3.02	3.01	2.80	2.80
2.86	2.61	4.50	4.00
6.34	6.00	7.60	7.20
8.50	9.91	11.40	12.40
8.13	8.43	10.00	10.30
6.25	6.69	8.40	10.30
1.43	3.02	3.30	4.10
1.19	0.98	2.00	1.70
4.04	---	6.70	---

tenant families and planted 16,000 acres to cotton. In 1969, cotton occupied only 7200 acres. The remainder supported soybeans, green beans, and pasture crops. Instead of croppers and mules tilling the land, Johnston's successors employed 510 wage laborers, no tenants, 150 tractors, 31 mechanical pickers, and 9 combines.[55]

The decline of agriculture, however, paled before the explosive expansion and diversification of industry. The South advanced in employment, value added, and productivity (Table 6-3). It became a leader in dynamic defense-related industries like aircraft, machinery, and electronics. The region outperformed the rest of the nation through both bust and boom seasons. The South withstood the recessions of the 1950s, displaying greater amplitudes of expansion and smaller contractions. During the 1958 recession, Luther Hodges boasted about a 33 percent increase in industrial expansions in North Carolina. Nationwide, investment in industrial facilities dropped 17 percent during that year. The region demonstrated similar resilience through the "deindustrialization" of the 1970s. Geographic dispersal of industries, industrial diversification of areas, and diminishing regional variations defined the prevailing patterns in American manufacturing from the mid-1950s until the 1980s. All these trends benefited the South. They led Walter P. Webb to conclude: "Industry is coming South because there really is no other place for it to go."[56]

What Webb, the great historian of the western frontier, witnessed was not merely the expansion of American business into the barren industrial frontier. The sunbelt boom was no simple regional shake-out, a product of the equilibrating flows of labor and capital. Southern and national capital-labor ratios, under the pressure of minimum wage legislation at one end and capital-intense defense contracts at the other, had already converged by 1960. "The new industries of the South," Rupert Vance noted with pleasure in the mid-1960s, "employ less unskilled labor and require more technical knowledge than those industries that formerly dominated the region." Mechanization proceeded faster in southern manufacturing than in the rest of the nation. To be sure the invisible hand of the market fueled regional growth, but the visible hand of the federal government accelerated the convergence of the southern and American economies. The fastest growing southern industries, the ones most responsible for narrowing the gaps in productivity and capital intensity, were the ones most closely tied to defense purchases.[57]

In fact, the region's outstanding growth in capital equipment and in productivity concentrated in a few key manufacturing industries which formed the South's "engine of growth." In the influential formulation of Walt Whitman Rostow, leading growth sectors, rich in profits and innovation, spur overall growth. These leading industries stimulate "supplementary growth sectors"—railroads stimulate coal, aircraft drives instruments and electrical machinery—and "derived growth sectors," which rise with the general volume of economic activity. The South had long lacked such leading sectors. As Franklin D. Roosevelt often lectured, the region's long dependence on low wage, low productivity industries provided neither the profits or the innovations to sustain expansion, nor the purchasing power for derived growth. Once federal policy had battered down the ramparts of that low wage economy, the national government helped provide the advanced industrial sector.[58]

The military contributed much to determining the nation's leading industries. So much so that the South benefited from national developments, even though the Pentagon intended no region-building effects. In the 1960s, electrical equipment, machinery, transportation equipment, and fabricated metals expanded faster than all other southern industries. From 1957 to 1979, the South scored comparative gains vis-à-vis the nation as a whole in electrical equipment, machinery, and instruments. All these lines of business relied heavily on defense demand. For the electrical equipment industry, for example, the military had accounted for almost two-thirds of aggregate sales growth in the 1950s. In transportation equipment, the South not only outpaced the old industrial belt but even pilfered the West Coast aircraft industry after 1968. Just as Hubert Humphrey had railed from the Senate caucus room, the Defense Department accounted in large part for the changing geography of those industries. Training his protest on the Florida aerospace boom, the Minnesota senator added "Only a few years ago it was sand dunes. Nothing." The migration of skilled workers and manufacturing facilities to that area or to any area, he explained, "just takes money."[59]

The existence of mini-Rustbelts in the post-1960 South demonstrated that regional growth was an industry-specific phenomenon. The old manufacturing towns of the South suffered decay not unlike that of Humphrey's home region. The region's once-thriving industrial cities—communities like Fayetteville, Tennessee, Muscle Shoals, Alabama, and Tupelo, Mississippi—suffered massive unemployment during the 1970s. In Fayetteville, 150 acres of a 170-acre industrial park stood vacant in 1981. The darkness of Muscle Shoals when contrasted with the sunlight of nearby Huntsville resembled the plight of Salinas, a struggling industrial town in the shadow of California's booming Silicon Valley.[60]

The defense state stimulated these leading industries. Even firms that held no Pentagon contracts benefited from military expenditures that built transportation systems or sustained overall demand. Federal expenditures brought in social overhead facilities—the airports and highways that altered the South's competitive position with other regions. Furthermore, defense contracts altered the production functions of industries like aircraft and sparked the sort of innovation that might not have appeared without federal sponsorship. The Pentagon created a de facto technological community in the scientifically undernourished South. From this the South gained industrial experience with automated factories earlier than what normally could have been expected.[61]

The military-industrial complex also encouraged the economic expansion of the nonmetropolitan South. The "rural renaissance," the rapid growth of nonmetropolitan areas and the slowdown in urban growth, constituted another national trend of particular benefit to the under-urbanized South. The spectacular growth of footloose industry, especially of electronics and service lines not tied to any geographic location, fueled this trend. Unlike heavy manufacturing, which was tied to supplies of raw materials or to cities with rail and water transportation, footloose firms could easily take advantage of the cheap land, surplus labor, warm climate, and amenities of the South. They could also settle in outlying communities, outside major urban areas. In a five-year period surveyed by the Advisory Commission on Intergovernmental Relations, southern manufacturing output actually expanded fastest outside the region's cities. Employment followed a similar pattern. Between 1963 and 1965, some 80 percent of the new manufacturing jobs in Tennessee were created outside standard metropolitan areas, as were 43 percent in Texas, and 64 percent in Virginia. Desire to avoid urban decay, racial strife, union constituencies, and regulations all contributed to the "rural renaissance."[62]

The defense state encouraged this development of nonmetroplitan industrial capacity. As Senator Humphrey noted, the Pentagon sometimes turned sand dunes into space centers. In the Tennessee Valley, defense spending concentrated in rural areas. Two-thirds of defense investment in the 1950s made its way into the region's 22 nonindustrial counties. Both the TVA's rural industrialization program and land availability played into those site selections.[63]

Far more important than occasional direct investment in nonmetropolitan areas was the development of regional transportation systems. One Alabama woman did not exaggerate when she told a reporter that if he was "looking for the New South, he'd do better to try some place closer to the interstate." Justified as national security programs, the interstate highway and federal airport programs brought isolated counties without access to markets or distribution centers into the national economy. The impact was greatest where existing systems were the poorest and resources to improve them the scantest. One southerner went so far as to declare: "The Federal Aid Highway Act of 1956 promises to have the most enduring effect on the South of any single piece of national legislation in United States History." The long southern freight rate battle had dramatized the age of railroads as the age of the North. The waterway development of TVA notwithstanding, it was the advent of truck transport that inaugurated the era of southern development. Interstate highways "opened up many parts of the South to manufacturing."[64]

The new accessibility of rural areas to manufacturers had profound consequences for the South. For every Hancock County, Mississippi, or Brevard County, Florida, rural sites transformed by NASA installations, there were hundreds of remote southern boroughs untouched by federal largesse, except for their attachment to interstate highways. This helped to reverse the traditional association of manufacturing with the city, so that the rural South became unusually dependent on manufacturing. In 1985, manufacturing accounted for 27 percent of the jobs in the rural South; it supplied only 18 percent in southern cities. Often one employer controlled a rural town, paying low wages and employing mainly unskilled workers. Poverty persisted in those areas outside the orbit of the wealthier cities. The workforce of the urban areas, with their concentration in service, professional, and highly skilled industrial employment, hardly resembled the rural manufacturing havens. The result was a divided economy and a divided labor force, a Sunbelt that cast long shadows.[65]

Although it left some areas and demographic groups unaffected, defense industry restructured the region's labor force as well as its industry mix. After all, in-migrants captured a large percentage of the new jobs in space and defense centers. Local residents received relatively few of the high-level positions. In 1965, only 21 percent of the labor force in Brevard County, Florida, site of Cape Canaveral, had worked in the county in 1960. The majority of employed persons had moved there from out of state during those five years. The same was true in Huntsville, and in Hancock, County, Mississippi, home of another federal installation. Some migrants worked in other industries, while some natives found space and defense jobs.[66]

But those lucky natives were the exceptions. Defense and space employers mined broad, national labor markets for technical and managerial talent. Between 1950 and 1980, non-native whites accounted for a larger and larger proportion of the southern population. In nine of the eleven Confederate states (all but Louisiana and Florida), the representation of non-native whites doubled. In Georgia and the Carolinas it tripled. In large measure, the new businesses recruited their best-paid workers from outside the South and hired locals for manual and custodial positions. Not surprisingly, one economist discovered in 1979 that the South's high in-migration areas were also its high out-migration zones. Different parts of the labor force responded to different labor market conditions. In 1986, SCI Systems, Inc., a Huntsville high-technology firm that had spun off from the space center, employed 1100 scientists and engineers and 3300 production workers. Many of the technical personnel had moved there from other regions. Almost all of the production workers, working in conditions SCI's onetime corporate counsel described as "just like a poultry plant," were longtime local residents. Siecor Corporation, a fiber optics firm, drew more than half of the scientific and engineering personnel at its facility in North Carolina's Research Triangle Park from other regions. The company credited its success in luring educated northerners in part to an aggressive national recruiting effort. In 1959, Atlanta fittingly celebrated its economic growth by honoring a Yankee newcomer. To represent the one-millionth Atlantan, the city Chamber of Commerce selected William Smith, a well-educated executive from Rochester, New York.[67]

Demographic upheaval proved basic to the reconstruction of the southern workforce. For half a century the South had served, in Rupert Vance's words, as "the seedbed of the nation." It cast off migrants, the richest southerners and the poorest, the best and the least educated, to the Northeast, the Midwest, and the West. That out-migration continued unabated in the 1950s, with the South suffering a net migration loss of over a million people in that decade. But around 1955, the flow changed direction. The South registered a small net gain between 1955 and 1960, and larger net gains thereafter. In-migration also became an increasingly large component of regional population growth. New arrivals accounted for 12 percent of the region's population increase between 1965 and 1970, 51 percent of the 1970 to 1975 jump. Between 1970 and 1976 the population of every southern state grew faster than the national average, and all but Mississippi, Tennessee, and Louisiana outstripped California's growth rate. In fact, during the 1970s, the land of Dixie attracted twice as many in-migrants as any other region, including the West.[68]

The characteristics of this migration revealed the impact of the defense boom. Blacks and poor whites continued to leave the region until the 1970s. Rural blacks tended to move to cities, both in and outside the South, that already possessed sizable black communities. What made this demographic shift a radical departure from historic migration patterns was not the characteristics of the out-migrants, but rather those of the in-migrants, many of whom were highly educated whites from the New England and North Central regions. These highly educated whites garnered the management and skilled positions in the aerospace, electronics, and chemical industries. The geographic direction of migration within the region bore out that pattern. Migrants headed overwhelmingly toward the new industries in the defense-related areas; they stayed away from the older industrial areas of the South. Since the new industries offered limited opportunities to the region's traditional (mainly black) out-migrants, sunbelt growth did not stem their exodus until comparatively late in the game.[69]

The labor market conditions of the 1960s accentuated the effects of migration patterns. Provoking much political and scholarly debate among contemporaries, this employment pattern was dubbed the "labor market twist." The twist referred to a long-run decline in demand for low-skilled labor and a rise in demand for highly skilled, highly educated labor in the 1950s and 1960s. It implied structural changes in the economy—new technology and changing consumption patterns—that altered the demand for labor more quickly than the labor supply could adjust to the new conditions. This produced rising unemployment and declining labor market participation rates for the less educated, and the opposite for the highly educated, even though the overall unemployment rate held fairly steady. The Department of Labor estimated that 2 million of the 2.5 million new jobs created between 1960 and 1963 required the highest education and skill levels. The "twist" loosened in the late 1960s for the young. Higher secondary school enrollments and the military draft reduced unemployment among unskilled youth. But it became even more pronounced for older workers. This pattern reinforced the development of an imported economic sector with a largely imported workforce at the highest rungs of the southern economic ladder.[70]

Alterations in the regional economy had, of course, lowered the demand for unskilled labor in the 1940s and 1950s. The federal government's assault on low wage southern employment dried up job opportunities for thousands of unskilled workers. The decline of farm tenancy exacerbated the problem. Large numbers of southerners without advanced educations or industrial skills had earned their subsistence in agriculture, however inadequate. Many of these displaced unskilled workers left the region in the decade after World War II. These migrants, many of them blacks barred by racial discrimination from most skilled positions in the South, swelled the impoverished neighborhoods of many northern cities. Poverty, at least as a matter of public debate, shifted with those travelers. As the federal government recognized in the 1960s, a predominately southern, rural problem became a largely northern, urban one.[71]

At the same time, the rise of the defense state advanced the twist at the other end of the educational scale by contributing both to the changing consumption patterns and the new technology that altered labor demand generally. Defense production also required a large supply of highly educated workers. Missiles, aircraft, and electronics all employed heavy concentrations of engineers and other high-skilled nonproduction workers. The military also influenced labor migration both directly and indirectly. Between 1946 and 1965, a strong correlation between defense spending and in-migration prevailed nationwide. Armed forces migration contributed directly to that correlation. Between 1965 and 1970, for example, retiring and active military personnel composed 14 percent of total interregional migration.[72]

Those patterns had particular impact on the South. Engineering employment expanded faster in the South and the West than in other regions. Defense demand steered many engineers to new homes in those sections. Military migration also redounded to the South's benefit. The largest block of armed forces migrants—both those entering and those leaving the services—settled in the South between 1965 and 1970. Defense activity also shaped southern wage patterns, creating engineering positions, but eliminating blue-collar jobs. In so doing, defense and other automated businesses nearly erased the sectional gap in salaries for technical personnel, but preserved regional differentials in many manufacturing occupations.[73]

Not only did the defense boom swell the flow of skilled labor to the South; it also introduced new types of labor management. Southerners had long bragged about the dedication and tractability of their labor supply. Henry Grady trumpeted the "Americanism" and hostility to radicalism of southern workers in the 1880s. In the postwar era, North Carolina's Luther Hodges often bragged about the "quality" of southern labor.

Hodges peppered his industrial recruitment speeches with anecdotes displaying the loyalty and discipline of southern workers. Audiences understood those boasts, however intended, as euphemisms for the cheap, tractable labor the region had long relied on and, increasingly, as code words for antipathy to organized labor. Indeed, despite ample contradictory evidence, the South's low union membership and anti-union environment received much of the credit for regional economic development.[74]

The vitality and virulence of southern anti-unionism no doubt contributed to that impression. Even Hodges, architect of the South's first state minimum-wage law, fought the scourge of unionism. Returning from a trip to the North and West, Hodges excoriated the control of those regions by "labor bosses" and their "gangsters." Sam Ervin, the North Carolina senator who later gained fame as the guardian of constitutional government during the Watergate scandal, co-sponsored anti-labor-racketeering legislation in the 1950s, and campaigned for national right-to-work laws in the 1960s. Union busting continued into the 1970s. The J. P. Stevens Company engaged in a brutal lockout and labor battle in 1976 and pioneered new techniques in "preventive labor relations."[75]

Not surprisingly, the South remained relatively union-free throughout the postwar era. In 1978, only Kentucky, with its large mining population, approached the national ratio of union members as a percentage of nonagricultural employment. The national standard stood at 23.6 percent; unions claimed 22.4 percent of Kentucky's laborers. The other southern states ranged from the nation's low of 6.5 percent in North Carolina, through Texas at 11.0 percent, to Alabama at 19.2 percent. As the region's share of manufacturing employment climbed, the unions, restricted to their old regional and industrial strongholds, lost membership and power nationwide. The rise of the South propelled this de-unionizing of America. General Motors, for instance, placed eight of the fourteen plants it opened between 1975 and 1980 in the Deep South, and one in Mexico. By the end of the 1970s, anti-unionism had practically replaced racism as the South's signature prejudice.[76]

The region's anti-union environment partly signaled its cultural and economic heritage. The inheritances of southern history—distrust of radicalism, a large pool of cheap farm labor, and racial conflict that inhibited worker unity—offered a harsh soil for labor organizers. The CIO learned that lesson in the immediate aftermath of World War II. The new anti-unionism of the 1960s and 1970s, however, in large measure departed from that legacy. First, federal contracts raised industrial wages. This practice undercut unions' leverage in wage negotiations. Second, military production's high concentration of nonproduction workers like engineers, and reliance on historically unorganized industries like aircraft, inhibited union growth. Third, in a few cases, the military's emphasis on performance to the neglect of cost allowed automation of aircraft, textiles, and other manufacturing lines. Such factories eliminated the skilled factory workers, the intermediaries between the unskilled and the engineers, which formed the core of union membership. The Defense Department, for example, financed research at the Spindletop Research Institute in Kentucky, which led to the automation of some southern textile mills. One Burlington Mills facility in Cordova, North Carolina, replaced virtually its entire labor force with new machinery. Where several workers had once tended each piece of equipment, one skilled worker controlled 25 machines. Fourth, Pentagon security regulations also braked union growth. By 1963, some 25,000 American private industrial firms fell under Defense Department ground rules. The regulations mandated all sorts of management surveillance, control, and supervision of employees. Since these rules were inflexible, they withdrew power from potential bargaining units. The Army, for example, blocked an investigation of factory safety at an armaments plant in Radford, Virginia. It denied the workers' petition for an inquiry by the Occupational Safety and Health Administration.[77]

An even more basic connection linked the rise of the defense establishment with southern anti-unionism. The same Cold War tensions which expanded military spending restricted union growth. As we have seen, the Taft-Hartley Act consolidated labor's wartime gains, but effectively barred unions from the industries and regions they had yet to organize. The anti-communism of the McCarthy era exacerbated the defensiveness of American labor organizations. National labor unions turned on their own radical wings, lost popular support, and concentrated on job security for their existing membership. They were never a force in the expansion of the aircraft, electronics, and electrical machinery industries. That political atmosphere compounded a deeper constraint. Nationwide, the 1970 level of union membership correlated best with 1880 urban population, that is, the extent of urbanization in 1880 predicted the strength of unions in a given area better than the 1970 urbanization levels. That correlation dramatized the association of unions with an earlier form and era of economic development. And just as the economy of 1970s and 1980s differed from the industrial economy of the nineteenth century, so did the styles of anti-unionism diverge. Notwithstanding the persistence of outfits like J. P. Stevens, the high-tech anti-unionism of the 1980s differed from the more brutal brand of earlier years.[78]

The labor environment of the sunbelt South rested then not on the region's historic anti-unionism and isolated labor market, but on a new sort of business climate. It depended also on the arrival of a new industrial economy—skilled workers, dispersed factories, automated plants. That new atmosphere included right-to-work legislation, but it also featured low taxes, unrestricted growth, and pro-business governments. The region consistently placed highest in rankings of business climates. A 1959 survey commissioned by the Southern Governors Conference denoted the South "the best business climate of any major region." In 1975 the Fantus Business Climate rankings placed seven southern states in its top ten. A 1986 study echoed those assessments: it rated the South near the top in business climate and rock bottom in "labor climate."[79]

In the minds of those surveyors, right-to-work laws, low taxes, and industrial recruitment programs formed the main components of the South's favorable business climate. By 1954, every southern state but Oklahoma had enacted anti-union legislation under section 14-b of the Taft-Hartley Act (Louisiana repealed its measure in 1956). In 1959 the Southern Governors' Conference celebrated that fact. The Conference communiqué announced that "Southern states have adopted and maintained the right-to-work laws and other legislation viewed as vital to industrial development." One economic analysis of industry migration went so far as to credit the lion's share of southern industrial growth to those laws. Low taxes on businesses complemented the restrictions on unions. In 1975, for example, effective state and local business tax rates stood well below the national average in every southern state. Alabama imposed the nation's lightest business tax burden. Aggressive promotion and special inducements for relocation rounded out the southern business climate. State governments offered subsidies ranging from free sites and publicly financed training programs to tax abatements and rent-free facilities. Even when states proffered no such gimmicks, the "fishing expeditions" of southern governors reeled in investment from other regions and overseas. In 1970, half of the annual foreign investment in the United States found its way below the Mason-Dixon line.[80]

The actual effects of the southern business climate fell short of its perceived stimulus to regional growth. Many studies have denied that right-to-work laws had any impact on union growth, the extent of industrialization, or wage levels. Even the economists who asserted the significance of anti-union legislation conceded that its effects decayed over time. The "selling of the South" likewise exerted only a modest impact on business location decisions, especially after state competition for industry grew so widespread as to render the states nearly indistinguishable in that respect. Business people and state recruiters could not even agree about which programs, if any, drew industry to the South. When they proved decisive, subsidies and tax inducements appealed chiefly to highly competitive, low wage, traditionally southern industries, and less to the dynamic, high wage lines that fueled regional growth. Thus, the selling of the South slowed rather than quickened the region's efforts at high-tech development and diversification. As for taxes, economic growth seemed to affect southern tax rates more than tax policy affected growth. The South maintained the lowest rates in the two decades after 1960, but that position represented slower increases from its former high tax position, rather than concerted efforts to reduce business taxes. In 1949, median corporate income tax rates in the South (including West Virginia) stood 1 percent above the 4 percent rate prevailing outside the region. By 1983, the South had crept from 4 to 6 percent, while the rest of the country jumped to 8 percent.[81]

Whatever its actual economic significance, the South's favorable business climate carried tremendous institutional and political influence. Local government policies accentuated migration patterns. During the 1960s, whites moved toward jurisdictions with low taxes and low welfare benefits, principally the South and Southwest. Blacks, on the other hand, traveled toward locales with high public spending levels and high welfare benefits. Furthermore, southern politicians felt bound by the dictates of the good business climate. Development programs, complete with subsidies, tax benefits, and right-to-work laws, created the appearance of aggressive recruitment of industry, if not the actuality. And with the electorate, the appearance of all-out effort often counted for more than actual results. These demands on southern politicians fostered a political environment hospitable to growth. Business-government partnership, flexible zoning, and environmental regulations, and celebration of leading businessmen as civic heroes became a common feature of the regional landscape.[82]

The absence of institutional barriers to economic growth has led influential economists like Walt Whitman Rostow and Mancur Olson to credit the rise of the South to the revolutions of economic cycles. According to Olson and Rostow, the relative maturity of the old industrial regions—their large, bureaucratic business units and well-developed unions and interest groups—created a kind of "institutional arthritis" that stifled economic growth. Long isolated from those developments by rurality, low wages, and racism, the South escaped that economic sclerosis in the 1960s and 1970s. During that period the nation's economic energy seeped into the unencumbered South. The recent resurgence of New England, the nation's first region to industrialize and its first to decline, and the slowing of sunbelt growth in the 1980s nourished such cyclic theories.[83]

Whether or not long-wave swings have determined the rise and fall of regions, the postwar upswing of the South required some agent to prime the region for growth. The South became a "blank slate" not because of a historical accident, but because political and economic forces wiped it clean. Before 1950, for example, the South was a high tax region; the search for industry and rising revenues inspired southern policymakers to keep taxes down. Federal policy aided southern state governments in that quest. The South preserved low taxes because the national government financed the social overhead necessary for industrial growth. While the northern states constructed their public infrastructure in an era of little federal support—in 1954, the federal government provided only 11.4 percent of state and local revenues—the South built its sewers, roads, and airports on the cheap. In 1974, 26.5 percent of state and local government revenues came from Washington. Moreover, the federal tax exemption on state industrial development bonds made such promotions attractive to industry and to southern recruiters. With Arkansas congressman Wilbur Mills at the helm of the House Ways and Means Committee, the South held off several attempts to remove that exemption.[84]

The federal government erased most of the "slate" that was the political economy of the South before 1950. Wage policy, farm and industrial development programs, and equalization grants assailed the colonial South of the 1930s and helped to erect a new political and economic order. The expansion of the military nourished those new institutions. Unique among American sections, the sunbelt South, like many foreign nations, has been a planned region. Pentagon procurement, the TVA, state industrial development commissions, and regional growth policy boards have all charted the region's economic future and channeled its progress. Whatever their relative success, they set the South on the same course, aiming it at an industrial and technological future free of the conflicts and social responsibilities of the North. The region, as Houston booster Marvin Hurley noted, pursued the research contracts of the Department of Defense, but resisted the welfare programs of the Department of Health, Education and Welfare.[85]

THE HIGH-TECHNOLOGY SOUTH

By the time Jimmy Carter entered the White House, research-based high-technology industry had become the watchword of the entire American economy. Carter preferred to bill himself as a plain-speaking Georgia farmer, but as a nuclear engineer and a retired naval officer he represented the nation's attachment to a high-technology future. It is fitting that the first Presidential embodiment of that faith was a southerner. The hunger for research and development funds had emerged early in the South. It owed ultimately to the influence of Howard Odum, who proposed research as the pillar of regional development, and to the intrusion of federal planners in the TVA and other agencies. In 1940, TVA chairman David Lilienthal named "inadequacy of industrial research and development" one of two barriers to the "restoration of economic equality among the regions of the United States." He noted that the region maintained only 7 percent of the nation's research facilities, a measly figure representing but a quarter of the South's share of America's population, and half its share of value added. Lilienthal's agency emphasized industrial research as the key to developing new industries which produced high grade consumer goods, paid high wages, and lifted the standard of living. Harnessing the potential of "technological wealth," one TVA report concluded, "constitutes the means of bridging the distance between the present situation in the South and a greatly improved Southern region."[86]

TVA and other southern planners demonstrated considerable prescience in that concern. They anticipated the post-World War II research revolution. Total expenditures for research and development jumped some fifteen times from 1940 to 1960, from $0.5 billion in 1939, to $2 billion in 1946, to $14 billion in 1960. By 1980, the total had reached $63 billion. Even after adjusting for inflation, United States R&D outlays increased three and a half times from 1955 to 1980. Economic growth centered in the research-based industries. Research dollars also acted as a magnet on highly trained manpower, rendering R&D funds an important source of institutional and regional development.[87]

The federal government, particularly the national defense establishment, supplied and channeled most of those research funds. In 1940, the national government financed less than half of the nation's research and development. Two decades later the federal government supported two-thirds of the research load, even though rising private expenditures leveled off that share after the mid-1960s. The Defense Department and the space program carried most of the research burden. After *Sputnik,* the scientific community, which had sought civilian control of all federal research at the end of World War II, lobbied for military control of even civilian research. Scientists realized that civilian programs promised a slow, uncertain stream of funds, while association with the Pentagon guaranteed large budgets undiluted by political debate.[88]

The federal government supervised American science by providing land for research installations and building laboratories across the country. The lion's share of state support, however, underwrote basic and applied research in universities. In fact, in the two decades after V-J Day, federal dollars became the largest component of the budgets of most major American universities. In 1963, Hubert Humphrey asserted that national authorities largely determined which universities prospered and which failed. "The Federal Government," Humphrey warned, "is making a vast intellectual wasteland out of America by having the R&D contracts concentrated as they are in limited geographic areas." Schools like Morehead State College in Minnesota could not compete for faculty with colleges which claimed Pentagon contracts and could thereby pay better salaries and fund expensive research projects. Humphrey predicted the "inevitable" decline of midwestern colleges "because no state legislature can appropriate the amount of money for a university that the Defense Department can make available."[89]

The military domination of scientific and industrial research dried up resources for research and development in nondefense industry. Even the U.S. Chamber of Commerce worried that government-financed R&D diverted too many resources from the private economy, sacrificing business prosperity for the "prestige value of space exploration, or overkill in military defense." In 1956, company-funded research and development expenditures increased 33 percent; in 1961, during a defense build-up, they grew only 3 percent. Competition for highly trained manpower led to a stockpiling of engineers. In order to bid for new contracts and to avoid labor shortages in peak "seasons," major military and aerospace contractors maintained large staffs in excess of current needs. Generous federal contracts covered the additional costs in the name of performance and reliability. In this respect, the South's emerging defense giants resembled the planters of the prewar years who signed tenants to year-long contracts to secure their harvest labor supply. But neither those landlords nor their tenants could look to Washington for fat pay envelopes.[90]

The South wasted no time in attaching itself to the research revolution. The region found comfort in both research's promise of rapid economic progress and its intimacy with the Pentagon. "The South's future," Luther Hodges lectured a commencement day audience at Wake Forest College in 1963, "depends upon its ability to move to the forefront of modern technology." Hodges and North Carolina quickly disseminated that faith across the region. The Southern Governors Conference affirmed that "Science and its applications offers a vast new area for industrial development in the South." The task would prove difficult. As Hodges warned, wooing research-based industry depended on the availability of sites with cultural and educational facilities. Winning research contracts required the strengthening of southern universities. The South could and did receive scientific facilities when the armed forces decided to build them from scratch, but the region lacked the graduate programs and college laboratories to attract basic research dollars and science-based industry. Recognizing that deficiency, the Southern governors committed their region to improving higher education so as better to woo high-technology industries.[91]

To do so, the southern states, the heirs of the state's rights tradition, pioneered the interstate compact, a regional pooling of resources to overcome state deficiencies. The covenant created the Southern Regional Education Board (SREB) in 1948. "The South stands to gain more than any other region from the technological revolution now sweeping American industry," asserted the SREB's manifesto "Within Our Reach" in 1961. But, the Board cautioned, "technological industrialization does not come automatically. It requires long-range planning and a high level of education." The SREB began operations in 1949, creating an academic common market in the South. Realizing the impracticability of developing graduate, professional and research programs in all fields, the Board directed funds to improving the strongest state programs in various fields. It encouraged talented southerners to take their graduate training in their home region and bestowed in-state admissions and tuition benefits to all students in the system. A Tennessean entering an engineering program in Alabama, for example, received the same preferment in admissions and paid the same resident's fees as an Alabaman. High-tech development formed the purpose of the common market. "Highly trained leaders and sound research," the SREB declared, "are indispensable to the progress of the southern states in utilizing their resources to the fullest and achieving higher levels of economic development."[92]

The Southern Regional Education Board's first objective was retention of the region's highly trained leaders. The Board sought to stem what it saw as the exodus of the region's most talented people. In the SREB's opinion, those men and women wanted to train and work in the region, but lacked the opportunities to do so. The Board achieved moderate success in its harvest of that talent. Engineering employment grew faster than the national average in the South during the 1950s, and so did the number of professionals. More important, since those figures reflected numerous in-migrants, the number of engineering enrollments and the number of doctorates earned grew faster in the SREB states (the South plus Delaware, Maryland, and West Virginia) during the 1950s than they did in the nation as a whole. The region registered those improvements even though the South's percentage of college-age youth in college remained well below the national average.[93]

Southern officials desired to "keep the talent home"; they also sought Defense Department research grants for their universities. In 1958 the South remained well below the national average in research and development personnel per capita. That deficiency focused the longing eyes of the Southern Governors Conference and the SREB on defense and atomic science. As recently as 1986, the Southern Growth Policies Board recommitted the region to seeking a larger share of federal research funds.[94]

Despite the immediate emphases on local talent and federal dollars, the region's quest to develop highly trained leaders and to enhance scientific research envisioned R&D as the spark to science-based industrial development and economic growth. Opposition to the minimum wage and chants of "segregation forever" waned in southern state capitals after 1960. But the belief that state-supported research and development would build prosperity never flagged. Luther Hodges argued that case when he commissioned North Carolina's Research Triangle in the 1950s. The Southern Growth Policies Board echoed him in 1986. It admonished that "Higher education should be given a high priority in any restructured state economic development strategy." The "root of research," that panel repeated, "spurs growth."[95]

That belief was so prevalent that southern officials did not limit their efforts to the Southern Regional Education Board. Southern state governments plunged directly into the promotion of research and high-technology industry. The efforts began in North Carolina in the 1950s. Impressed with the results of privately funded research complexes like California's Silicon Valley and New Jersey's Princeton Research Park, Luther Hodges viewed a similar installation as the answer to North Carolina's "Bread and Butter" problem. He believed that R&D facilities packed "tremendous economic wallop," that they likely would prove "the most important single factor in advancing productivity." His hopes bore fruit in 1958. Then, after three years of planning, the first research firm moved into the Triangle described by the state's three great research universities—the University of North Carolina in Chapel Hill, North Carolina State University in Raleigh, and Duke University in Durham. Situated near Interstate 40 and the Raleigh-Durham Airport, Research Triangle Park represented, in Hodges's words, "the heart and hope of North Carolina's future." In 1959 the Research Triangle Institute (RTI) was formed. It bought out the private investor who had helped finance the state's initial venture and began conducting contract research. RTI received its first contract from the Atomic Energy Commission in 1960. Tarheels celebrated the park as "the marriage of North Carolina's ideals for higher education and its hopes for material progress."[96]

Hodges's departure for the Commerce Department in 1961 did not diminish his fondness for Research Triangle Park. As Secretary, he used his influence in Washington to woo an IBM facility to the Triangle. Technitrol, an IBM supplier, moved to the Park shortly after IBM arrived in 1965. Two federal government research installations followed soon thereafter—the National Center for Health Statistics, and the Park's largest employer to date, the National Environmental Research Center. By 1973, the Park housed twenty-two facilities. Over the next ten years, the Triangle expanded, almost keeping pace with the rapid growth of Silicon Valley. While the Stanford-San Jose area created 30 new jobs for every 100 workers in its 1982 labor force between 1973 and 1982, the Triangle counties created twenty-six. And unlike Silicon Valley or Boston's Route 128 area, the Research Triangle Park never received much in Defense Department dollars.[97]

If Research Triangle Park was only an incidental beneficiary of the explosion of military-funded research, the South's other research facilities and research parks drank directly from the Pentagon's fountain. Such facilities included the Air Force's Arnold Engineering Development Center in Tennessee and the Air Force Proving Ground Center in Florida. The Army operated the Jet Propulsion Research Center of the Army in Texas and the Rocket and Guided Missile Research Center of the Army in Alabama. The National Advisory Committee on Aviation Laboratories was headquartered in Virginia. The South also housed the Atomic Energy Commission Complex in Oak Ridge, Tennessee, and NASA's extensive program. Secretary Hodges did not exaggerate the impact of space and defense research during a visit to Huntsville: "Northern Alabama," he intoned, "has been made over in the past decade by research, backed mainly by the Federal Government."[98]

The state-supported advanced technology incubator in Atlanta exemplified the South's continued dependence on defense-inspired research and development in the 1970s and 1980s. Associated with the Georgia Institute of Technology, the state's Advanced Technology Development Center (ATDC) evolved slowly. It began during World War II, when for the first time, Georgia Tech's Engineering Experiment Station garnered the major part of its budget from defense industry and the federal government. Tech remained primarily an undergraduate institution through the 1950s. But over the next decade, the university began pursuing contract research and planned a model industrial park. At the same time, the Engineering station concentrated on electronics and drew an increasing share of its support from the Pentagon. In October 1971, Georgia Tech named as its eighth president Joseph M. Pettit, Dean of Engineering at Stanford University and a disciple of Silicon Valley architect Frederick Terman. Pettit's selection represented a commitment to remold Georgia Tech on the Stanford model. The Trustees hired Pettit to enhance the institution's graduate and research programs. At Stanford, Pettit had staunchly defended research funded by the Defense Department against considerable opposition. He found Atlanta and the South a more hospitable environment in which to pursue such grants.[99]

In the late 1970s the university and the state government coordinated their efforts in the Advanced Technology Development Center. This incubator for high-tech companies opened in 1980. In 1984 it graduated six companies and housed thirty more. Most ATDC firms made computer software, but the enterprises ranged from biotechnology to computer applications for textile mills. Following the objectives of the Southern Regional Education Board, the ATDC furnished a means to keep entrepreneurial talent and Georgia Tech graduates in the region.[100]

Just as the South pioneered the Regional Education Board in the 1940s and the state-supported research park in the 1950s, so did it launch the new high-tech ventures of the 1970s. Georgia and North Carolina led a trend away from site-specific industrial recruitments toward efforts to create statewide high-tech business environments. Rather than wooing industry from other regions, these programs aided development of local high-technology businesses. North Carolina sponsored a second-generation microelectronics center with $88 million in state funds and only $20 million from the participating companies. In addition to the ATDC, Georgia also established a Microelectronics Research Center. By 1985, forty states had some program to support research-based industry, but the South led the way. The region's thirteen states, never known for interventionist government, housed seven of the nation's technology incubators and nine of the twenty-seven research parks. Among the southern states, only South Carolina did not sponsor such a program. With four separate programs, Mississippi became the nation's fastest growing research center in the early 1980s.[101]

The regional, indeed national, consensus that high-tech research facilities guaranteed future growth obscured their immediate economic impact. Research jobs offered little to the region's legion of poor, low-skill workers in traditional industries. They even exacerbated the region's traditional inequality by creating a two-tiered employment system with elite professionals at one end and low paid production and clerical positions at the other. In 1985, those nonscientific personnel received wages amounting to only 57 percent of the average unionized steel or auto worker's. Furthermore, whatever its future prospects, high-tech industry has provided very little employment in the South. During the 1980s, only Texas and Virginia exceeded the national average in high-tech firms per resident, and in no southern state did R&D personnel reach the standard share of nonagricultural employment. North Carolina, with all its boasting about the world's highest concentration of Ph.D.s within the Research Triangle, only ceded its position as the nation's low wage state in 1984. The low-wage mantle fell, instead, on Mississippi, despite its incubator, research park, and technology-transfer program.[102]

Those dark spots failed to dampen the South's research and development craze. Even the clear-eyed 1986 report of the Southern Growth Policies Board reasserted the need to "build bridges between Academia and the Private Sector" and to "increase the South's capacity to generate and use technology." Those pronouncements, of course, merely echoed the calls of the TVA in the 1940s and the southern governors in the 1960s. But the contrast between the South's aggressive development of higher education and its lukewarm stand on federal aid to public education is instructive. Federal aid to public education emerged as a program of southern liberals in the 1930s; it promised the equalization of educational resources without interference with the tradition of local school supervision. Accordingly, it won wide support among southern politicians in the 1940s. After 1950, however, the efforts of northern congressmen to apply federal aid as a lever against segregation reversed the regional alignments on the issue. Frank Graham, the mentor of the original aid bills, felt compelled to oppose such a measure in 1950. While the southern New Dealers had championed federal aid to public schools, their Whig successors instead emphasized federal support for the region's universities and research programs. For higher education, the SREB states spent more than the U.S. average per capita and per student. At the same time, investment in elementary and secondary schooling fell far short of the national standard. Such an approach steered federal funds to southern schools with little divisive debate over race or class. It tapped into the seemingly inexhaustible and unrestricted resources of the defense establishment without recourse to the national welfare state. It supported regional development without demands for equity. With its patrons in the defense establishment, the South could oppose the restrictions and programs of the welfare state. Joseph Pettit's move to Georgia Tech demonstrated that the South did not quarrel with the Pentagon restrictions that so exercised other regions. The South's heritage of conservative educational reform, going back to Progressivism, persisted into the 1980s.[103]

The emphasis on R&D might ultimately prove an economic boon to the South. The rest of the nation certainly followed the South down the high-tech path. But regardless of its economic prospects, the research boom offered a development strategy peculiarly compatible with the political economy of the sunbelt South. First, it nurtured Whig politicians, offering avenues for aggressive industrial development programs without worries about urban blight or labor conflict. It sanctioned government action in the business of industrial promotion, without corresponding demands for equity or social welfare. Second, it allowed intimacy with the national defense establishment that underwrote research and development expenses and provided union-free manufacturing growth. From the defense state, the South received all of the benefits of federal largesse with a minimum of regulation and restriction.

The apparent paradox was not lost on the surviving southern liberals of the New Deal generation. At a 1968 convocation celebrating the twentieth anniversary of the Southern Regional Education Board, one conferee saw the need to remind the development-hungry audience that "we should expect more from education than economic growth." And Jonathan Daniels noted: "We have come almost to the indissoluble combination of promoters and professors—both performing in a sort of mutual hypnosis by the wonderful tools they require." Research, Daniels said, "is not only respectable; far more important, it is profitable. Learning is praised on billboards, and corporate recruiters come, hat and cash in hand, to enroll the young erudite on their payrolls." No institution, Daniels added, seemed "to be more the symbol of our opulent society than our expanding universities. But beside them poverty remains."[104]

Beside them poverty remained. Beside the best state-financed high-tech facility, the nation's lowest wage rates prevailed. Beside the rapid growth, lived a population that still trailed the rest of the nation in income. In 1980, southern per capita income reached 88 percent of the national average, a long climb from 60 percent in 1940, but leaving a long road still to travel. The region contained the same number of impoverished people in 1980 as it did in 1965. Notwithstanding all of the national defense activity, the region drew little aid from the federal area redevelopment and antipoverty programs of the 1960s and 1970s. Despite the need for butter, the South wanted and received mostly guns. In 1986 the Southern Growth Policies Board recognized the persistence of southern poverty, the maldistribution of much of the new-found wealth, the shadows on the Sunbelt. The South, after its long journey from plantation and sweat shop to space center and research park, remained, in the words of the Board's chairman, Arkansas governor Bill Clinton, "halfway home and a long way to go." The Southern Growth Policies Board even warned the region's leaders to "defend" federal welfare programs. "Under the twin pressures of growth at the top and poverty at the bottom," the Board's Commission on the Future of the South concluded, "the South cannot underestimate the worth of some federal infrastructure and social service programs." That proved a difficult lesson to learn.[105]

Chapter 7 ❂ "Shadows on the Sunbelt"

Roads—*Tobacco Road,* "Dixie Highway," "Crossroads Blues," "On the Road Again." Highways have long claimed a place of distinction in southern culture. They have also dominated political debate and concentrated economic effort. The region, Mississippi senator Pat Harrison admitted in the 1930s, had an obssession with "climbing out of the mud." Pavement symbolized progress. For visitors to the region, nothing so altered the face of the South as the construction of the National Defense Interstate Highway System in the late 1950s. Federal-aid roads paved the South, laying more road below the Mason-Dixon line in three years than southerners themselves had cleared in the years from 1789 to 1930.[1]

The construction of Interstate Highway 95 opened a new gateway to the South in the 1960s. From Washington, D.C., source of so much that transformed the South, the expressway's broad, smooth pavement ribboned to Richmond, capital city of the old Confederacy. Just beyond Richmond, the highway split. One branch pointed southwest, through the Virginia countryside and across the North Carolina border. That road skirted the Research Triangle, the parcel of land boasting the world's highest density of Ph.D.s, and pressed on toward Atlanta, Huntsville, Houston.

The other fork continued due south. Few motorists exited this express route into the poor, rural areas of the eastern Carolinas. Across the South Carolina border, the main highways sometimes bled into dirt paths. Travelers, tourists bound for Florida, or tractor-trailers laden with manufactures, ignored those trails. "It's a funny thing what the automobile has done," one witness told a Congressional committee. "The automobile takes you past these houses. It doesn't take you in them, and it takes you past these people. Most white people have not seen the inside of those shacks ever." But during the 1960s, Donald Gatch, a transplanted Yankee doctor, ventured down those dirt paths. He discovered, according to one southern newspaperman, that "they led to huts of poverty, to people who don't go to doctors, to hovels where breakfast for a shack full of kids is grits and lunch is grits with a bit of fatback added—if there is lunch." No one had computed the number of Ph.D.s per square mile in that area.[2]

Those two highways connected the sunbelt South with the still poor southern region it overlayed. The roads exposed, in the words of the Southern Growth Policies Board, the "twin pressures of growth at the top and poverty at the bottom," gripping the South of the 1970s and 1980s. "A flash flood of change," the Board lamented in 1986, "stranded" many southerners amid the rising tides of the sunbelt boom. The torrent swept some into prosperity, primarily immigrants and educated, middle-class natives. But poor, under-educated southerners found themselves "high, dry, and unemployed."[3]

The persistence of southern poverty led many commentators to detach the label "Sunbelt" from the South, or at least to specify that such advanced prosperity only reached the region's fringes—Florida in the southeast, oil-producing Oklahoma in the northwest, Texas at the region's southwest corner, the northern Virginia suburbs in the northeast. Others portrayed the modern South as a collection of oases scattered about the old South—Atlanta, the North Carolina Triangle, and the Texas cities.[4]

The two Souths of the past four decades prove difficult to distinguish from one another. Economic development was not confined to urban areas. The hinterlands of middle Tennessee flourished under the TVA, and later as part of the Tennessee Technology corridor. Other nonmetropolitan areas—northwest Arkansas, north Georgia, and coastal South Carolina experienced substantial growth. Moreover, boom and bust often inhabited the same landscape. In 1980, North Carolina, the South's most industrialized state, with its research center, its reknowned state university system, and its progressive reputation, had the lowest average wage in the nation. No wonder the journalist Ferrel Guillory observed that "the farther you get from North Carolina the more progressive it looks." In 1970, the sunbelt South's growing cities still maintained higher inner city subemployment rates than the decaying metropolises of the Northeast.[5]

In Houston, "the golden buckle of the Sunbelt," 99 large firms and hundreds of smaller suppliers opened for business between 1971 and 1978. Half a million new arrivals moved to the city to claim jobs in such new establishments. But, the *U.S. News and World Report* noted in 1978, "left behind in Houston's headlong flight toward growth and economic success," lurked "an estimated 400,000 people who live in a 73-square mile slum that, says a college professor, has an infant mortality rate 'that would have embarrassed the Belgian Congo.'" The city boasted a modern highway system, inner and outer loopways clogged with traffic, but had yet to pave more than 400 of its streets. New condominiums and apartment complexes sprang up at a frightening rate, many never filling, while a quarter of the city's poor population lived in substandard housing.[6]

The coexistence of severe poverty and spectacular growth perplexed many observers during the 1980s. It manifested itself in what the U.S. Department of Labor termed the "puzzling lag in southern earnings." As one Labor Department economist observed, "Business booms but average earnings remain relatively low in the South." Despite rapid economic and population growth, average southern earnings remained 17 percent below the national standard between 1973 and 1978, only a slight gain from the 20 percent differential that Victor Fuchs calculated in 1959. Earnings in southern trade and service establishments more closely approached the rest of the nation than did manufacturing wages. No cost of living index accounted fully for this large gap, even though energy costs enhanced the South's cost of living advantage during the oil price shock of the 1970s.[7]

Economists have furiously debated the nature and intensity of the regional wage differential. Some deny its existence. They emphasize not only the South's lower living costs but also its heavy concentration of low-paying occupations, the small size of southern cities, and the legacy of discrimination borne by the region's large black population. Others argue the persistence of the regional wage gap. They contend that the regional occupation mix and the wage rate reflect the same economic factors. Low wage employers locate in the South because it is the low wage region, rather than the other way around. Moreover, even after adjusting for variables like cost of living, city size, and racial discrimination, these studies have uncovered real regional wage gaps, especially in the blue-collar occupations. In any case, low earnings distinguish the South, whether the "puzzling lag" truly constitutes a regional differential or merely stems from the region's peculiar combination of other demographic and economic traits.[8]

The wage gap did in part reflect the region's continuing reliance on low wage industries. Despite the remarkable convergence of the southern and national industrial structures, the South continued to devote more of its manufacturing capacity to textiles, lumber, and apparel than the rest of the nation (Table 6-3). In 1978, one-third of southern factory workers labored in industries with earnings below the national average. Only one-fifth of their counterparts outside the South worked in those low wage lines. And such low wage industries were always among the most aggressive shoppers for the subsidies and tax abatements advertised by the South. Nonetheless, the composition of southern industry accounted for only a small segment of the wage differential.[9]

The wage differential also functioned as an indirect index of rurality. Southern manufacturing, especially low wage manufacturing, concentrated in nonmetropolitan areas. That explained much of the unusually large regional differential in factory wages (25 percent in 1978). Most trade and service employees, on the other hand, resided in cities, where their earnings climbed within 10 percent of the national average. The rurality of southern manufacturing also inhibited unionization, further exacerbating the earnings gap. Furthermore, southern poverty remained a rural phenomemon, even if the majority of poor southerners lived in cities. In 1970, some 25 percent of rural Texans lived in poverty. Only 17 percent of the state's city dwellers faced similar economic hardship. Farm laborers and unskilled factory operatives, both rural residents in the

Table 7-1
Race and the Regional Earnings Gap, 1959–1978
Ratios of Census South/Northeast* Weekly Wages for Male Workers, by Age and Race

	White Males			*Non-white Males*		
Age	*1959*	*1969*	*1978*	*1959*	*1969*	*1978*
20–24	.871	.884	1.108	.765	.685	1.071
25–34	.869	.967	.947	.694	.766	.786
35–44	.872	.859	.931	.672	.717	.825
45–54	.861	.872	.989	.672	.693	.747
55–64	.882	.857	.996	.680	.668	.637

*"Census South" indicates South plus West Virginia, Maryland, Delaware, and the District of Columbia. The "Northeast" is New England and the Mid-Atlantic states (New York, New Jersey, Pennsylvania).

Source: Robert J. Newman, *Growth in the American South* (New York: New York University Press, 1984), 145–46.

South, suffered the highest incidences of poverty in the Lone Star State. "Cotton is no longer King," one observer of the southern countryside remarked in 1968, "but its harsh reign lingers on." Nearly two decades later, the Southern Growth Policies Board lamented the tenacity of that rural poverty: "The sunshine on the Sunbelt has proved to be a narrow beam of light, brightening futures along the Atlantic Seaboard, and in large cities, but skipping over many small towns and rural areas."[10]

The "widely publicized new jobs at higher pay," that 1986 document noted, "have been largely claimed by educated, urban, middle-class Southerners." Many of whom, the report neglected to mention, had recently migrated south for those jobs. Indeed, skill and education differentiated the "two Souths." Defense plants, automated industries, and research parks required highly skilled employees. Moreover, skilled personnel in the South—scientists, engineers, machinists, professionals—earned wages near the national standards for their occupations. Throughout the postwar period, the greatest regional differentials prevailed at the unskilled level. "Twenty years ago, when the national unemployment rate was near four percent," the Southern Growth Policies Board explained in 1986, "people with grade school educations could still find jobs. Today the will to work must be matched by the skill to work."[11]

Southerners, however, remained the nation's most poorly skilled people. During the Korean War, the South posted the worst record in Armed Services entrance examinations. Southerners were twice as likely as other recruits to score in the lowest two mental groups—outright rejection and unsuitable for advanced training. The region repeated that poor performance during the Vietnam War. With the exception of Oklahoma, every southern state exceeded the national average in the rate of educational rejections. Poor school systems prepared citizens no better for private enterprise than for the military. Despite the region's growing commitment to public education, the South still spent less per pupil than the national average in 1986 and paid America's lowest teacher salaries. In percentage terms, fewer southern children completed secondary education and fewer high school graduates went on to college. As a result, the region possessed a smaller percentage of high school graduates in its population and a markedly higher adult illiteracy rate.[12]

The failure to improve education or to provide training programs ensured that poor southerners would hold few of the new, high paying jobs. Employers had to lure sufficiently qualified migrants to fill these positions, so that while the South imported new industry and a new labor force, it exported or left in poverty many of its citizens.

Southern blacks bore the brunt of this dislocation. As whites entered the region in the late 1950s and 1960s, blacks continued to move out. After 1970, they trickled back, with many black senior citizens returning to their birthplaces during the 1970s. That reverse migration slowed by 1980 and never matched the flood of whites into the region during the 1970s. Regional wage differentials continued to be greater for blacks than for whites (Table 7-1). Only white wages narrowed the North-South gap between 1959 and 1978. And whites claimed almost all of the region's attractive new jobs.[13]

The black-white income gap remained widest in the South. In 1969, for example, per capita income of black Mississippians made up only 35 percent of the lowly total for white residents. In North Carolina, blacks earned just 46 percent of per capita white income. In 1970, racial wage differentials also remained widest in the South. By the late 1960s, many civil rights leaders had turned their attention to the economic distress of black southerners. Blacks demanded renewed federal economic intervention targeting the people southern economic development had left behind. Echoing FDR's original pronouncement on southern underdevelopment, one commentator declared in 1975 that "economic discrimination against blacks is the nation's number one economic problem."[14]

Leading proponents of regional economic development shared that view. They identified the failure of blacks to share in the sunbelt boom as an impediment to southern economic progress. In 1967, TVA director Frank Smith conceded that "for more than a decade, the South has been surging forward with renewed economic growth." But, he cautioned an audience of Huntsville business leaders, "no region can hope to sustain this progressive pace unless all of its people are allowed to make their full contribution to the economic productivity of that region." Commerce Secretary Luther Hodges concurred: "For the South, especially, the matter of better opportunities for our Negro citizens is an urgent economic necessity as well as a duty imposed by our sense of fairness and right."[15]

Unfortunately, the welfare of southern blacks did not prove to be an urgent economic necessity for the region's white political and economic leadership. The sunbelt boom proceeded largely without black participation in the new prosperity. Regional economic development opened the national labor market to southern employers and put a premium on skills generally possessed by whites. Businesses found it unnecessary to promote southern blacks or to lobby for improved training programs. Only in the industries where blacks had previously built up experience did employers hire or promote them. In one sample of Deep South employment growth during the 1950s, blacks received 21 percent of the new jobs in the region, even though they made up 43 percent of the population.[16]

New industry concentrated in predominately white areas, restricting opportunites for black southerners. In the nonmetropolitan South, heavily black counties suffered employment declines in the 1950s and 1960s. At the same time, white areas—northwest Arkansas, northern Alabama, coastal Mississippi—experienced significant employment growth. Relatively abundant skilled labor attracted mechanized plants to white communities. So did the ability to recruit skilled whites from outside the South, many of whom refused to relocate in heavily black neighborhoods. By opening in white areas, employers also hoped to avoid racial conflict or to evade affirmative action strictures keyed to the area's level of black population. They also believed black workers were more likely to join unions.[17]

Regional economic development bypassed southern blacks geographically. It also circumscribed them occupationally. First, blacks remained concentrated in traditionally black occupations—unskilled labor and custodial work. Even the southern aerospace industry, an industry dominated by federal regulations, hired few blacks in well-paying positions. Blacks filled less than 2 percent of the industry's white collar positions in 1966 and less than 7 percent of the blue collar jobs. At the same time, blacks represented more than 40 percent of the unskilled laborers and more than one-third of the service workers. Furthermore, as a result of the civil rights movement, blacks also moved into what the Civil Service Commission termed "new traditional Negro jobs." These opened new lines of work to blacks—the police force, fire fighting, social work in anti-poverty programs. But blacks in those positions served only their communities' black populations. Black police officers seldom patrolled white neighborhoods. Black caseworkers rarely aided impoverished white families.[18]

Second, black employment growth was centered in agriculture and declining industries. Blacks found positions only in the unskilled, low paying lines that regional development was eliminating. During the 1960s, blacks made up more than one-third of the southern labor force in agriculture, sawmills, and personal services. But blacks held only 10 percent of the positions in the South's seven rapidly expanding industries—electrical machinery, transport equipment, rubber, apparel, machinery, paper, and metals. Blacks also found limited entry into the trades and financial services—the region's fastest growing nonmanufacturing sectors. New industry and white laborers migrated to the South; they built a new regional economy. Southern blacks who remained in Dixie continued to toil in sawmills and on farms, in the remnant of the old southern economy.[19]

In 1972 Maynard Jackson, soon to become Atlanta's first black mayor, reflected on the condition of black southerners. "About one hundred and ten years ago, of course, I was selling for around twelve hundred dollars," Jackson reminded his audience about the grim legacy of chattel slavery. But "today," despite billions in federal spending and a generation of progress, Jackson noted ruefully, "more likely than not I would find my life not only not worth twelve hundred dollars, but maybe not worth a dime."[20]

National policy shaped the prospects of southerners, black and white. It molded the South's peculiar dual economy. Federal intervention had ignited growth at the top, but had neglected the poverty smoldering at the bottom. For a variety of reasons, growth-oriented federal programs flourished in the South, but welfare agencies accomplished little. In part, that mixed record reflected the overweening national concern with northern urban poverty in the 1960s and 1970s. In part, it reflected southern attitudes. In his home state of South Carolina, Senator Ernest Hollings explained, many farmers received annual sums in excess of $40,000 from the federal government for not planting crops. Yet no one, according to Hollings, doubted that those farmers were "as red-blooded, capitalistic, free enterprising, and patriotic as ever before." Welfare, however, represented another matter altogether. "Give the poor, little hungry child a 40-cent breakfast," Hollings complained, elucidating the regions's prevailing attitudes, "and you've destroyed his character. You've ruined his incentive."[21]

Southern politicians brought to Washington this hostility to welfare programs. Congressional committee membership reflected the region's priorities. The South dominated the Armed Services and Aeronautics Committees which funneled military largesse into the region, but relinquished influence on social policy. In 1937, Alabama's Hugo Black, author of the Fair Labor Standards Act and a leader of the assault on the traditional southern economy, chaired the Senate Committee on Education and Labor. Three other southerners served with Black on that panel, including Claude Pepper, the leading southern liberal in Washington. In 1970, when the Senate Labor and Public Welfare Committee convened to deliberate an unprecedented expansion of welfare payments, not one southerner sat on the panel.[22]

Nonetheless, southern politicians did not ignore social programs. From their positions of influence in the Congress, southern legislators led the opposition to federal welfare programs in the 1960s. They did so while approving increased defense spending and safeguarding agricultural subsidies. They did so even though the South stood to benefit from such programs, to receive far more in outlays than it would contribute in taxes. In 1972, Maynard Jackson noted the same contradiction that irked Hollings, but apportioned the blame to specific southern politicians, not general southern attitudes. "One of them," Jackson fumed, "is Senator Eastland of Mississippi, a man who can somehow reconcile taking thirteen thousand dollars a month not to plant food when there are kids running around within the shadow of his estate in Sunflower County with stomachs bulging from malnutrition and worms."[23]

Like Eastland, most southern leaders embraced federal welfare for business, but remained suspicious of national anti-poverty programs. In the 1960s and 1970s, the South welcomed all development-oriented initiatives. Southern politicians sponsored and supported many of them. But federal efforts aimed at uplifting poor people, rather than at developing poor areas, won little support or cooperation below the Mason-Dixon line. The new Whigs demanded improved universities and increased defense spending, not the higher wages, better nutrition, and literacy programs the southern New Dealers had desired. And where northern liberals sought public housing in the 1960s, southern leaders constructed football stadia and interstate highways. Welfare and training programs might have led some of the South's impoverished people into the stream of economic growth, but the region proved altogether inhospitable to such efforts. The political and economic institutions of the South, shaped in large part by four decades of federal intervention, deflected the few attempts to ameliorate poverty.[24]

This antipathy to welfare was deep-seated. Public opinion polls showed that southerners overwhelmingly opposed welfare; one survey of Georgians revealed a widespread tendency to blame the poor for their poverty. White southern politicians linked national welfare programs with spendthrift government, encroachment on state's rights, and the encouragement of idleness and immorality. One congressman, E. C. Gathings (D-Ark), feared that a federal war on poverty would draw away needed farm labor, enticing rural southerners to cities with promises of lucrative handouts. "We cannot get help when we need these people so badly on farms to cultivate and harvest our crops," Gathings complained about people drawing food assistance. "Our farmers would like to have some of these people." A New Orleans *Times-Picayune* journalist captured white southern attitudes on federal welfare programs in 1963: more military spending, foreign aid, and space exploration "may be necessary," he allowed, but domestic spending was wasteful, unnecessary, and a trespass on states' prerogatives.[25]

Dallas, Texas, dramatized this pattern. With strong ties to the Pentagon, "Big D" became a high technology center after 1960. Its "Metroplex" neighbor—Fort Worth—was a major military aircraft producer. But under the watchful eye of the city's Citizens Council, an exclusive group of corporate executives who dominated Dallas affairs, the city adamantly refused federal welfare funds until the mid-1970s. The Citizens Council believed that federal dollars weakened local control and shifted the city's priorities away from economic growth. Dallas refused public housing programs in the 1950s and urban renewal in the 1960s. The city even shunned participation in the food stamp program. Dallas's leadership finally relented during the Nixon administration. Exasperation at the loss of federal funds encouraged the reversal, while Nixon's reforms of the grant system made the revised course more palatable. Under Nixon's "new federalism," a program of revenue sharing and federal-local partnership, Dallas could expend federal funds as it wished. It chose to spend the money on sewers rather than day-care centers, to emphasize business development, and to ignore questions of economic justice in the disbursement of federal funds.[26]

Dallas represented the limit of southern intransigence. It was almost alone in its refusal to accept previously authorized funds. Most southern communities, however, shared Big D's peculiar relationship with the federal government. Southern elites welcomed the federal programs that promised industrial development. They accepted funds earmarked for highways or downtown shopping centers and enthusiastically supplied matching funds for such programs. At the same time, they rejected the anti-poverty programs which enfranchised the poor, empowered black communities, or steered resources away from business development. The region's misery had, of course, animated the original federal effort to rebuild the southern economy. Sourthern poverty had inspired the *Report on Economic Conditions*. But by 1960 the economic and political forces the *Report* had set in motion no longer concerned themselves with the plight of the region's impoverished people.

Ironically, during the 1960s and 1970s, national attention focused on the dispossessed. The federal government directed its efforts at the impoverished, the illiterate, the chronically unemployed. In the minds of national policymakers, these pockets of desperation did not float up with rising economic tides. Such people seemed impervious to the general Keynesian economic stimuli that the federal government had relied on in previous decades. To attack that "impacted" poverty, the Kennedy administration launched a distressed area redevelopment program in 1961. The Johnson administration followed with an all-out "War on Poverty."[27]

The coincidence of renewed concern for the bottom third of the nation and the South's lingering poverty offered the region a propitious opportunity to uplift its poor. It suggested renewal of the New Deal's original assault on the "Nations's No. 1 Economic Problem." After three decades of Washington's beneficence, the South of the 1960s seemed more fertile soil for federal welfare programs than had the suspicious, recalcitrant region of the Depression years. But unlike the War on Poverty, the New Deal had portrayed poverty and underdevelopment as alternate sides of the same coin. They appeared as the twin scourges of a colonial economy, collateral symptoms which concerted federal action could treat simultaneously in all-encompassing regional program. That view informed the *Report on Economic Conditions* and animated late New Deal policy in the

South. The Fair Labor Standards Act, for example, promised the modernization of southern industry and the eradication of oppressive working conditions, more efficient factories and higher pay for workers.

The policies of the 1960s, however, dissociated growth and welfare programs. In the minds of many policymakers, persistent poverty and chronic unemployment demonstrated that federal policy could not at once promote equality and efficiency. Two influential books, John Kenneth Galbraith's *The Affluent Society* and Michael Harrington's *The Other America,* sharpened this focus on the paradox of poverty in the midst of plenty. Beginning with the recession of 1949, federal relief efforts had concentrated on areas of persistent unemployment, mostly old industrial towns. With these initiatives evolved new theories about "structural unemployment"—pockets of joblessness immune to general economic stimulation. Those theories competed with the regional emphasis of New Deal programs like the TVA. "A generation ago, when the geographic South was a national economic problem," Frank Smith explained in 1965, "many aspects of that problem could properly and beneficially be attacked on a basis of their regional characteristics." But, he added, "Very few economic problems in the South today are susceptible to regional solutions, and very little attempt is being made to achieve any." Indeed, the federal programs of the 1960s demanded a choice between accelerated regional growth and the amelioration of southern poverty. Southern authorities chose the former and neglected the latter.[28]

WAR ON POVERTY IN THE SOUTH I: AREA DEVELOPMENT VERSUS COMMUNITY ACTION

The contest between growth and welfare programs remained undecided when the Kennedy administration fired the first salvos of what would become the War on Poverty. The battle opened with a diverse series of area relief programs. Some concentrated on structural unemployment in the old industrial belt. Others, drawing on the experience of the TVA, pursued industrialization in underdeveloped areas. Among the latter remained the Tennessee Valley Authority, the grandfather of social welfare programs, one originally aimed at sourthern, rural poverty. The TVA operated throughout the postwar era. According to David E. Lilienthal, the TVA represented "more than a Federal Program." It was a concerted, popular project to help a poor people rebuild their own home—"grassroots democracy on the march." And in 1964, Frank E. Smith, one of Lilienthal's successors on the TVA Board of Directors, informed the director of President Johnson's Office of Economic Opportunity that "TVA regards the war on poverty as part of its basic mission."[29]

The TVA did ameliorate economic distress in the Southeast. Its development program spearheaded the industrialization and economic diversification of the Tennessee Valley. In the 1940s, under Lilienthal's direction, the Authority had shifted its efforts from preserving the rural economy to fostering the growth of manufacturing. During the 1950s the TVA abandoned the headlong search for factories and concerned itself with the quality as well as the quantity of the industrial jobs it sought for the region. The Authority dedicated itself to diversifying the regional economy by pursuing employment in the trade and service lines. In so doing, the TVA maintained its independence, its ties to the "grass roots." The Authority condemned the use of tax abatements and business subsidies for industrial promotion. It even criticized the Tennessee Division of State Information for exploiting the TVA in its advertisements.[30]

The animosity of the Eisenhower administration forced retrenchment on the TVA. Budget cuts curtailed TVA activities, reducing it, one commentator lamented, "to a glorified Power Company." At the same time, the shining symbol of New Deal reform attempted to insulate itself from political attack by becoming a defense agency. The Authority had lured war plants to the Valley during World War II. After the war, the TVA pursued that association with national defense by linking its expansion to the needs of the region's energy hungry AEC and DOD facilities. The agency devoted most of its funds to increasing power production, while spokesman stressed the TVA's contribution to the struggle against communism.[31]

At the same time, political controversy forced the agency closer to its patrons within the southern political structure. Despite Frank Smith's pledge, the Authority retreated from the war on poverty and dedicated itself instead to the South's battle for industry. By 1960, regional business interests both championed and controlled the TVA, confining it to the Whig program of cheap power, R&D, and industrial recruitment. The Authority boasted about its "partnership" with local elites. It gradually abandoned Lilienthal's promises of "grassroots democracy" and dissociated itself from the persistent poverty of the Tennessee Valley. The Authority's advocacy of nuclear power and bitter disputes with environmentalists further distanced it from the grassroots. One cautious TVA official privately warned a colleague not to "expect too much in the way of results and lessons from TVA's experience."[32]

Despite its many travails, the TVA served as a model for the area development programs of the 1960s. All through the 1950s the Congress considered programs to relieve unemployment in labor-surplus areas. Just as the persistence of one senator, Nebraska's George Norris, bore fruit in the creation of the TVA, so did similar fortitude on the part of Illinois Democrat Paul Douglas culminate in the Area Redevelopment Act of 1961. Douglas's various proposals focused on mature industrial communities like Pittsburgh, Pennsylvania, offering assistance to cities harboring large populations of unemployed blue collar workers with dim job prospects. These were generally urban areas of average per capita income, which suffered slow economic growth or even economic decline. Indeed, Douglas's 1958 proposal so skewed its resources toward the the North that nearly half of the southern Democrats in Congress opposed the measure. The bill, eventually vetoed by President Eisenhower, won southern support primarily from the the states with old industrial towns—Tennessee, Alabama, and Texas. Finally, in May 1961, President Kennedy signed the Area Redevelopment Act, a four-year program "to alleviate conditions of substantial and persistent unemployment in certain economically distressed areas." Administered by the Area Redevelopment Administration (ARA), the program disbursed $507 million in the form of low interest business loans, grants and loans for public facility construction, and technical assistance for local planning efforts. The bill also called for training programs, but little was expended for that purpose.[33]

The ARA accomplished little and had almost no effect on the South. Its vague mandate had ensured Congressional approval, but it also prevented the agency from determining a workable definition of a distressed area. The ARA divided scarce resources between the low income South and the high unemployment areas of the North. Most of the aid found its way to New England and the Great Lakes states. Of the businesses launched by ARA loans, only about one-fifth opened in the South, less than the region's share of the nation's population and well below its share of the nation's poor areas. In any case, the ARA so fragmented its programs that more that 1000 counties participated, nearly one-third of the nation's total. The Public Works Acceleration Act of 1962 partly remedied the depletion of resources. It appropriated $900 million to supplement the public works portion of the ARA.[34]

Even with the additional public works funds, the ARA proved insufficient. When the agency's four-year grant expired, the Johnson administration responded with the Public Works and Economic Development Act of 1965. The act established an Economic Development Administration (EDA) in the Department of Commerce to encourage business development. It lodged the responsibility for project initiation with state and local governments and emphasized creation of new employment opportunities. For a five-year program, the Act authorized $3.3 billion and specified seven criteria for eligibility. The list included low median family income, but unemployment rates over 6 percent qualified the greatest number of areas. The law also promised aid to Indian reservations and assured that each state would receive distressed area funds.[35]

The EDA continued operations after exhausting its original appropriation. Over one five-year period in the 1970s, the South received 29 percent of EDA projects, and only 21 percent of EDA project funds. Those figures represented a slight gain from the original Area Redevelopment Act. Nonetheless, the Commerce Department concluded that the programs had little impact on the South or the nation as a whole. Furthermore, southern governors disliked the EDA program. Even though local authorities initiated all projects, the federal government decided which projects received funds. State governments could not veto local projects of which they disapproved, nor could they steer the money toward favored highway and public building projects in cities and "growth areas."[36]

The EDA's economic and political failure was highlighted by the success of a similar agency created in the same year, the Appalachian Regional Commission (ARC). In the early 1960s, bitter labor conflict in the coal fields of eastern Kentucky attuned national media attention to the plight of Appalachia. Terrible floods in the winter of 1963 dramatized the inadequacy of the ARA, especially in the central Appalachian region of eastern Kentucky, southern West Virginia, western North Carolina, and eastern Tennessee. To assuage the mountaineers' troubles, President Kennedy, motivated perhaps by his tour of Appalachia during the 1960 campaign, appointed the President's Appalachian Regional Commission (PARC) in 1963. Many commentators compared JFK's interest in Appalachia to FDR's concern for the South during the 1930s. The commission consisted of emissaries from each Appalachian state, designated by their respective governors, and representatives of the relevant federal agencies. PARC debated two approaches to the economic difficulties of Appalachia. The so-called centralist position reasserted the views which had governed the 1938 *Report on Economic Conditions of the South.* It blamed the distress of Appalachia on the failure of state and local government and recommended direct federal action. The second or "federalist" position acknowledged the states' fiscal weakness, but desired "partnership" in the allocation of federal funds.[37]

The latter view won its way into the 1964 PARC report and eventually into the 1965 Appalachian Regional Development Act. Steered through the Congress by West Virginia senator Jennings Randolph (D-WVa), the bill received crucial support from Senator John Stennis (D-Miss). As the reward for his cooperation, Stennis exacted the inclusion of northeast Mississippi in the program, including his home county. The Act created the Appalachian Regional Commission and designated as its bailiwick the mountain region stretching from New York to Stennis's Mississippi. In 1971 the Commission won new appropriations from the Congress, despite the opposition of the Nixon administration. Enthusiastic support from the region's governors ensured renewal of the ARC. In 1981, the commission submitted a five-year termination proposal.[38]

The ARC owed its popularity in southeastern statehouses to two unusual features. First was its so-called partnership. The commission included fourteen voting members, one representative from each state and the federal co-chairman. Representatives could propose projects only within their own states' borders and they were loath to reject proposals from other members lest their own states' projects receive similar treatment. With the national government relegated to one vote, the board became an unobstructed pipeline of federal money to the states. Second, the Appalachian Regional Commission (ARC) elaborated the "growth strategy." Unlike the EDA, which directed funds to the most distressed low income and high unemployment areas, the ARC dispersed its funds to communities with high growth potential. The Appalachian Regional Development Act gave priority to the areas which the states deemed to have "a significant potential for future growth, and where the expected return on federal dollars invested will be the greatest." The ARC never provided jobs, training, or welfare to the region's needy. It instead offered development assistance to southern Appalachia's wealthiest communities. The Commission dedicated itself to highway construction, technology development, and technical training, all regional favorites. While Appalachia's poorest settlements received little aid, many less impoverished areas, including part of the Atlanta metropolitan area, received grants.[39]

Indeed, the ARC became mainly a highway program, but its road construction differed from other highway systems. The Commission designed highway corridors to break down the isolation of Appalachia and serve as a foundation for development, rather than to connect existing cities. And it built an extensive system. Between 1965 and 1980, highways received between 60 and 65 percent of ARC funds. Land stabilization and water drainage projects garnered the bulk of the remaining money. People-oriented programs—health care, housing, education—received little.[40]

The southern Appalachians certainly deserved a development program, whether or not it included direct aid for the region's impoverished. Welfare programs for Appalachia, however, would never have won such acclaim from southern leaders. In the 1930s, federal relief payments built a constituency for federal intervention in the South which eventually undermined the Bourbon elite. But the ARC never interposed federal administrators between southern governments and their citizens, never built a constituency for direct federal intervention as the TVA and relief programs had. Schooled in the manipulation of federal programs, the Whig successors to the southern leaders of the 1930s accepted federal funds without incurring such risks.

Whatever the contribution of ARC highways, Appalachia registered considerable economic progress in the 1970s. After five decades of slow population growth and heavy out-migration, the region recorded a healthy population increase. For the first time in three centuries, immigration added more population than natural increase. The southern Appalachians, especially Georgia and the Carolinas, experienced the greatest surge in population and income. Southern leaders attributed the turnaround to the ARC. Most governors agreed that it was the best federal program in their states. They thought the highway program invaluable, not only for developing Appalachia but also for easing mountain crossings and expanding the commerce of Nashville, Louisville, Raleigh-Durham, and Atlanta. "Under the traditional patterns of transportation development," Luther Hodges observed, "states like North Carolina have had no access to trans-mountain industrial areas like Ohio and Pennsylvania." A completed trans-Appalachian highway system would end that isolation. It would "connect Charlotte and Western North Carolina with Cleveland . . . connect Knoxville and Tennessee with Dayton and Detroit."[41]

The Appalachian Regional Commission did not develop such a fond following in Washington corridors. The ARC conflicted with the Economic

Development Administration and clashed with the other Great Society agencies. Those disputes revealed the South's tenuous relationship with the national welfare state. The ARC and the EDA collided in the South because the EDA rejected the "growth stategy." The EDA targeted needy areas and small business. As a wholly federal agency, it served patrons and clients quite different from those of the state-run ARC. Attempts to coordinate efforts in Kentucky, Tennessee, and Georgia met with repeated failure.[42]

Those disputes paled in comparison with the ARC's bitter conflicts with the Office of Economic Opportunity (OEO). Established by the Equal Opportunity Act of 1964, the OEO embodied the centralist perspective that the President's Appalachian Regional Commission had rejected. The OEO not only assumed the inadequacy of state and local anti-poverty efforts, but blamed local power structures for the misery of their constituents. The Office strove to create community action agencies with maximum popular involvment. It administered Vista, the Job Corps, the Neighborhood Youth Corps, and the Community Action programs. Poor people were the OEO's overriding concern in the South. It rejected the economic development plans, especially locally administered development which denied benefits to the least fortunate. These views put the OEO in direct conflict with the ARC. They also alienated southern leaders across the region. Mississippi congressman Jamie Whitten, an influential member of the House Appropriations Committee, curtailed OEO activities in his Appalachian home district. He also blocked passage of a new domestic anti-hunger program in 1968. Meanwhile, the state governments of the Southeast petitioned that the ARC receive precedence over the OEO in their jurisdictions.[43]

In fact, disputes over the methods and objectives of federal welfare policies permeated the entire policymaking community during the 1960s and 1970s. Bureaucrats, academics, and the Congress debated the proper criteria for welfare programs: Should an area's median income matter if the programs targeted only the poor within a community? Would development programs uplift entire regions and break the cycle of dependency? Should the federal government raise income in an area without reducing inequality in its distribution? Should jurisdictions receive funds or should they go directly to the poor inhabitants who qualified the location for aid in the first place?

Federal policy, indeed liberalism itself, has been growth-oriented since the 1940s. The eclipse of the southern liberals in the Roosevelt administration, the gradual shift from New Deal to New Economics, had changed the direction of federal intervention in the South, and in the nation. But during the 1960s a group of reformers challenged those priorities. Centered in the Office of Economic Opportunity, these young officials hoped to mount a frontal attack on poverty; they planned to circumvent local institutions and governments and to empower the poor directly. "And that is why the war on poverty is more than a program to help individuals—

one by one," OEO chief Sargent Shriver explained. "It is a program to help institutions and entire communities view and treat poor people in a different way." At the same time, the Great Society also identified new targets for federal action—not just public assistance, but Head Start, school lunches and breakfasts, compensatory education, and others. By 1965 the national government devoted a greater share of its budget to domestic programs than it had in 1954, and less to national security.[44]

OEO officials worried particularly about state governments obstructing the war on poverty. The states, most emphatically the southern states, had long restricted social security programs like Aid to Families with Dependent Children (AFDC), and welfare officials hoped to avoid such problems. The OEO supervised many programs: Adult Basic Education, Work Experience, Rural Loans, Job Corps, Neighborhood Youth Corps. Most threatening to southern state governments and most controversial were the Community Action programs. The OEO described them as "a unique partnership between local communities and the federal government," unique in that they bypassed state and local government. "Community Action," the OEO explained, "lies at the heart of the War on Poverty. It means focusing the total resources of the community on the problem of poverty. It means direct involvement by a cross-section of community members—especially the poor." These programs hardly promised the growth-oriented area development so dear to southern Whigs.[45]

But the Economic Opportunity Act, a product of Congressional compromise, was itself vague about state participation in the anti-poverty fight. It endorsed community action, including the "maximum feasible participation" of the poor. But in order to win Congressional approval, the Johnson administration accepted amendments to its original proposal offered by Senator George Smathers (D-Fla). Smathers added safeguards for the states, granting governors the power to veto Jobs Corps and Community Action agency contracts in their states.[46]

This set the stage for bitter fights between southerners and the OEO, although the agency also fought with state and local governments in the North. OEO claimed to be fomenting a "quiet revolution," and, for a time, it seemed to succeed. The War on Poverty launched major campaigns in the South. The first two adult literacy programs were established in Kentucky and in North Carolina. Within a year, they had enrolled 7000 adults in free classes. Southerners who had never received adequate educations, 60 percent of them black, learned to read and write in North Carolina. In Mississippi, Project Star, a Community Action program, trained illiterates in reading and job skills. One student, a mother, borrowed a book of fairy tales to read to her children for the first time.[47]

The Rural Loan Program also waged battle in Dixie. Designed to help rural people "'caught in a box'—too old or unskilled for retraining and too young for Old Age Assistance," the program provided low interest loans and financial advice. Jesse Davis, a fifty-seven year-old East Texas farmer who could not meet his mortgage, was saved from foreclosure by a loan. With the money he refinanced his mortgage, purchased livestock, planted food crops to make his family self-sufficient, and, with the aid of government advisers, expanded and diversified his holdings.[48]

These incursions, modest as they were, nonetheless enraged many southern leaders. George Wallace took exception to federal programs providing so many services to blacks, programs which he believed trampled state's rights and hastened integration. More moderate leaders were also alarmed. In Florida, Governor Haydon Burns objected to OEO-supported programs which bypassed state agencies. He tried to seize control of all local Community Action and anti-poverty programs and threatened to veto all direct grants from the federal government to local agencies.[49]

The fears proved unjustified; the bite of Great Society programs never matched their bark. Eventually, Congress itself reined in the OEO and sharply curtailed the independence of Community Action agencies. But long before Congress enacted those restrictions, southern leaders had tamed the most threatening initiatives of the Great Society. Atlanta, one of the capital cities of the sunbelt South, ran one of the most prominent and controversial Community Action programs. Avoiding "maximum feasible participation of the poor," local elites controlled the Atlanta project. It provided modest services to individual poor people and favorable publicity for Atlanta, without empowering the poor or doing much to relieve poverty. OEO officials criticized the project and recommended terminating its funding. The Atlanta CAP, however, won such praise from the press and the Congress that OEO director Sargent Shriver continued federal support for it.[50]

The most threatening prospect of the Great Society—political organization and empowerment of the poor—never materialized. The impoverished in general and poor blacks in particular were under-represented on the boards of nearly every Community Action program in the south. "The concept of including the poor in the planning and in the operation, of course, is a new concept, and the people are having difficulty assimilating it," conceded Representative Sam Gibbons (D-Fla), head of the Southern Task Force investigating the anti-poverty program. Black rights activists and reformers were not so sanguine about the South's hostility toward Community Action. New York congressman Adam Clayton Powell charged: "In the South, black people have been excluded from planning boards, committees, staffs, and, in turn, enjoy the most minimal per capita participation in the poverty program." Marian Wright of Jackson, Mississippi, complained that conservative local officials had commandeered the federal government's war on poverty: someone "in Mississippi and in Alabama and in Arkansas and rural Georgia really cannot tell the difference between the State and Federal Governments." The programs, Wright concluded, "are not reaching the poor."[51]

Marian Wright did not exaggerate. Administration really mattered during the 1960s War on Poverty—which bureaucracy controlled a program, what its ideology and objectives were, whether state or federal officials implemented it—these determined the reception a program would receive from southern politicians and from black southerners. They also determined the effects of the program. It was no trifle when a southern state official petitioned Congress to assign all OEO educational programs to a more conservative federal agency, nor for a reformer like Dr. Gatch to seek broader powers for the OEO.[52]

WAR ON POVERTY IN THE SOUTH II: FOOD STAMPS AND FEDERAL AID TO EDUCATION

The Food Stamp program, for example, was administered to the specifications of southern Whigs. President Johnson signed the Food Stamp Act in 1964, but the program's long pedigree stretched back to 1939, when the federal government operated a small, four-year long experimental program. This pilot program shut down in 1943, and although every Congress for the next two decades considered a permanent Food Stamp plan, none was enacted. The Eisenhower administration strongly opposed all efforts to revive food stamps. Still, Ike continued the practice of distributing the surplus commodities in government storehouses. Under this program, the federal government packaged and delivered surplus food to state welfare authorities. Many southern states received these surplus commodities during the 1950s; the program cost the states nothing and helped local planters by sustaining demand for farm produce.[53]

The modern Food Stamp program took shape in the 1960s. President Kennedy launched a small pilot program in 1961 and, in 1964, the Food Stamp Act made the program available nationally on state option. In return for a deal on tobacco subsidies, southern members of Congress generally supported the bill.[54]

The Food Stamp program's conservative features appealed to southern politicians. First, food stamps were in-kind assistance not cash benefits—as Secretary of Agriculture Orville L. Freeman put it, the stamps were a food and agriculture program, not a general welfare program. As such, the program did not encourage the immorality and idleness many southerners ascribed to welfare. Eligible families could buy stamps which increased their purchasing power for food. Stamps could be redeemed only for food; they could not be squandered on drink or other idle pleasures. And, since one needed to purchase the stamps, food stamp users, unlike recipients of cash benefits, could not avoid work. This purchase feature, Secretary Freeman noted, was "difficult for many people to immediately understand and accept, including welfare officials and the families themselves." Not surprisingly, Dr. Donald Gatch opposed the purchase requirement and the Food Stamp program in general as too niggardly, while Georgia senator Herman Talmadge, a foe of the Great Society, applauded it.[55]

Second, the program was controlled by the South's favorite federal bureaucracy—the Department of Agriculture. Southerners relied on the Department to defer to the interests of big planters and to ensure that food assistance never siphoned away rural labor. Also Agriculture was unlikely to join HEW in cutting off funds to states that practiced racial discrimination. The Department of Agriculture, Senator Talmadge proudly proclaimed in 1969, "has never attempted to cut off Federal food assistance funds as a means of social engineering."[56]

Third, the Food Stamp Act provided generous federal funds with a minimum of supervision and regulation. Washington covered the full cost of the benefits and a substantial part of the administrative costs, but let the states run the programs and establish the eligibility requirements. The USDA itself set the benefit levels, but it even made a special concession to the South on this score. There were two scales of food stamp benefits, a parsimonious one for the South, another, more generous one for the rest of the nation. The Agriculture Department claimed that this practice geared "the program to economic needs of the participating areas and families." In fact, it merely ensured that benefits would remain low in the South.[57]

Not surprisingly, then, the conservative and business progressive leaders of the South generally welcomed food stamps. In the late 1960s, Mississippi was one of only six states in the country with a federal food assistance program in every county. According to the Atlanta *Journal and Constitution,* food stamps were "catching on" in Georgia as well. Still, few southerners and few Americans in general participated in the program during the 1960s. The restrictions were too stringent.[58]

This cozy arrangement, however popular with southern state governments, did little to relieve hunger or poverty in the South, and liberals sought reforms. In 1968, "Hunger, USA," the report of the private Citizens Board of Inquiry into Hunger and Malnutrition in the United States, focused attention on federal food assistance programs, especially in the South. CBS News followed up on the report with a television documentary, "Hunger in America," featuring poverty and malnutrition in Mississippi. The film and report galvanized the Congress to liberalize the Food Stamp program. The Senate immediately established a Select Committee on Nutrition and Human Need, chaired by Senator George McGovern (D-SD). In 1969, McGovern, Senator Walter Mondale (D-Minn), and Senator Joseph Montoya (D-NM) each introduced bills expanding the program. Herman Talmadge countered with a more modest alternative plan. After heated debate, Congress amended the Food Stamp Act several times during the early 1970s. The regional benefit scale was eliminated and a single, higher national standard was applied to the South. Purchase requirements were reduced and control over eligibility passed from the states to the federal government. As a result of these reforms, total benefits payments soared.[59]

These changes dampened the enthusiasm of southern politicians for the program. Southern state governments stepped up their investigation of citizens for fraud and other abuses. The South led the nation in such prosecutions, and some observers believed that such practices frightened people away from the program and harassed participants. Still, despite the reforms, food stamps remained a conservative food and agriculture program; it never became a general welfare or income maintenance plan. It neither uplifted the South's impoverished nor forced state governments to address their problems. Food stamps, in fact, proved less effective in reducing poverty in the South than in other parts of the United States. This salvo of the War on Poverty never challenged the South's Whiggish political economy.[60]

Neither did federal aid to education, another centerpiece of Johnson's Great Society. Like food stamps, federal education subsidies had a long, contentious history stretching back to the Great Depression (and even to the nineteenth century). Aid to education persistently fueled Congressional debate from 1937 until 1965. When the nation first turned to the problems of the South in the late 1930s, none had loomed larger than the region's inferior educational systems, a liability which promised the indefinite continuation of southern backwardness. In 1940 the South trailed the nation's other regions in every index of educational resources. As the *Report on Economic Conditions* noted, 1500 Mississippi schools lacked even buildings, forcing students to attend classes in "lodge halls, abandoned tenant houses, country churches, and in some instances, even in cotton pens." The South set the lowest standards in the nation in school attendance and literacy and, with the single exception of Texas, the southern states ranked below the national average in median school years completed.[61]

Southerners portrayed the poverty of southern public education as a national, rather than a regional problem. Echoing the claims of Rupert Vance and Howard Odum, southern liberals described the region as the "seed-bed of the nation." Lister Hill warned that "The educational level of the people of the United States is not a state problem alone; it is a national problem. With the increasing mobility of our people," he averred, "it cannot be otherwise."[62]

With the political momentum for equalization in the late 1930s, federal aid to education began its long, tormented legislative history. In various sessions of Congress during the 1930s and 1940s, sponsors included southern liberals such as Hugo Black, Claude Pepper, and Lister Hill, but the motivating force behind the measure was none other than Mississippi's Pat Harrison. This fiscal conservative and opponent of federal interference in the South gave his name to the 1938 version, a measure which promised some aid to every state in the Union, but devoted a substantial portion of federal resources to "equalizing educational opportunities." This translated into favoritism for the South. Title I of the 1938 Harrison-Fletcher bill, for instance, earmarked more than 60 percent of its funds for the southern states. Republican Senator Robert Taft led the early opposition to the Harrison bill, decrying it as "a bill for relief of the South."[63]

The controversy raged for ten years, during which time the Congress recommitted six separate bills without final action. The debate revolved around the "three R's" of federal aid to education—not reading, 'riting, and 'rithmetic, but region, race, and religion. Region as an economic issue was the primary "R" during the 1940s. Congress would enact no bill to relieve the South without some benefit for the rest of the nation, so all federal aid proposals devoted some funds to general support of all states and others to equalizing resources in poor states. But even such largesse could not assuage some northern legislators. Henry Cabot Lodge assailed the unfairness of several of the bills to his home state of Massachusetts, which paid taxes into the federal treasury, but received little in return under the terms of an equalization bill. Other northerners echoed Lodge's argument. This alignment of interests explained the relative success of federal aid measures in the Senate, where southern support supplied more than half of the necessary votes for passage, and the utter failure of the bills in the House of Representatives.[64]

Race—the second "R"—gradually replaced region as the major roadblock to federal aid to education. During the 1930s and early 1940s, the specter of "federal control" and meddling with state's rights had seldom reared its head. Proponents of federal aid explicitly wrote restrictions against federal control into the text of the bills. And Pat Harrison, whose credentials as an advocate of state sovereignty no one could challenge, even amended his 1938 bill to strengthen such guaranteees. In addition to the prohibitions against federal interference, proponents of federal aid cited a long tradition of national support for public education, dating back to the land ordinances of the Continental Congress and the land-grant colleges of the nineteenth century. More recently, they noted, the FERA, WPA, and RFC had established education programs on a small, local scale. Such financial support formed "a policy as old as the hills"; it marked no departure from the American tradition of local education. It would be aid without interference.[65]

The perverse political alignments around the issue of federal aid, the irony of southern Democrats defending a federal program against the state's rights objections of northern Republicans, righted themselves by 1950. With both political parties courting the black vote, the sponsors of the 1943 bill acceded to an amendment which forbade the states' trimming their contributions to Negro schools. Even with that provision, Republicans charged that the South would never devote its federal grants to black education. Nothing, Colorado senator Eugene Millikin maintained, prevented the southern state legislatures from disbursing their entire allotment of federal funds to white and none to needy blacks. Soon thereafter, the Republicans, with the aid of some northern Democrats, attached a rider to the education bill prohibiting the distribution of funds to segregated schools. To no one's surprise, that move killed the measure. Such riders found their way onto succeeding bills, often at the behest of House Democrats from northern cities. Since even as committed a southern liberal as Frank Graham could not accept such a restriction, a legislative stand-off emerged. There were sufficient votes to defeat any federal aid proposal with or without such a rider.[66]

The religion issue—whether private and parochial schools should receive federal aid—further bollixed the legislative works. In the absence of a national consensus on federal aid to education, then, Congress settled for a few limited, piecemeal measures. During the Korean War, aid for so-called federally impacted areas was enacted; these laws helped areas with military (and other federal) installations build and operate schools. Federal facilities swelled local school populations with the children of soldiers and bureaucrats and occupied land without contributing to the tax base. Impacted area aid compensated the affected districts.[67]

Once again, a long desired social program was justified as a military necessity, implemented piecemeal without its reformist objectives. It sought simply to offset the financial "burdens" the federal government had placed on these localities, even if the "burdens" had been greedily sought by the affected communities. Impacted area aid particularly benefited the South. During the 1952–53 school year, the South gorged itself with 41 percent of the aid for school construction and 35 percent of the aid for operation and maintenance.[68]

In 1958, in the wake of *Sputnik*, Congress expanded the hitherto modest aid program with the National Defense Education Act (NDEA). It appropriated federal funds for a variety of educational programs, especially for strengthening the teaching of science, mathematics, and modern languages. This measure too made itself felt below the Mason-Dixon line. In 1962 the Arkansas Commissioner of Education credited the NDEA for vast improvements in his state's schools. But while any district in the nation was eligible for NDEA funds, the law provided neither general aid to education nor special assistance for school districts in poor areas. In 1959–60 the federal government spent $15 billion on public education, a trifling 4 percent of the total public schools bill.[69]

Finally, in 1965, the Johnson administration broke through the legislative logjam and secured passage of the Elementary and Secondary Education Act (ESEA). With the support of the liberal Congressional majority, LBJ steered federal aid to education through the rapids of the three R's. Johnson had long desired such legislation; he had entered national service in the 1930s as Texas state director for the NYA, one of the earliest federal educational programs, and he had fathered the NDEA. Johnson considered the ESEA the centerpiece of his Great Society. He flew to Johnson City, Texas, to sign the bill, to the very one-room schoolhouse where he began his own schooling. The ESEA was far more ambitious and liberal than any of the proposals LBJ had supported in the 1940s and the 1950s. It not only aided poor areas like the South but underwrote a program of "compensatory education" designed to uplift impoverished people. Johnson saw education as a passport out of poverty, not just for backward regions, like his Texas hill country, but for individual poor people as well.[70]

The ESEA funneled money to regions, states, and district facilities, but, at its center, was Title I, special programs for the educationally deprived. Policymakers had long noticed the correlation between income and education—the more years of schooling a person completed, the higher her or his likely income. Impoverished children, for a variety of reasons, tended to perform poorly in school, and therefore to leave school early. This, the architects of Title I believed, trapped them in poverty. Massive federal aid, they assumed, if targeted for programs to improve the performance of impoverished children, would raise their grades, keep them in school, and lift them out of poverty. This program of compensatory education allocated funds to districts on the basis of the number of poor students they served. Title I, then, was envisioned as an anti-poverty weapon, not merely a tool to equalize educational resources from state to state. Furthermore, under Title VI of the Civil Rights Act, this infusion of federal funds would be withheld from segregated school systems. In 1965 the Office of Education published guidelines requiring segregated schools to submit a desegregation plan or forfeit ESEA funds.[71]

The ESEA seemed to confirm the worst fears of conservative white southerners and Whig business progressives. Instead of financial aid with no strings attached, southern schools had to devote resources to special programs for impoverished students and to dismantle dual school systems immediately. But this policy, which ostensibly pitted a liberal national administration against a conservative South, actually played into the hand of southern Whigs. Title I allocated funds according to the number of poor students in an area, but it entrusted the districts to spend the money on compensatory education. Many districts, in the South and across the nation, used Title I funds for routine expenditures. Quiet noncompliance allowed southern states to enjoy federal largesse without meeting disagreeable federal regulations.[72]

The desegregation guidelines similarly redounded to the advantage of Whig politicians. The rules encouraged the tokenism and quiet obstructionism which Whigs commonly termed "moderation." The Johnson administration hardly desired confrontation on this issue—LBJ refused to allow the race issue to block the educational aid he had so long sought for his native region. With Johnson's prodding, the U.S. Office of Education ensured that nearly every segregated southern district would submit a plan before the 1965–66 academic year began. The office approved so-called voluntary integration schemes, plans that minimized integration. They allowed a few blacks to enter white schools, but forced no whites to attend black schools. The federal government, one black southern educator lamented, had given the South a slew of clever tricks with which to avoid desegregation while continuing to collect ESEA funds.[73]

Accordingly, the peculiar racial moderation of pro-business politicians became the order of the day in Dixie. Overt resistance would sacrifice federal funds; full-scale desegration would inflame recalcitrant whites. Either extreme would spark visible conflict and stall economic development. Eventually, the federal courts interceded and forced integration, but the early guidelines eased and slowed the process. By the time southern schools submitted to the courts, they had become thoroughly hooked on federal funds. Federal aid became a still more powerful argument for moderation on the race issue.[74]

What, then, were the effects of the ESEA, one of the centerpieces of the Great Society? It did rescue southern schools. Indeed, southern schools rapidly became dependent on federal aid. R. E. Hood, the Brunswick, Georgia, superintendent of schools, did not exaggerate when he told a Congressional hearing that ". . . we would never have survived had there not been an up here [Washington] to appeal to." In 1970–71 the South grabbed more than 40 percent of NDEA and ESEA funds. While the average state relied on federal aid for just 7 percent of its education budget, the southern states banked on funds from Washington for between 9 and 22 percent of their total schools bill (the median percentage was 12.6). Federal aid to education assisted the southern states, financed new buildings, and established new programs, but it did not issue poor southerners a passport out of poverty, end racial discrimination, or threaten the probusiness political order. The ESEA allowed the South to upgrade its dismal schools without raising local taxes or devoting more resources to education. In that same 1970–71 year, per pupil expenditures on education in the South nowhere matched the U.S. average, and only Virginia, Florida, and Louisiana reached 80 percent of the national standard.[75]

No Great Society initiatives met a warmer welcome from southern leaders than aid to education and food stamps. None, perhaps, influenced southern life as much. But they did not sufficiently nourish the poor or lead them out of poverty. In 1965 the Atlanta *Journal and Constitution* published a cartoon, celebrating the promise of the Great Society. It depicted LBJ carrying a teeming horn of plenty. The horn bore the label "The Great Society," and out of it trailed such fruits as education, food, and medicine. But the cornucopia proved as beneficial to southern elites and politicians as to the region's poor. That hardly worried Lyndon Johnson. He hoped to reward local politicians as well as relieve poverty, and he saw little conflict between those goals. The Johnson administration, despite the efforts of the OEO and the rhetorical commitment to an all-out war on poverty, hardly challenged the political economy of the emerging sunbelt South. Throughout the 1960s and 1970s, Washington never forced southern states to reorder priorities from business development to poverty fighting. Nor did the federal government abandon its own emphasis on economic growth for a truly redistributive economic policy. Both Lyndon Johnson and his successor remained faithful to the Keynesian creed.[76]

In 1965, Leon Keyserling, architect of "Growth Keynesianism," ridiculed the War on Poverty. Even with a tremendous infusion of additional funding, Keyserling asserted, federal anti-poverty programs would remain ineffective; they were low-caliber weapons for an all-out war against such an intractable foe. Keyserling insisted that the "dominant aspects of getting rid of poverty in the United States relate to programs and efforts that are entirely apart from the Economic Opportunity Act." Only sustained economic growth could eradicate poverty. Keyserling was enunciating the fundamental priority of American economic policy, in the 1960s and 1970s as well as the 1940s and 1950s. Presidents Kennedy, Johnson, and Nixon, despite their various redistributive and anti-poverty initiatives, never abandoned the emphasis on growth. The early War on Poverty, the Great Society, and Nixon's welfare reforms only reinforced the South's drift toward a dual economy and a Whiggish political order.[77]

The actual contours of national policy reflected this theoretical and political confusion. Throughout the postwar era, federal public welfare programs remained fragmented. Just as the ARC and EDA competed in Appalachia, so national welfare policy embraced a mélange of categorical, uncoordinated programs, often with overlapping objectives. Varying degrees of local participation—voluntary community enrollment, direct payments, block grants, sliding scales of state matching requirements—further complicated the system. In contrast, the Defense Department, despite the legendary competition between the services, represented a model of single-minded efficiency. Still, federal welfare policy displayed three consistent features. First, the federal role expanded throughout the postwar era. In 1950 the national government contributed 46 percent of the nation's total social welfare expenditures. The federal share grew steadily to 50 percent in 1960, 55 percent in 1970, and 61 percent in 1980. Second, the nation preferred social insurance to public assistance. The former, principally social security benefits, catapulted from one-fifth of federal welfare expenditures in 1950 to almost two-thirds in 1980. Public assistance, even with the War on Poverty, held fairly steady. It accounted for 11 percent of federal welfare outlays in 1950, 13 percent in 1970, and 16 percent in 1980. Third, Congress expanded federal welfare spending in fits and starts, depending on economic conditions and political fortunes. Social spending reached its peaks when the greatest number of Democrats from outside the South sat in the House of Representatives.[78]

The implementation and impact of these programs reflected their perverse formulation. The distribution of need, measured either by numbers of poor families or areas of low income, never matched the pattern of disbursement. Inter-regional transfers of federal welfare payments benefited the West, the nation's second richest region, almost as much as the South, the nation's poorest. In 1970, the two regions registered approximately the same net surplus of income gains over tax costs for the welfare system. Other regions also received greater per capita benefits. In fiscal 1975, the Middle Atlantic and Western states garnered larger per capita federal welfare payments than the South.[79]

THE SOUTH AND FISCAL FEDERALISM

The South received less federal welfare support by choice. Southern governments set standards in order to extract the maximum possible federal aid with the minimum possible state and local investment. Because southern authorities allocated so little, the federal government contributed a larger share of the South's public assistance costs than that of any other region. In 1970, for example, Washington supplied a national average of 52 percent of state welfare costs. Of the southern states, only one (Virginia at 62 percent) received less than 70 percent of its welfare budget from the national government.[80]

Even with that federal commitment, southern welfare benefits remained well below the national average. For the major assistance programs—Aid to Families with Dependent Children, Aid to the Blind, Aged, and Disabled, and general assistance—none of the average monthly payments in the South approached the national standard. In 1970 the national average monthly AFDC benefit was $187. The South ranged from Virginia's regional high of $179 to $46 in Mississippi. Besides Virginia, no southern state awarded as much as three-quarters of the national average. Over the 1970s, southern benefits deteriorated even further. In 1985 the Southern Growth Policies Board calculated that in the United States as a whole food stamps plus AFDC provided 68 percent of the poverty line income. In the South, they added up to just 51 percent of that meager figure.[81]

Local control allowed such regional inequalities. Under both AFDC and Aid to the Blind, Aged, and Disabled, the two largest support programs, the states set eligibility requirements and benefit levels. The national government matched state contributions on a sliding scale. The first few dollars—a basic benefit—were underwritten entirely by the federal government. States received matching funds if they augmented the basic payment, but the higher the benefit, the greater the required state contribution. Keeping benefits low, as did southern states, minimized state expenditures for public welfare. Many southern states further discouraged their citizens from seeking assistance by establishing rigorous eligibility standards. From 1950 to 1969, AFDC case loads inched up more slowly in the South than in any other region, despite the South's intense poverty. Those in need either left the region or did without.[82]

Ordinary southerners faced the same choice with regard to the Great Society's manpower training programs. If any federal initiatives offered the region an opportunity to enlist its poor, unskilled youth in the sunbelt South of aerospace research and high-technology manufacturing, it was those training programs. But the region's leaders refused. Southern governors routinely vetoed the construction of Job Corps training centers in their states. The program therefore contributed little to the South, although a large contingent of young southerners left the region to enter the program. The Neighborhood Youth Corps fared only slightly better.

Many southern communities objected to the requirement that Corps workers receive the federal minimum wage for their labor and avoided the program. In 1962, Congress amended the Social Security Act to include the Community Work and Training program. CWT supplied training and work experience to AFDC recipients in order to foster economic independence. The federal government provided half of CWT's administrative costs and three-quarters of its social service costs. Only one southern state, Kentucky, participated in the program. Two years later, under Title V of the Economic Opportunity Act, Congress replaced CWT with the Work Experience and Training program. The new plan targeted the same people, but Washington picked up the total cost. Under those generous terms, every southern state but Alabama received some aid, but only Arkansas and Kentucky fully enrolled in the program.[83]

In general, the South preferred outright grants to matching programs. Per capita federal spending for welfare services reached the highest levels in both rural southern and urban northern communities. The types of spending, however, varied greatly between the two regions. In fiscal year 1975, for example, spending for both matching and non-matching programs stood about even outside the South, and slightly higher for matching programs in the Northeast. In the South, non-matching programs provided nearly double the funds of the matching programs. Southern officials listed preservation of low tax rates, antipathy to cumbersome regulations, and strained fiscal resources as their reasons for spurning matching programs. They wanted to maintain the sunbelt South's favorable business climate.[84]

No such scruples inhibited southern states from appropriating matching payments and accepting federal restrictions on development-related programs. Strapped southern treasuries managed maximum involvement in the federal-aid highway and airport programs. In 1960, all but two southern states devoted a larger share of their budgets to highway construction than the national average. Every southern state but North Carolina exceeded the U.S. level of highway expenditures per thousand dollars of personal income in 1965. Mississippi spent twice the national ratio. In 1970, per capita public welfare expenditures remained below the national average in twelve of the thirteen southern states. Yet five of those states exceeded the U.S. standard for per capita highway outlays and three more almost matched it. The bias became more pronounced by 1980. And just as southern state governments mustered the matching funds for highways, so they tempered their antipathy to federal regulations. Once urban redevelopment programs offered aid for mammoth football stadia and modern civic centers, Atlanta, New Orleans, Houston, and Tampa gladly submitted to the cumbersome regulations.[85]

State fiscal policies reinforced the emphasis on development-oriented ventures and the antipathy to public welfare spending. Highways drew more substantial investments from southern governments than public assistance. The southern states also devoted larger than average shares of their budgets to industrial development programs. Higher education received considerable support. In 1957, only Louisiana and Oklahoma reached the national average for per capita spending on higher education. In 1982, six southern states exceeded it. As expected, that list included the region's most developed states—Texas, North Carolina, and Oklahoma. But the South so desired improved universities that the region's poorest states—Mississippi, South Carolina, and Arkansas—also made that grade. On the other hand, 1982 welfare expenditures remained below the national average, whether measured on a per capita basis, as a percentage of state budgets, or even as a percentage of personal income. These priorities led the Southern Growth Policies Board's 1986 Commission on the Future of the South to insist that the South "Change welfare eligibility and benefits to more closely reflect the national average."[86]

The Southern Growth Policies Board demanded that southerners reverse their attitudes toward the federal welfare state. It asked southern state governments to embrace federal social service programs. Such a request must have been awkward for the Growth Policies Board—an organization dedicated to economic development. Twelve southern states had formed the Board during the 1970s in response to the political organization of the Frostbelt. Its purpose was the coordination of southern efforts in the interregional competition for defense contracts, grants, and industry. Still, the Board felt compelled by the persistence of southern poverty to condemn southern attitudes toward the welfare state.[87]

Enthusiastic participation in federal welfare programs would not have eliminated southern poverty. The programs, after all, had no such success in the Northeast. But opposition to them did shape the economy of the sunbelt South. The South's alliance with the federal government regarding regional development and its resistance to national welfare programs put a premium on luring educated whites. It pushed down or pushed out less educated blacks and whites. In the mid-1960s, one development official in Yazoo City, Mississippi, admitted that he searched only for jobs for whites. Employment opportunities for blacks, he feared, might stem out-migration and lead to a black political takeover of Yazoo City. Other southern leaders recognized that low AFDC benefits not only saved money but helped drive welfare recipients to the high benefit regions. One recent study of the migration effects of local fiscal policies demonstrated that blacks moved toward jurisdictions with high levels of social spending while whites relocated to low spending, low tax communities.[88]

BLACK SOUTHERNERS AND THE GREAT SOCIETY

As it had since the New Deal, national policy reinforced the racial biases of southern economic development. During the Great Depression, southern economic interests had acceded to the AAA and NRA because they could largely shift the costs of these policies to blacks. Jobs for blacks, like women and children's work, were sacrificed to the modernization of southern industry and agriculture. Meanwhile, southern authorities attempted to exclude blacks from New Deal relief programs and resisted federal relief when they could not do so. When World War II finally ended the Depression in the South, racial discrimination barred blacks from many jobs in war industries.

Racisim similarly fueled southern antipathy to federal welfare programs in the 1960s and 1970s. Unlike defense spending and growth programs, which mainly benefited whites, blacks were to be the principal beneficiaries of Great Society anti-poverty efforts. Black politicians and their constituencies formed a solid bloc of electoral support for the liberal welfare state. Local administration of federal programs, however, allowed many southern officials to ignore black poverty. Civil rights laws prohibited racial discrimination; they even threatened cut-offs of federal aid to offending states and localities. But federal authorities never limited highway funds to states which provided decent health care or maintained a minimum level of welfare benefits. Often, such fiscal decisions disguised subtler forms of racism. In 1981 a southern Republican party leader noted that a pro-growth, anti-welfare platform appealed to the racial resentments of many white southerners. Obvious racist appeals no longer succeeded in the 1970s. Instead, politicians like this Republican used fiscal conservativism to evoke racial fears. "You're talking about cutting taxes, and all these things are totally economic things and a by-product of them is that blacks get hurt worse than whites. And subconsciously maybe that is a part of it."[89]

Certainly, the War on Poverty and the civil rights revolution were linked in the minds of black and white southerners, and Americans in general. Even in the South, the battles for equal opportunity and against poverty overlapped. The school lunch program, for instance, was geared toward black southerners. Special federal funding underwrote school lunches for predominately black southern schools during the 1960s. In Wilcox County, Alabama, according to one official, only blacks received free lunches. Whites refused to participate. They viewed school lunches as a program for blacks.[90]

There were also more direct links between the War on Poverty and federal anti-discrimination efforts. In 1961, President Kennedy established the Equal Employment Opportunity Commission (EEOC), a federal agency charged with assuring that government contractors did not practice racial discrimination. The EEOC gained permanent statutory authority under the Civil Rights Act of 1964. It would theoretically force southern employers to comply with the Civil Rights Act or lose lucrative government contracts. In general, the EEOC deferred complaints to state bodies; only in the South, which lacked such state agencies, did the EEOC conduct its own investigations. In 1967, twelve states lacked Fair Employment Practices regulators of their own—eleven of them in the South. And the two southern states with anti-discrimination laws on their books—Oklahoma and Tennessee—lacked adequate enforcement procedures. The EEOC was poised to eliminate racial discrimination in many southern businesses.[91]

The Commission, however, lacked enforcement powers. Not until the 1970s did the Congress, against the objections of most southern representatives and the Nixon administration, allow the EEOC to sue offending parties in the courts. Still, even with expanded powers, the EEOC chalked up few victories in the South. It rarely forced southern employers into more than token compliance with the anti-discrimination laws.[92]

Similarly, Title VI of the Civil Rights Act—the provision calling for denial of federal funds to segregated institutions—disappointed black rights activists. Eventually, this policy forced the desegregation of hospitals, schools, and public accommodations, but only slowly. Only constant agitation from black activists and a flurry of lawsuits effected change.[93]

In general, white southerners continued to place themselves and their restrictions between the federal government and black southerners. "The white man," as ex-sharecropper Ned Cobb explained, ". . . tried to keep the government hid from the colored man and the colored man hid from the government." Local regulations kept blacks off the public assistance rolls in many southern communities. In Sunflower County, Mississippi, only "responsible citizens" were entitled to relief and that standard was interpreted by white officials. Some county governments even used the welfare system to suppress black political activity. In Leflore County, Mississippi, authorities cut off food relief when civil rights demonstrations expanded. Restrictions, harassment, and low benefits all signaled to the poorest black southerners that they seek elsewhere for the Great Society. Many did so.[94]

One legacy of the original New Deal assault on the South, the national minimum wage, played an ongoing role in this process. The Fair Labor Standards Act largely accomplished its original intention. It speeded the South's integration into the national labor market. It weakened the grip of low wage industry on the region. It helped raise family incomes in the South and across the nation. The national minimum also accelerated mechanization and skill upgrading and redirected growth to high wage, capital intense lines. But different demographic cohorts carried the burdens and realized the benefits of those developments. Like the sharecroppers of the 1930s, young unskilled blacks lost job opportunities. And without access to education or training programs, they had little chance to reclaim them. Many of them departed the region in the 1950s and 1960s.[95]

Poverty migrated along with many of those emigrants. Largely blacks from the mills and fields of the South, they headed toward northern cities where opportunities were scarce and jobs for teenagers non-existent. The minimum wage, of course, only partially accounted for this migration. Declining employment in agriculture, the promise of relief from segregation, the attractions of settled black communities, all pushed and pulled blacks northward. As these impoverished people moved north during the 1950s and 1960s, poverty metamorphosed from a rural to an urban problem, from a southern to a northern one, from an affliction of places to one of people. As one federal official explained in 1976, the large northern cities had become "the nation's new cottonfields. . . . The rural poor of the South and elsewhere now reside in the cities of the Northeast." The South remained the nation's poorest region in the 1970s, but it advanced as the North declined. The North suffered more from unemployment, decrepit housing stock, population decline, and slow income growth.[96]

In 1976 the Coalition of Northeastern Governors looked to Washington to remedy the desperate conditions of their cities. "As the nation in the past recognized the development needs of the Western frontiers and the Rural South," the governors argued, "so now the nation must acknowledge a similar commitment" to the Northeast. That announcement and others like it touched off the so-called Frostbelt-Sunbelt controversy of the 1970s, a competition for federal aid which *Business Week* dubbed "the Second War Between the States." The northeastern states, beset by fiscal difficulties and high federal tax burdens, attacked favoritism in federal assistance for the South in 1975. And the South did enjoy a more favorable balance of payments with Washington. Not suprisingly, military spending accounted for almost all of the South's advantage. The North received more in welfare spending.[97]

Of course, preferential federal treatment for the South was hardly novel in 1975. Some northern leaders had even complained about federal aid to the South in the 1930s. But before World War II, northern politicians generally supported federal development efforts in the South. National regulations promised an end to competition from low wage southern industry, and regional development offered federal contracts to firms in the industrial Northeast. In the 1930s and 1940s, for example, the TVA purchased a considerable volume of goods from the Middle Atlantic states. The underdeveloped South of the 1930s could not supply that demand. By the 1970s, southern firms met those needs.[98]

The improving fortunes of the South and the decline of the Northeast, however, obscured the regions' actual economic status. The South had narrowed the economic gap between itself and the rest of the nation, but still remained the nation's poorest region. Its cities contained a higher percentage of residents below the poverty line. Crime, disease, and malnutrition festered in the cities and the hinterlands of the sunbelt South, at the worst rates in the nation. And try as it might, the region could not export all of its impoverished, white or black.[99]

Indeed, a 1980s update of Howard Odum's opus would have shown the South with the nation's highest infant mortality rate, lowest income, and highest percentage of people below the poverty line. As outgoing Mississippi governor William Winter observed in 1986, "There remains that other South, largely rural, undereducated, underproductive, and underpaid that threatens to become a permanent shadow of distress and deprivation in a region that less than a decade ago had promised it better days." Winter's invocation of "two Souths" emerged as a familiar motif during the 1980s. The emergence of "two Souths" fueled political debate in that decade. Liberal southerners published critiques of southern growth such as "The Divided South" and "Shadows on the Sunbelt."[100]

In this respect, however, the South no longer differed from the rest of the nation. New York governor Mario Cuomo electrified the 1984 Democratic National Convention by describing the United States as a "tale of two cities." Other commentators adopted Cuomo's version of two Americas—"one prospering, the other suffering." At the same time, the South's rejection of the welfare state and embrace of the defense establishment found echoes throughout the nation. The South emerged as a national political and economic leader. In 1980 the South even repudiated its native son, Jimmy Carter. Among southern states, Carter won only his home state of Georgia in the 1980 presidential election. On the strength of his showing in the South, Ronald Reagan, friend of the Pentagon and foe of "big government," won in a landslide. Reagan also broadened Republican influence in the traditionally Democratic South.[101]

When Carter first entered the White House in 1977, many southerners, according to *Time*, compared him to Franklin D. Roosevelt. Certainly, Carter's election symbolized the South's reunification with the nation, the entry into the American mainstream that FDR had sought. Roosevelt might even have recognized the proliferation of defense plants and the expansion of southern higher education which accompanied that integration. No doubt he would have gestured approvingly with his cigarette holder at the news of automated southern factories and mechanized southern agriculture. But Roosevelt never could have imagined the revolution in race relations that erupted after his death. Nor could he have imagined the transformation in politics that his program ignited. Ironically, a program designed to nurture intelligent and liberal Democracy in the South would ultimately yield conservative Republicanism.[102]

Chapter 8 ✪ Conclusion: Place Over People

In 1939, David Lilienthal, chairman of the TVA and devoted New Dealer, took stock of his own and his President's recent actions in the South. "How could TVA help change the income level, not only of the section," Lilienthal wondered, "but of the low income groups—how could we increase the income which remains in the hands of the great masses of miserable people, most of them darned good stuff, too?" But policies aimed at aiding poor people seldom materialized. Instead, from 1940 until the 1980s, federal economic policy stressed regional development rather than human development. Originally designed to secure an "American standard of living" for the southern people, federal policy increasingly concentrated on ensuring an American standard of industrial progress for southern places.[1]

The triumph of place over people stemmed not from the fading of liberalism or the waning of federal action after 1940. On the contrary, federal activity in the South and around the nation broadened beyond the wildest expectations of the most ardent New Dealers. And liberal reform, while suffering occasional setbacks at the polls, also flourished from the 1940s to the 1970s. Postwar America witnessed a transmutation of federal intervention in the South rather than a retreat from it, a transmogrification of the liberalism that had nourished the New Deal rather than its eclipse.

The civil rights struggle contributed to these transitions. New Deal liberalism retained a strong southern component, with many young white southerners enlisted in Roosevelt's ranks. Committed to economic reform, they firmly believed that ameliorating southern poverty would relieve racial tensions, but they remained wary of tampering with segregation itself. As one southerner put it years later, ". . . people who get decent jobs don't stay rednecks for very long. That's the hope of the South, of course." It was, at least, the fervent belief of the southern liberals of the 1930s who planned to reform the South without directly challenging white supremacy. These attitudes carried the day in the Roosevelt administration, but after World War II, northern liberals increasingly embraced the civil rights cause—isolating the southern New Dealers and the programs they championed.[2]

As liberals tied themselves more closely to the black struggle for freedom, they slowed their drive for economic reform. Northern liberals embraced Keynesian economics, resting their hopes for social justice on sustained economic growth, not on redistributing wealth or restraining "economic royalists." Even during the 1960s War on Poverty, Keynesian principles held sway.[3]

In so doing, American "Big Government" followed the international trend. Even in Western Europe, where liberals held stronger electoral majorities than their American counterparts, Social Democrats and Laborites traded reformist Marxism for Keynesianism. They too came to conceive of strong economic growth as their top priority—as a prerequisite for social security or economic justice.[4]

But American Keynesianism proved harsher and more conservative than its European counterparts. The United States concentrated a much larger share of its government spending in the military—military Keynesianism with its peculiar political corollaries and economic consequences. The U.S. also relied more heavily on borrowing and less heavily on progressive taxation to finance its spending, and it never established what welfare state theorists term the "institutionalized state"—a guaranteed minimum income for all citizens. American liberalism, despite the promises of Franklin Roosevelt and Lyndon Johnson, never really attempted to ensure an American standard of living for all citizens, nor did it remove the social stigma from Americans receiving such assistance. Nowhere was this more apparent than below the Mason-Dixon line.[5]

To some extent, the emergence of federal policies which raised the section, but not its "great masses of miserable people," represented a failed promise of the New Deal era. The Roosevelt administration had attempted to uplift the South because it harbored a large share of the bottom one-third of the nation, and because it was a colonial economy subservient to other regions. A generation and a half later, in the midst of the sunbelt boom, many of the region's poor remained ill-housed, ill-clad, and ill-nourished, and the region depended as much as ever on outside economic interests. Many old southern liberals lamented those continuites. They found defense dollars not only ineffective but morally tainted, and they doubted that research could sustain southern economic development. The region's economic growth and racial progress, for example, could not console Jonathan Daniels. He felt betrayed by the modern South. He despised the region's new leadership, which he labeled a perverse partnership of professors and promoters.[6]

It is tempting to echo Daniels, to bemoan the unrealized promise of federal intervention in the South and to downplay the region's progress. Federal policy did not do as much to alleviate southern poverty as liberals wished; it did not do as much as it could have. But even though different national policies might have produced a broader, more equitable southern boom, such policies might never have mobilized sufficient political support. And, whatever its economic effects, federal intervention in the South certainly catalyzed vast changes in politics and relations.

RACE, POLITICS, AND ECONOMIC DEVELOPMENT IN THE CONTEMPORARY SOUTH

No change proved more startling than the civil rights movement and the rise of black political power. Although the region's economic upturn had largely bypassed them, black southerners had won a long struggle against segregation and disfranchisement through protests, litigation, and political action. Federal policy helped to arm southern blacks in their battle for freedom. The federal courts ordered the desegregation of the nation's schools, the Congress enacted civil rights and voting rights legislation, and the executive branch sent election examiners and troops to enforce those policies.

Regional development also influenced the southern civil rights movement, even if it did little to advance blacks economically. Civil rights leaders adapted their tactics to the peculiar conditions of their communities. In places like Selma and Birmingham, they forced dramatic confrontations. In other southern cities, they negotiated with local whites, who agreed to desegregate public accommodations or register black voters in return for racial peace.[7]

In 1958, Harry S. Ashmore argued that economic development would lead the South away from racial conflict. "It is not that the bustling gentlemen at the local Chambers of Commerce or the State Industrial Development Commissions are particularly concerned with race as a moral problem," the Arkansas journalist explained. Rather, those business interests realized that "sustained racial disorder would be fatal to their effort to lure new industries and new capital from the non-South, and that the existing level of tension isn't doing their handsomely mounted promotional campaigns any good." Cities that only lightly felt government-sponsored economic development and population growth—Memphis, Birmingham, Augusta, Saint Augustine—experienced explosive racial conflict. In booming cities like Atlanta, Dallas, and Hunstville, however, the civil rights struggle followed a more moderate course. In every case, the business community, by either its action or its apathy, governed the white response to black protests.[8]

In the early 1960s that pattern was not yet apparent, and southern history cautioned against the too easy association of economic development and racial progress. The southern business elites of the early twentieth century, as C. Vann Woodward noted in an influential 1961 article, had both pressed for new investment and established Jim Crow. To assume that business boosterism ensured racial moderation, Woodward alerted his readers, was to be hoodwinked by the "New South fraud." Woodward's point was well taken, but not entirely applicable to the post-World War II era. Unlike the first New South campaign, postwar economic development altered the outlook of southern white leaders, who never advocated far-reaching reform and generally preferred segregation. Many even shared the prejudices of the racist White Citizens Councils. But white leaders desired civic harmony, federal aid, and continued economic advance so strongly that, if necessary, they would accede to racial change in order to secure them. And black southerners soon forced those concessions.[9]

In several crucial respects, then, the postwar crusade for industrial development differed from the New South movement of the late nineteenth and early twentieth century. First, black southerners stood in a stronger position vis-à-vis the region's ruling class. The protestors of the 1950s and 1960s had the law of the land on their side. Furthermore, the civil rights movement had the benefit of a half-century of experience. Those cities where blacks had established civil rights groups before 1954 adjusted fastest to desegregation.[10]

Second, a vanguard of white southerners led the South away from segregation and toward racial justice. These courageous writers, journalists, and clergy publicly lobbied for racial change. Even in politics, many white southerners risked electoral defeat to speak out for racial justice. Among others, Senator Frank Graham of North Carolina, Governors James Folsom of Alabama and Leroy Collins of Florida, and U.S. Representatives Brooks Hays of Arkansas and Frank Smith of Mississippi fought courageously for racial change. As a senator, Lyndon B. Johnson refused to sign the "Southern Manifesto" protesting the *Brown* decision. As President, Johnson signed the Civil Rights Act and the Voting Rights Act. The white South of the 1950s and 1960s no longer presented a united front against black demands for integration.[11]

Third, in the earlier period, southern leaders maintained an economic stake in segregation. White supremacy blocked black access to skilled positions in industry and to land ownership in agriculture. In so doing, it maintained a reservoir of low-paid black workers that kept wages down and allowed planters and mill owners to exploit both black and white laborers. But, from the New Deal forward, federal policy eroded the South's economic investment in segregation. The South became less dependent on cheap labor. Employers, whatever their racial views, no longer needed white supremacy to safeguard their labor supply.

Indeed, national policy substituted new and powerful incentives to acquiesce in integration. Before 1930 the South neither desired nor received much economic assistance from Washington. But by 1960 the loss of federal aid because of noncompliance with desegregation orders posed a genuine threat to southern leaders. Similarly, southern employers of the 1960s, unlike their New South counterparts, relied on outside sources of labor and investment. A reputation for racial bigotry frightened away northern investors and workers. The region's new-found dependence on federal largesse and interdependence with the national economy supplied the civil rights forces with a powerful weapon. The South could no longer resist racial change without severe economic costs.[12]

Where racial moderation prevailed, it manifested itself not so much in the liberalization of white racial attitudes as in the triumph of southern Whiggery. Before 1960, economic development offered southern politicians an alternative to the restrictive shackles of the race issue. After 1960 the rising intensity of civil rights activity and the region's growing economic dependence on the federal government, national businesses, and northern labor supplies, further loosened those manacles. With the exception of Alabama's George Wallace, the southern governors most dedicated to development were racial moderates. Politicians like Florida's Leroy Collins, South Carolina's Donald Russell, North Carolina's Terry Sanford, Georgia's Carl Sanders, and Arkansas's Winthrop Rockefeller avoided fruitless race-baiting by making economic development the number-one political issue of their campaigns.[13]

Florida's Leroy Collins embodied the politics of moderation. Born in 1909, Collins, like most residents of what was then the South's least populous state, grew up in the state's northern panhandle. Collins witnessed the miraculous transformation of peninsular Florida from uninhabited swamp into one of America's richest and fastest growing regions. He established close relationships with Florida businessmen and became renowned as an industrial recruiter. When Collins took office in 1955, he was determined to promote industrial progress and lobbied for political reform consonant with economic growth.[14]

A basically conservative man of fervent religious belief, Collins at first resisted innovation in racial matters. In the immediate aftermath of the *Brown* decision, Collins, like most southern "moderates," pledged to use all legal means to preserve segregation. He even volunteered to plead Florida's case before the Supreme Court. But, as governor, Collins rejected defiance of the Court and looked forward to a time when "acceptance of non-segregation is developed in the hearts and minds of people."[15]

Events eventually forced Collins into action. In the 1956 election a virulent segregationist challenged Collins for the Democratic gubernatorial nomination. During the bitter campaign, Collins reasserted his fealty to segregation, but defused the race issue by trumpeting his contribution to Florida's economic development. He warned that "nothing will turn . . . investors away quicker than the prospect of finding here communities hopped up by demagoguery and seething under the tension and turmoil of race hatred." The lyrics of Collins's victory song boasted that "He made the legislature legislate/ He put roads in places that help the state/ He went travlin' North, met the V.I.P.s/ And brought back millions in new industries." A few months before the election, Howard Hughes announced plans to build a research installation in Florida, and Collins's campaign song immediately incorporated the news. "Floridians, here's a governor we must choose," the ditty concluded, "or we'll lose that project with Howard Hughes."[16]

Collins won the election by the largest margin in the history of the Florida Democratic party, sweeping the booming southern portion of the state, with its large cities and immigrant population. After the election, Florida's opposition to integration intensified, but Collins became more moderate. In his 1957 inaugural address, Collins warned Floridians to expect inevitable changes in race relations. Later that year, he "spanked" Orville Faubus in a speech at the Southern Governors Conference after the Little Rock crisis. When Tampa blacks launched a sit-in movement in 1960, Collins appointed a biracial committee to settle the dispute. After finishing his term in 1961, Collins accepted positions in the Kennedy and Johnson administrations.[17]

Collins might have been an unusual southern Whig in his genuine commitment to racial change—a moral impulse that went a little further than protecting the state's reputation in federal government offices and northern corporate boardrooms. North Carolina, on the other hand, provided the unalloyed model for the subordination of racial conflict, and to some extent racial justice, to economic growth. Another southern governor who would serve in the Kennedy and Johnson White Houses, Luther H. Hodges, presided over North Carolina's school crisis. Hodges, the archetypal Whig politician of the postwar era, proclaimed industrialization "the number one goal" of his administration and relentlessly pursued that objective. Governor Hodges established a development credit corporation, personally hunted for industry in Europe and across America, and reformed and expanded the state highway department. His administration created the state's board of higher education and established its community college system. Hodges even recognized that the state's abundance of low wage manufacturing threatened economic progress. "North Carolina was dependent on and satisfied with these industries," he recalled in his memoirs, "but other industries that paid higher wages were needed." To get them, Hodges built the Research Triangle Park.[18]

Nimbly casting himself as the voice of moderation, Hodges managed to pursue his economic program in the midst of the conflict over school desegregation. At the same time, he stifled debate by portraying all opponents, even blacks seeking full compliance with the Supreme Court, as dangerous extremists. "I was besieged on both sides," Hodges claimed, criticizing both Orville Faubus and the NAACP in the same breath. With that approach, Hodges steered a complicated local option plan into law. The plan allowed limited integration in communities that desired it, but postponed desegregation in most of the state. Nonetheless, Hodges successfully billed it as the only moderate option and discredited all other proposals. In so doing, Hodges achieved all of his goals—to forestall integration, to prevent racial conflict, to protect the state's progressive image, and to prevent race from derailing economic expansion.[19]

Hodge's successor in Raleigh, State Senator Terry Sanford, followed in the same vein. The state's leading liberal and an early supporter of John F. Kennedy, Sanford promised Tar Heels a "New Day," and declared that the "South would rise again, not with bayonets, but with textbooks." On segregation, Sanford preached racial justice, but he went little further than Hodges. Sanford based his integration campaign on voluntarism and a statewide network of good neighbor councils. Despite his liberal reputation, Sanford also had the support of the state's business leaders. "Because there were Terry Sanfords who handled desegregation problems in an astute fashion," one North Carolina black official reflected in 1976, "North Carolina didn't get the attention, and the pressure was not brought to bear as greatly as on other Southern states."[20]

Political moderation muted racial conflict and sustained economic growth. Its effects on the black struggle for justice, however, proved less salutary. Progressive communities could stop at limited, token integration. Moderate states also erected all sorts of "non-racial" procedural and administrative barriers to postpone desegregation. North Carolina's pupil assignment plan was the most ingenious of these. Not until the 1964–65 academic year did as many as one pecent of North Carolina's black students attend integrated schools. In 1971, Greensboro had still not complied with federal integration guidelines, and that same year, Charlotte was the district involved in the *Swann* decision that upheld court-ordered busing.[21]

Racial moderation effectively rewarded southern politicians for doing nothing. Collins, Hodges, and their ilk won praise not by fighting for justice, but by not fighting vociferously against it. As Whitney Young of the National Urban League put it, praise for such moderation was like "rewarding a child for not having a tantrum." To divert attention from the race issue, Whig politicians needed to recast desegregation as a white issue—open schools and economic development versus closed schools and economic stagnation. In so doing, they often ignored pleas for racial justice.[22]

But if the politics of moderation delayed the black struggle for freedom in some cases, it nourished it in others. The concern for civility headed off the viciousness that reigned in Mississippi and Alabama. Although many black southerners were courageous enough to risk their lives in their cause, the absence of such severe threats permitted demonstrators to organize and protests to flourish. After all, the sit-in movement, as the *New York Times* noted (with misplaced irony), began in North Carolina communities with liberal reputations and token desegregation policies.[23]

With a few exceptions such as Dallas, the moderate approach also enhanced black political power. A dozen years after Atlanta blacks negotiated the desegregation of downtown, they controlled the city. By the mid-1970s, Atlanta had sent black representatives to the state legislature (Senator Julian Bond), to the mayor's office (Maynard Jackson), and to the U.S. Congress (Representative Andrew Young). Young won his seat in a white majority district, outpolling the national Democratic ticket in each of the city's white precincts. Even with minimal support from white communities, blacks used city government to press for power in the late 1960s and 1970s.[24]

Indeed, if the politics of moderation were effective in the early 1960s, they became necessary after 1965. After that date, blacks became a force to be reckoned with in the southern electorate. The federal Voting Rights Act of 1965 proved to be the crucial ingredient in the enfranchisement of southern blacks. It outlawed discriminatory regulations that had kept blacks from the polls and sent federal examiners to monitor elections in seven southern states. Despite uneven enforcement (federal officials preferred voluntary compliance and limited active involvement whenever possible), the Voting Rights Act expanded black political participation and spawned a new stage of civil rights activity.[25]

The Act also withered the political appeal of segregationism. Gradually, militant segregationists faded from the southern political scene. Active supporters of integration won elections and strident opponents either lost, like Orville Faubus, or modified their views, like George Wallace. In 1971, Julian Bond returned to a Mississippi town where several activists had been murdered in the early 1960s. Bond and his companions received the red carpet treatment and a police escort from local authorities. Shortly thereafter, John Lewis of the Student Nonviolent Coordinating Committee remarked that "the 'closed society' had begun to open its doors to the knock of the black ballot."[26]

Black politicians pursued different electoral strategies just as they had adjusted tactics to local conditions in the desegregation battles. In Mississippi and Alabama, the states where conflict was most pronounced, blacks built separatist political organizations. In the other southern states (and eventually in Mississippi and Alabama), blacks eschewed separatism and entered coalitions with moderate white Democrats. That strategy ended segregationist domination of state Democratic parties and elected candidates more responsive to black aspirations. Many of the southern Democrats elected by black-white coalitions in the 1970s proved demonstrably more liberal than their predecessors or colleagues.[27]

Such coalitions initially empowered southern blacks, but eventually compromised black political strength. Tied to the fortunes of the Democratic party, black voters had nowhere to turn when white Democratic candidates moved to the right in order to broaden their support with appeals to segregationists or economic conservatives. Sometimes in such circumstances, black southerners split their tickets. Arkansas blacks, for instance, cast 91 percent of their ballots for the Republican Winthrop Rockefeller in 1968, while voting overwhelmingly for Democrats Hubert H. Humphrey and J. William Fulbright in the presidential and senatorial elections. Similarly, Georgia blacks denied Herman Talmadge reelection in 1980. Even though Talmadge's Republican opponent hardly appealed to black voters, they could not countenance returning one of the South's legendary racists to the Senate. "Talmadge's defeat," Atlanta's mayor Maynard Jackson asserted after the election, "demonstrates that Afro-American voters cannot any longer be taken for granted in Georgia—you cannot spit in our eye and tell us it's raining."[28]

Jackson's prediction proved premature; those cases remained exceptions to the rule of straight-ticket Democratic voting by southern blacks. They became the region's new "yellow dogs"—the epithet for southerners who would vote Democratic even if the party placed a yellow dog on the ballot. Across the South and the nation, black politicians increasingly felt themselves and their constituencies taken for granted. Such frustration led Jesse Jackson to seek the Democratic presidential nomination in 1984. "The fundamental relationship between blacks and the Democratic Party," Jackson asserted, "must be renegotiated."[29]

Black political fortunes required renegotiation in the 1980s because the events of the 1960s and 1970s not only expanded black political activity, but also restructured the political loyalties of southern whites. Civil rights agitation and voting rights legislation brought millions of black southerners to the polls. But at the same time, population growth, increased partisan competition, and the final repeal of the poll taxes swelled white registration across Dixie. For if the South, in its odyssey from "Economic Problem No. 1" to sunbelt power, witnessed any political development more stunning than black registration, it was the growth of white Republicanism. Presidential Republicanism had established itself in the South in the 1950s, when Eisenhower captured the electoral votes of Florida, Tennessee, Texas, and Virginia, and it sank roots deep in southern soil. Over the next two decades, Republicans also won statewide office in every southern state, increasing their share of southern gubernatorial and senatorial seats from less than 3 percent of the total in 1960 to more than 40 percent in 1980. Republican registrations also skyrocketed. By 1980, nearly the same percentage of white southerners in the ex-Confederate states (25 percent) identified themselves with the GOP as did whites nationwide (26 percent).[30]

Republican gains proved the ironic harvest of the economic transformation of the South. Most analysts credited the GOP surge to Democratic in-fighting over the race issue, an internal struggle which drove segregationist whites into the Republican camp. That view had some merit. Barry Goldwater's Deep South landslide in 1964 manifested rejection of the Democratic civil rights program. So did George Wallace's strong showings in the 1968 and 1972 presidential races. But, for the most part, broader socioeconomic developments nourished Republican gains. Florida and Texas, the South's economic success stories, led the way to southern Republicanism. Furthermore, after 1975, when the race issue had abated, a far larger percentage of voters registered Republican in both Florida (a growth state) and Louisiana (a Deep South state) than had during the late 1960s and early 1970s.

Three separate strains of southern Republicans emerged between 1950 and 1980. First, migrants arrived from outside the region. The same businessmen, professionals, and skilled workers who filled the new positions in defense firms and research laboratories swelled the Republican rolls in the South. The region imported Republicanism much as it had imported new industry and a new technical workforce. During the 1970s, half of the grass-roots Republican activists in Virginia were immigrants to the Old Dominion. In 1980, Georgia senator Herman Talmadge labeled his Indiana-born opponent a "carpetbagger." With an impressive showing in the state's suburban counties, where "carpetbaggers" resided in strong numbers, ex-Hoosier Mack Mattingly became Georgia's first Republican U.S. Senator since Reconstruction.[32]

Native urban businessmen and southern migrants from small towns joined the transplanted Yankees to form Republican strongholds in the South's economically vital urban and suburban communities. The GOP flourished in Texas and Florida, especially in the boom areas. Republicanism thrived in Florida's "urban horseshoe," a solidly Republican suburban region which voted for Nixon on 1960. In Texas, the Houston and Dallas areas contained the strongest Republican counties. Moreover, the GOP fared best among younger white southerners. In the 1980s, older white voters, those over sixty years of age, maintained their traditional allegiance to the "party of the fathers." The youngest cohort, white southerners between 18 and 39 years of age, registered Republican. In the 1980 presidential election, white voters over sixty in the eleven ex-Confederate states gave Jimmy Carter a small plurality, while the 18- to 39-year-olds went overwhelmingly for Ronald Reagan. Older white southerners voted more strongly for Carter, and younger white southerners more strongly for Reagan, than their counterparts outside the region.[33]

The Goldwaterites, the renegades on the race issue that defected to Goldwater in 1964, constituted the third source of southern Republican growth. Barry Goldwater pioneered the vaunted Republican "Southern Strategy." In 1961 he told an Atlanta audience: "We're not going to get the Negro vote as a bloc in 1964 and 1968, so we ought to go hunting where the ducks are." That signaled an appeal to segregationist southern whites, a promise to consign the task of integration to state authorities. Goldwater followed up that pledge with Senate votes against the Civil Rights Act of 1964. In the November election he won 87.1 percent of the vote in Mississippi, 69.5 percent in Alabama, and majorities in South Carolina, Louisiana, and Georgia.[34]

Most of those Goldwater supporters voted for Wallace in 1968 and for Nixon in 1972. In the 1980s they rallied to the standard of Ronald Reagan. But these mainly poor, rural whites hardly feathered the Republican nest. Goldwater could not carry the urban, Republican strongholds of Florida and Texas or the traditional GOP voters of Tennessee. And below the presidential level, the southern strategy failed the Grand Old Party. The Republicans never organized the rural areas, the bastions of Wallace and Goldwater support. Victories in national and state elections have rarely translated into local political power. Most hard-line segregationists, both office seekers and voters, remained registered Democrats. Indeed, in the 1970s and 1980s, photogenic, mealy-mouthed Democrats dominated state and local government across the South. They pasted together illogical "night and day coalitions" of traditionally Democratic conservative, rural whites and liberal, urban blacks. Only against the region's most liberal Democratic candidates did Republican mobilization of the Goldwaterites contribute to victory.[35]

Nonetheless, the Goldwater candidacy indirectly augmented Republican power in the South. Goldwater initiated a southernization of the national Republican party, a gradual takeover of the GOP leadership by its sunbelt wing. As southern and western interests captured the Republican party, they largely transformed the GOP from the party of Wall Street bankers to the voice of cowboy entrepreneurs and evangelical Christians. Out of these Republican conflicts of the 1960s and 1970s evolved a populist conservatism. The tax revolt of the 1970s, for example, represented a genuine populist movement with a real economic grievance—the increasing share of the federal tax burden borne by middle and low income taxpayers. But instead of a liberal movement designed to shift the burden back to the rich and the corporations, the tax revolt took the form of an attack on government itself. In the South, the Republicans' rejection of welfare programs appealed to middle-class migrants from the North and poor, rural whites alike. Opposition to welfare dovetailed with the economic development proclivities of urban businessmen, the economic grievances of poor workers, and the racial bitterness of southern whites against blacks who were the principal beneficiaries of welfare spending. Across the nation, that union of traditional business Republicanism with populist conservatism energized the GOP in the 1970s and 1980s.[36]

In 1949, V. O. Key predicted that with the race issue resolved, the South, because of its poverty, would embrace economic liberalism. That prophecy, already dubious when Key propounded it, rang even more hollow by 1980. The passing of forty years had all but eliminated racial demagoguery. But, at the same time, conservative, pro-business, anti-government politics developed in the region. "Public relations," one southern critic admitted in 1980, had replaced "racism as the compelling southern vice."[37]

This changing political landscape explained the region's paradoxical attitudes toward the national state. Black southerners generally applauded the federal government's role in race relations and the economy. At the same time, black leaders from Julian Bond and Martin Luther King, Jr., in the 1960s to Andrew Young and Jesse Jackson in the 1980s, often questioned the national defense establishment. For white southerners, the relationship proved more ambiguous. Because the New Deal found its greatest potency and exerted its deepest economic impact below the Mason-Dixon line, the region remained wedded to some of FDR's legacies. Middle Tennesseeans praised the TVA, tobacco farmers depended on the farm subsidy program, elderly southerners embraced social security. Notwithstanding those attachments, white southerners generally decided in the 1970s that Washington (not counting the Pentagon) had grown too strong. The region still relied heavily on the activities of the federal government. But, largely because it excepted defense and growth-oriented programs from its rejection of federal initiatives, the South earned a deserved reputation for hostility to the national government.[38]

By the late 1970s the prevailing themes of southern politics dominated political debate across the country. "Liberal" had become a naughty word. In 1976, Republican challenger John Grady attacked Florida senator Lawton Chiles as "a liberal who is soft on defense and spendthrift on social programs." Four years later, John East, a little-known protégé of Senator Jesse Helms, unseated Democratic Senator Robert Morgan in North Carolina. East echoed Grady, labeling his opponent a "big spender on social programs but tight-fisted when it came to defense." Ronald Reagan, of course, swept the South and most of the nation, except for his opponent's home state, with similar campaign rhetoric.[39]

The Republican rise in the South, and the economic changes that engendered it, ensured the triumph of Whig politics in the region. With few exceptions, development-oriented politicians ruled both parties in the South after 1960. In 1967, Mississippi governor Paul Johnson agreed to a tax-exempt $130 million bond issue to secure Litton Industries' "Shipyard of the Future" for Pascagoula. Litton threatened to withdraw if the bond issue became politically controversial. Governor Johnson convened all the contenders for statewide office. On tape, each one pledged not to debate the bond issue. They kept their promises. Even in Mississippi, no politician wanted to put at risk an automated, modern facility with 12,000 high-paying jobs.[40]

In fact, at both ends of the political spectrum, southern politicians labored to create environments amenable to economic growth. Black politicians joined in the chorus, often tabling welfare and relief programs to court their cities' economic elites. Mayors Andrew Young of Atlanta, Harvey Gantt of Charlotte, and Richard Arrington of Birmingham, the *New York Times* reported in 1985, sought investment and touted their cities like other southern boosters. Cooperation with business proved such a political and economic necessity for southern officeholders that black politicians disappointed many civil rights leaders. In the zeal for investment, Julian Bond saw indifference to the needs of the poor. According to Bond, the first generation of black politicians had been "loud and frequently lonely voices speaking for the dispossessed." But by 1979, Bond lamented: "Many of the region's black elected officials have turned out to be only slightly better than the white officials whose places they took." For good or ill, Bond's diagnosis has proved accurate.[41]

White Democrats exhibited the same fealty to Whig principles. John West and Richard Riley in South Carolina, David Pryor and Bill Clinton in Arkansas, Jimmy Carter in Georgia, Reubin Askew in Florida, William Winter in Mississippi all campaigned for improved education, road construction, and high-tech development. Republicans, of course, sounded the same pro-business tunes. These business progressives retired the staunch segregationists; they also drove out some of the region's few remaining liberals.[42]

The Whig spirit caught on even in Mississippi, heart of the old Cotton Belt. With deals like the Litton Industries bargain, Mississippi increased its share of defense contracts more than twenty-five times between fiscal year 1951 and fiscal 1978. In the 1970s the state government opened a Research and Development Center in the Jackson suburbs to spur economic development in the state. The state increased investments for education, higher education, and transportation systems. It elected William Winter to the governorship, a candidate with strong commitments to education and economic development. In 1982, in expenditures per capita and as a share of state budget outlays, Mississippi outpaced the national average expenditures for higher education and public schools. Meanwhile, per capita income in the Magnolia State remained lowest in the country and the infant mortality rate second highest. In 1987 a policy analyst at Silicon Valley's SRI International called Mississippi's high technology development program the "hottest" in the country. Yet a larger percentage of Mississippians lived below the poverty line than did residents of any other state. In exaggerated relief, the Mississippi of the 1980s demonstrated what forty years of federal intervention, forty years of economic transformation, had wrought.[43]

In 1976, one of the South's great historians encapsulated the odyssey of the rising South. He charted economic gains and accompanying changes in the racial and political systems. Waxing lyrical over regional progress, George B. Tindall caught himself. "But before this begins to sound like Henry Grady warmed over and spiced up with a dash of pollyanna," Tindall warned, "let us not forget that if experience is any guide, the South will blow it. We will have to make the same mistakes all over again, and we will achieve the urgan blight, the crowding, the traffic jams, the slums, the ghettos, the pollution, the frenzy, and all the other ills that modern man is heir to." Thinking perhaps of Texas or Florida or Georgia or his own home of North Carolina, Tindall concluded: "We are already well on the way."[44]

Where Tindall erred in that compelling vision was in his view of the South as mere follower. Such a portrait fit the early twentieth-century South which pursued northern capital and railroads. It captured the region of the 1930s and 1940s; then, national policymakers set in motion the region's economic transformation in order to remake the South in the image of the rest of the nation. In recent times, however, the tables have turned. The South, for good or ill, has been bidding the nation to follow its lead. The South has passed on much to other regions—a fondness for high technology, a craving for defense industry, a suspicion of unions, a divided economy, an antipathy to welfare, an uneasy accommodation between black urban leaders and white business conservatives. The South has also picked Presidents. In the 1980 election, southern whites supported Reagan more enthusiastically than northern whites. And among blacks, even though few voted for Reagan, southern blacks gave him a slightly higher percentage than northern blacks. In 1981, Virginia governor Charles Robb, a prospective Democratic presidential candidate, returned the Democratic party to power after a long hiatus in Virginia. Robb resembled Reagan far more closely than he did any recent Democratic President, including his father-in-law, Lyndon B. Johnson. As early as 1962, the New Orleans *Times-Picayune* had predicted that the South's blend of development, urbanization, and "innate conservatism" would transform national politics: The South would find "its interests no longer sectional, its needs no longer marginal, and its outlook no longer blurred by ancient grievances."[45]

FROM COTTON BELT TO SUNBELT

In the two decades after 1962, the South seemed truly to become an integral piece of the national polity and economy, a part of the main. The contemporary South, many commentators maintained, skipped the intermediate stage in America's passage from rural to urban to suburban nation, from agrarian to industrial to service economy, but rejoined America's post-industrial mainstream. Long-term economic processes—the mechanization of the cotton culture, the steady swelling of cities, the high-technology boom, the ever-narrowing income gap between the regions, the northward migration of poverty—knit the regions together and brought the South a two-party political system, even if those processes by-passed large segments of the South.[46]

Certainly, such "natural" processes played a large role in the South's postwar economic transformation. The oil economy, with its cycle of boom in the 1970s and bust in the 1980s, dominated the fortunes of the Southwest. The long, sustained period of postwar prosperity produced what economists call "spread effects." That is, continuously strong demand stimulated the expansion of plant, while other factors encouraged that new capacity to locate away from the old manufacturing belt. This industrial de-concentration especially benefited the South in the 1950s and 1960s.[47]

The postwar era also witnessed the rise of so-called footloose industries. For these firms, unlike the old heavy industries, geographic proximity to supplies of raw materials was unnecessary. They could readily take advantage of the South's appealing business climate. Footloose industries included the high-tech lines, but also many highly competitive, traditional southern industries like textiles. But, of course, high technology produced few jobs for local residents, and footloose low wage plants moved frequently. Many southern communities found themselves just as dependent as ever on the local mill, and just as vulnerable when that employer departed for another locale.[48]

Tourism and retirement also emerged as major postwar businesses, flooding the warm South with tourists and retirees. Florida led the way in both respects, but by the 1970s, hillside resorts and retirement condominiums had emerged even in Appalachia. Tourism development, however, provided little direct benefit to southerners. Aged southerners were less likely than elderly Americans to collect social security benefits, and aged black southerners the least likely to do so because domestics and agricultural workers did not participate in social security before the 1960s.[49]

But these basic economic processes—like the many others analyzed above—owed much of their potency to the federal government. Macroeconomic policy sustained postwar prosperity and pushed industry toward the South. Tourism and the retirement business depended on social security, interstate highways, and airports. Pipeline development and tax policies nourished the oil patch.

"Cotton is going West, Cattle are coming East, Negroes are going North, and Yankees are coming South." That popular southern quip, Frank Smith noted in 1965, captured "a great many of the changes through which the South is passing in these middle years of the twentieth century." The national government figured prominently in all of those epic migrations. The AAA reorganized southern agriculture, reducing cotton production in the Deep South and diminishing the tenancy that sustained it. At the same time, the revised economics of southern agriculture and the vicissitudes of the farm program encouraged the cultivation of livestock. Along with the demise of sharecropping, federal wage policy catalyzed a reorganization of southern manufacturing. The effect was to dry up low wage, unskilled employment and send black workers to northern cities. At the same time, war and space industry and high-tech spin-offs of the defense state attracted educated Yankees below the Mason-Dixon line. They fueled the rise of business-oriented politics. In the 1970s the South's booming cities captured the imagination of the American press as its poverty had in the 1930s. Tales of Sunbelt prosperity replaced stories of Cotton Belt woe in the national consciousness. Even as the Sunbelt boom slowed in the 1980s, the southern economy continued to outpace the national average in employment growth and other indicators, and journalists wondered at the reverse migration of blacks toward the region so many had once fled.[50]

The region's boosters believed the South's new politics consonant with its economic development. They numbered the end of one-party Democratic rule as just another instance of southern progress. That might have been the case. Even the liberal Democrats of the 1930s looked forward to a future South of two-party competition. The shadows on the Sunbelt, however, exposed the folly of unqualified celebration of southern progress. The South, for all its changes, remained a place apart in the 1980s. Poverty persisted, especially in the rural areas. Education still lagged. Public health facilities remained inadequate. Blacks had yet to enter the economic mainstream and the tax structure remained notoriously regressive.

The Southern Growth Policies Board was wise to concede in 1986 that the South had journeyed only halfway down the highway to economic prosperity.[51]

The federal government largely paved the way. The South's odyssey from Cotton Belt to Sunbelt, however tortuous and incomplete, depended on national policy. It also dramatized the evolution of postwar national policy, the shift from economic programs to defense-related expenditures. Aid to underdeveloped regions gave way to a competitive scramble for the fruits of federal spending.

In the 1930s, FDR had punctuated nearly every speech about his hopes for regional development by lecturing southerners on their need to truly join the nation. He had echoed Howard Odum and the Depression-era southern liberals, calling for an end to sectionalism and the birth of a South with closer economic and political bonds to the rest of the nation. Luther Hodges picked up Roosevelt's theme when he arrived in Washington in the 1960s: "We Southerners don't want to saw off any part of America, and we are sick and tired of hearing people talk as if they wanted to saw us off, to isolate us from the mainstream of American progress and prosperity." Roosevelt's dreams of a vitalized southern economy were realized but, in the perverse way of dreams, the newest New South appeared not so much as FDR had envisioned it, but as Luther Hodges had.[52]

❂ Essay on Selected Sources

What follows is a discussion of selected sources. A complete listing of sources with full citations is presented in the notes. Statistical information can be found in several regular U.S. Government publications. The U.S. Bureau of the Census provides essential data in the *Statistical Abstract of the United States,* the *Census of Manufactures,* and the *Current Population Reports.* Consult also the regular *Bulletins* of the U.S. Bureau of Labor Statistics, and the Bureau's monthly journal, *Monthly Labor Review.* Especially useful is the 1947 Bureau of Labor Statistics report on labor in the South. See Bureau of Labor Statistics, *Bulletin No. 898.* Private economic studies also provide a treasure trove of information on the southern economy. First among these is Howard W. Odum's *magnum opus, Southern Regions of the United States.* See also the works of Odum's student Rupert Vance, especially his *All These People;* and *Economic Resources and Policies of the South,* a 1951 study by Calvin B. Hoover and B. U. Ratchford.

I. U.S. Government Archives and Publications

The records of most U.S. government agencies are housed in the National Archives (NA) in Washington, D.C. The records of the Department of Labor (RG 174), the Office of War Mobilization and Reconversion (RG 250), and the War Production Board (RG 179) are all available there. The index for the Labor Department archives, an unwieldly color-coded file of index cards that bedevils colorblind researchers, resides across town in the library of the Department of Labor, in the Frances Perkins Building. An especially rich source for students of the southern economy, of federal policy, and of all American industry in the late 1930s, is RG 155, the Records of the Wages and Hours and Public Contracts Division of the Department of Labor. Entry 16, the records of the Industry Committees established by the Fair Labor Standards Act, contains abundant material on the organization of American industry on the eve of World War II.

Two important collections, the Records of the National Recovery Administration (RG 9), and the Records of the National War Labor Board

(RG 202), are scattered among several repositories. Most records remain at the National Archives in Washington, as do, for some inexplicable reason, the records of the NWLB's southwestern field office. The agencies' records pertaining to the Southeast, however, are held at the Federal Archive and Record Center for Region IV, located at East Point, Georgia (FARC). That regional center also contains, *in toto,* the Records of the Tennessee Valley Authority (RG 142). Especially valuable are the Records of the TVA's Commerce Department.

The National Recovery Administration, the Federal Emergency Relief Administration, and the Agricultural Adjustment Administration all published regular reports on their operations. In 1938, the National Emergency Council issued as a pamphlet the *Report on Economic Conditions of the South.* The *Report* contains President Roosevelt's charge to the Conference on Economic Conditions. The Federal Writers' Project, an enterprise of the Works Progress Administration, conducted interviews in the rural South during the 1930s. Their eloquent testimony of the attitudes of ordinary southerners is collected as *These Are Our Lives.*

On wartime developments, the *Termination Report of the National War Labor Board* supplements the records of the National War Labor Board, the War Production Board, and the Office of War Mobilization and Reconversion. See also the publications of the U.S. Fair Employment Practices Committee. Postwar adjustments are described in the publications of the Civilian Production Board, and in the *Postwar Economic Studies* of the Federal Reserve Board. The Department of Defense and the National Aeronautics and Space Administration describe their own activities in the South between 1945 and 1980. The Department of Defense's regular publication of *Prime Contract Awards, By State* is the essential record of defense spending in the postwar South.

The U.S. Advisory Commission on Intergovernmental Relations, established during the 1950s, remains the principal chronicler of American federalism. Its publications detail the federal grant system, report state and local fiscal policy, and assess regional development programs. Two of the commission's publications—*The Role of Equalization in Federal Grants* and *Multistate Regionalism*—are particularly valuable. On highways and airports, consult the reference works of the Department of Transportation—the Federal Highway Administration's *America's Highways 1776–1976,* and the Federal Aviation Administration's *FAA Historical Fact Book.*

The Economic Development Administration (EDA) of the Department of Commerce disseminates information on the federal government's regional economic development programs. It also has published valuable studies of the impact of research and development on economic growth. Consult also the publications of the EDA's predecessors, the Area Redevelopment Administration and the Office of Area Development. *Appalachia,* the organ of the Appalachian Regional Commission, describes that agency's activities.

The *Congressional Record* transcribes the relevant debates on Capitol Hill. More valuable, however, are the hearings and reports of the Congressional committees considering legislation of particular relevance to southern economic development. Four sets of such Congressional documents deserve special attention. The Tolan Committee of the House of Representatives, officially known as the Select Committee to Investigate the Interstate Migration of Destitute Citizens before Pearl Harbor and as the Select Committee Investigating National Defense Migration after December 1941, compiled an unparalleled collection of testimony and data on wartime migration, morale, economic performance, human experience. In 1949, the Joint Committee on the Economic Report published a follow-up study to the *Report on Economic Conditions of the South,* entitled *Impact of Federal Policies on the Economy of the South.* In the early 1960s a subcommittee of the Senate Select Committee on Small Business conducted a series of hearings on the economic impact of defense spending. Under the direction of Minnesota senator Hubert Humphrey, that committee uncovered the role of defense spending (and of the welfare state) in postwar American social and economic development. It also reveals and discusses the nation's attitude toward its government's various functions. Finally, the Committee on Education and Labor of the House of Representatives and its Senate counterpart, the Committee on Labor and Public Welfare, conducted hearings on all aspects of the 1960s War on Poverty. Most valuable is the House Committee's 1965 *Examination of the War on Poverty Program.*

II. Personal Papers, Memoirs, and State Archives

At the University of North Carolina in Chapel Hill, the Southern Historical Collection houses the most valuable collection of personal manuscripts on the modern South. The Southern contains the papers of Frank Porter Graham, the leading southern liberal of his generation. Graham also served as chairman of the Conference on Economic Conditions of the South, Industry Committee chairman, member of the National War Labor Board, and briefly, as United States Senator from North Carolina. The papers of Jonathan Worth Daniels, another influential southern liberal, who served as a White House adviser to Franklin Roosevelt, also reside in Chapel Hill. So do the Luther Hartwell Hodges Papers. The Hodges personal papers include the Governor's reports from his "fishing expeditions" for new industry. They should be supplemented with the official Governor's Office Papers at the North Carolina State Department of Archives and History in Raleigh.

In the Manuscript Division of the Library of Congress, in Washington, D.C., researchers can profitably consult the papers of leading southern legislators. The manuscripts of Hugo Lafayette Black, the architect of the Fair Labor Standards Act, and of Thomas Terry Connally, the Texas senator, are particularly useful. The Library also contains the papers of Jesse

H. Jones, the Houston magnate and onetime Secretary of Commerce, Harold Ickes, FDR's Secretary of Interior, and of Chattanooga editor George Fort Milton.

The Governor's Office Records in the Tennessee State Library and Archives provide insight into that state's relationship with the Tennessee Valley Authority and the Atomic Energy Commission. The state archive also maintains complete sets of the records pertaining to the various multistate southern organizations. The proceedings of the Southern Governors Conference and the Southern Regional Education Board can be found in the Governor's Office Records in Nashville.

Published collections of papers are also valuable, particularly those of the Presidents of the United States. Of these, the thirteen volumes of *The Public Papers and Addresses of Franklin D. Roosevelt* contain the most material on the interaction of national policy and regional economic development. In 1933 and 1934, Lorena Hickok, motored around the nation and reported on the progress of the New Deal and the condition of the nation's people. Her letters to her supervisor, Harry Hopkins, and to her friend, Eleanor Roosevelt, are collected as *One Third of a Nation: Lorena Hickok Reports on the Great Depression.* Edited by Richard Lowitt and Maurine Beasley, Hickok's letters describe both the objectives and the impact of the New Deal in the South. From the 1930s through the 1960s, the sociologist Rupert Vance remained an insightful observer of southern life. John Shelton Reed and Daniel J. Singal have edited a collection of Vance's essays under the title *Regionalism and the South: Selected Papers of Rupert Vance.*

Many crucial actors have published memoirs or autobiographies. Of varying trustworthiness, these include the reminiscences of southern politicians like Jimmy Byrnes, Tom Connally, Jesse Jones, Claude Pepper, and Luther H. Hodges. The administrators of federal agencies have also recounted their experiences. The most important of these are David Lilienthal's *Journals* (see also his book *TVA*), and Donald Nelson's *Arsenal of Democracy,* a memoir of Nelson's experience as chairman of the War Production Board. In 1946, Ellis Arnall published his vision of the South's future as *The Shore Dimly Seen.* Two liberal southern Congressmen, both of whom served as directors of the TVA after losing their House seats, have each written two volumes of memoirs. Brooks Hays's *A Southern Moderate Speaks* and Frank E. Smith's *Look Away from Dixie* are the more noteworthy volumes.

Two extraordinary memoirs relate the experiences of black southerners, before and during the Great Depression. They are the reminiscences of the farmer Ned Cobb (alias Nate Shaw), *All God's Dangers,* compiled and edited by Theodore Rosengarten; and *Lemon Swamp and Other Places* by Charlestonian Mamie Garvin Fields, with the assistance of Karen Fields.

III. Secondary Sources

The American South between Reconstruction and the Great Depression has been the subject of much superb scholarship and even more acrimonious debate. For an overview of this scholarship, see the essays in John B. Boles and Evelyn Thomas Nolen, eds., *Interpreting Southern History: Historiographical Essays in Honor of Sanford W. Higginbotham.* C. Vann Woodward pioneered the field in 1951 with his *Origins of the New South 1877–1912.* Extending and revising Woodward's analyses of the twentieth-century South are George B. Tindall's monumental *The Emergence of the New South 1913–1945* and Gavin Wright's stimulating *Old South, New South.*

The post-Civil War reconstruction of southern agriculture has generated considerable attention. Works cited in this study include Gerald D. Jaynes, *Branches Without Roots;* Jay Mandle, *The Roots of Black Poverty;* Roger L. Ransom and Richard Sutch, *One Kind of Freedom;* and James L. Roark, *Masters Without Slaves.* On the evolution of southern industry in the nineteenth century, see Dwight B. Billings, Jr., *Planters and the Making of a "New South";* David L. Carlton, *Mill and Town in South Carolina;* James Fickle, *The New South and the "New Competition";* Jacqueline Dowd Hall et al., *Like a Family;* Joseph Pratt, *The Growth of a Refining Region;* and Jonathan Wiener, *Social Origins of the New South.*

On politics and public policy in the South between 1880 and 1930, see Blaine A. Brownell, *The Urban Ethos in the South 1920–1930;* Paul D. Escott, *Many Excellent People;* Richard Sylla, "Long-Term Trends in State and Local Finance," in Stanley L. Engerman and Robert Gallman, eds., *Long-Term Factors in American Economic Growth;* J. Mills Thornton, "Fiscal Policy and the Failure of Radical Reconstruction in the Lower South," in J. Morgan Kousser and James McPherson, eds., *Race, Religion, and Reconstruction: Essays in Honor of C. Vann Woodward;* Peter Wallenstein, *From Slave South to New South;* and two works by J. Morgan Kousser, *The Shaping of Southern Politics* and "Progressivism—For Middle-Class Whites Only," *Journal of Southern History 46* (May 1980): 169–94.

For reflections on the historiography of the postbellum South, see James C. Cobb, "Beyond Planters and Industrialists," *Journal of Southern History* 54 (February 1988): 45–68; and Woodward's memoir, *Thinking Back: The Perils of Writing History.*

A collection of essays edited by James C. Cobb and Michael Namoroto examines *The New Deal and the South.* President Roosevelt's personal relationship with the region is described by Frank Freidel in *F. D. R. and the South.* The wide-ranging work of Pete Daniel, Jack Temple Kirby, Warren Whatley, and Lawrence J. Nelson investigates the ramifications of New Deal farm policy in the rural South. At this writing, their work is most accessible in Daniel's *Breaking the Land* and Kirby's *Rural Worlds Lost.* Cambridge University Press will soon publish a revised version of Whatley's doctoral dissertation, "Institutional Change and Mechanization in the Cotton South." See also Arthur Raper's classic study, *A Preface to Peasantry.* On New Deal relief programs, see Donald S. Howard, *The WPA and Federal Relief Policy;* Howard B. Myers, "Relief in the Rural South"; and Michael S. Holmes, *The New Deal in Georgia.* Early TVA policy is discussed in three anthologies: Erwin C. Hargrove and Paul K. Conkin, eds., *TVA:*

Fifty Years of Grass Roots Bureaucracy; Roscoe C. Martin, ed., *TVA: The First Twenty Years;* and John R. Moore, ed., *The Economic Impact of TVA.* See also the stimulating essay by Jane Jacobs, "Why TVA Failed," which is adapted from her full-length study *Cities and the Wealth of Nations.*

NRA policy is described by Leverett S. Lyon et al., *The National Recovery Administration,* Charles F. Roos, NRA Economic Planning; and Leon C. Marshall, *Hours and Wages Provisions in N.R.A. Codes.* Michael M. Weinstein assesses the impact of the recovery program in *Recovery and Redistribution Under the NIRA.* Raymond Wolters's *Negroes and the Great Depression* and Raymond Gavins's *The Perils and Prospects of Southern Black Leadership* analyze the reaction of black southerners to the NRA and the New Deal in general. See also on this subject Michael S. Holmes, "The Blue Eagle as 'Jim Crow Bird'"; Harvard Sitkoff, *A New Deal for Blacks;* and Nancy Weiss, *Farewell to the Party of Lincoln. Manufacturers' Record,* the organ of the Southern States Industrial Council, illustrates the attitudes of low-wage southern manufacturers toward the NRA and toward later federal policies. Michael Bernstein assesses the experience of American industry in general in his provocative *The Great Depression: Delayed Recovery and Economic Change in America, 1929–1939.*

On southern liberals in the 1930s and 1940s, the most valuable sources are John T. Kneebone, *Southern Liberal Journalists and the Issue of Race, 1920–1944;* Daniel J. Singal, *The War Within;* and, especially, Morton B. Sosna, *In Search of the Silent South.* Thomas A. Kreuger chronicles the Southern Conference for Human Welfare in *And Promises to Keep.* The works of Howard Odum, Rupert Vance, and Walter P. Webb are also available in their own right. See also the writings of prominent southern liberals like Jonathan Daniels's *A Southerner Discovers the South;* Claude Pepper's "A New Deal in Reconstruction"; Brooks Hays's *Politics is My Parish;* and Clark Foreman's "Decade of Hope."

Orme Phelps has recorded *The Legislative Background of the Fair Labor Standards Act.* Steve Davis analyzes the origins of the *Report on Economic Conditions of the South* in "The South as 'the Nation's No. 1 Economic Problem.'" A. Cash Koeniger, "The New Deal and the States," Robert A. Garson, *The Democratic Party and the Politics of Sectionalism, 1941–1948,* James T. Patterson, *Congressional Conservatism and the New Deal,* and Samuel Lubell, *The Future of American Politics* place the 1938 "purge" in the context of long-term national and Congressional politics.

On the economic theory underlying the minimum wage law, consult J. Ronnie Davis, *The New Economics and the Old Economists;* and Robert Lekachman, *The Age of Keynes.* John Moloney assesses "Some Effects of the Federal Fair Labor Standards Act upon Southern Industry." In his unpublished doctoral dissertation, David E. Kaun ably and comprehensively analyzes the "Economics of the Minimum Wage." On southern labor, see Allan R. Richards, *War Labor Boards in the Field,* F. Ray Marshall, *Labor in the South,* and Nelson Lichtenstein, *Labor's War at Home.* Horace C. Hamilton, "Educational Selectivity of Net Migration from the South," Jack

Temple Kirby, "The Southern Exodus, 1910–1960," Joseph J. Persky and John F. Kain, "Migration, Employment and Race in the Deep South," and Elizabeth M. Suval and Horace C. Hamilton, "Some New Evidence on Educational Selectivity in Migration to and From the South" all consider aspects of interregional migration.

During World War II, John Dos Passos toured the South to gather material for his *State of the Nation.* A native of Dixie, Birmingham journalist John Temple Graves, published his reflections on the region in wartime as *The Fighting South.* Frederick L. Deming and Weldon Stein report on southern defense industry in *Disposal of Southern War Plants.* Frederic C. Lane's *Ships for Victory* describes the activities of the U.S. Maritime Commission in the South. On wartime transformations in agriculture and their long-term implications, see the works by Pete Daniel, John Temple Kirby, Warren Whatley, and Gavin Wright cited above. See also Richard Day, "The Economics of Technological Change and the Decline of the Sharecropper"; Gilbert C. Fite, "Southern Agriculture Since the Civil War"; Harry S. Fornari, "The Big Change: Cotton to Soybeans"; and Charles R. Sayre, "Cotton Mechanization Since World War II." On industrialization, consult James C. Cobb, *Industrialization and Southern Society, 1877–1984;* Victor R. Fuchs, *Changes in the Location of Manufacturing in the United States Since 1929;* and the 1949 National Planning Association Report by Glenn McLaughlin and Stefan Robock, *Why Industry Moves South.*

William Nicholls's classic 1960 study exposes the conflict between *Southern Tradition and Regional Progress.* Two anthologies investigate the federal grant system—*Fiscal Federalism* and *Grants-in-Aid,* edited by Peter Mieskowski and William Oakland, and *Essays in Fiscal Federalism,* edited by Richard A. Musgrave. Mark H. Rose describes the birth of the national interstate highway system in *Interstate: Express Highway Politics, 1941–1956.*

V. O. Key's *Southern Politics* remains the starting point for analysis of that subject. Three studies have updated Key's state-by-state findings. They are Jack Bass and Walter DeVries, *The Transformation of Southern Politics;* William C. Havard, ed., *The Changing Politics of the South;* and Alexander P. Lamis, *The Two-Party South.* See also the collection edited by James F. Lea, *Contemporary Southern Politics;* and, on the same subject, Earl Black and Merle Black, *Politics and Society in the South.* Numan V. Bartley has contributed much to the history of postwar southern politics. See his *The Rise of Massive Resistance* and *From Thurmond to Wallace.* On national politics during the 1940s, see Alonzo L. Hamby, *Beyond the New Deal;* Harvard Sitkoff, "Harry Truman and the Election of 1948"; and the *Senate Journal* kept by Alan Drury.

Walter A. McDougall's masterful . . . the *Heavens and the Earth* analyzes the space program. See also *The Political Economy of the Space Program* by Mary A. Holman. The impact of defense spending is assessed in Roger E. Bolton, *Defense Purchases and Regional Growth;* Bolton, ed., *Defense and Disarmament;* James L. Clayton, ed., *The Economic Impact of the Cold War;*

Roger W. Lotchin, ed., *The Martial Metropolis;* John E. Lynch, *Local Economic Development After Military Base Closures;* Roger Riefler and Paul Downing, "Regional Effect of Defense Effort on Employment"; and Bernard Udis, ed., *The Economic Consequences of Reduced Military Spending.* See also *Our Own Worst Enemy: The Impact of Military Production on the Upper South,* an investigative report prepared by the Highlander Research and Education Center; and "Southern Militarism," the special first issue of the southern liberal periodical *Southern Exposure,* published in 1973. For a profile of L. M. Rivers, see Marshall Frady, "The Sweetest Finger This Side of Midas," *Life* 27 February 1970. David F. Noble, *Forces of Production,* and Nathan Rosenberg, *Inside the Black Box,* consider the military's effects on industrial technology.

The Southern Regional Education Board disseminated considerable material about its activities, and about southern education in general. See especially the Board's *The Future South and Higher Education.* Robert C. McMath et al., *Engineering the New South,* charts the history of Georgia Tech from the nineteenth century to the opening of the Advanced Technology Development Center. On education in general, see Diane Ravitch, *The Troubled Crusade;* and Joel Spring, *The Sorting Machine.* Edward H. Kolchum, "Georgia High Tech Facility Is Key To Growth"; Marc Miller, "The Lowdown on High Tech"; John W. Moore, "High Tech Hopes"; and Stuart Rosenfeld, "A Divided South," all consider the development of high-technology industry in the region.

The contours of postwar southern economic growth are sketched by David Goldfield, *Promised Land: The South Since 1945;* Robert J. Newman, *Growth in the American South;* and Bernard Weinstein and Robert E. Firestine, *Regional Growth and Decline in the United States.* Richard Gillam, ed., *Power in Postwar America;* Godfrey Hodgson, *America in Our Time;* and Alan Matusow, *The Unraveling of America,* examine the evolution of liberalism and of federal policy in the postwar era. On the nature of the modern American state, consult Leonard Krieger's seminal essay "The Idea of a Welfare State in Europe and the United States" and William E. Leuchtenburg's stimulating address "The Pertinence of Political History." Recent scholarship on this subject is summarized and collected in Peter Evans et al., eds., *Bringing the State Back In.* Literature on the Sunbelt begins with Kevin P. Phillips's discovery of *The Emerging Republican Majority* in 1970. See also Phillips's *Post-Conservative America,* and *Power Shift* by Kirkpatrick Sale. Two anthologies investigate various issues regarding contemporary southern urban and economic development. They are *Shades of the Sunbelt,* edited by Randall M. Miller and George E. Pozetta, and *The Rise of the Sunbelt Cities,* edited by David C. Perry and Alfred J. Watkins. C. Vann Woodward criticizes the concept of the Sunbelt in *Thinking Back: The Perils of Writing History.* The Sunbelt-Frostbelt debate attracted national attention in the mid-1970s with two articles in the popular press:

"The Second War Between the States," *Business Week,* 17 May 1976; and "Federal Spending: The North's Loss is the Sunbelt's Gain," *National Journal,* 8 (26 June 1976).

The "Sunbelt-Frostbelt" controversy is ably summarized in James C. Cobb's *The Selling of the South,* a masterful study of state economic policy in the modern South. It should be supplemented with Richard M. Bernard and Bradley R. Rice, eds., *Sunbelt Cities;* and David C. Perry and Alfred J. Watkins, Jr., eds., *The Rise of the Sunbelt Cities.* Alan M. Huber is at work on "A Fiscal History of the South Since 1940." The Southern Growth Policies Board recently analyzed the region's prospects and problems in *Halfway Home and a Long Way To Go.* That report and the other publications of the Board's Commission on the Future of the South chart the successes and failures of government policy in the modern South.

The nature and extent of the regional wage differential is debated in Don Bellante, "The North-South Differential and the Migration of Heterogenous Labor"; Michael Bradfield, "Necessary and Sufficient Conditions to Explain Equilibrium Regional Wage Differentials"; Philip R. P. Coehlo and Moheb A. Ghali, "The End of the North-South Wage Differential"; H. M. Douty, "Wage Differentials: Forces and Counterforces"; Robert S. Goldfarb and Anthony M. J. Yezer, "Evaluating Alternative Theories of Intercity and Interregional Wage Differentials"; and George Stamas, "The Puzzling Lag in Southern Earnings." The migration effects of government policy are assessed in Richard J. Cebula, "Local Government Policies and Migration."

The War on Poverty is beginning to attract considerable scholarly attention. For a variety of perspectives, consult Marvin E. Gettelman and David Mermelstein, eds., *The Great Society Reader;* Marshall Kaplan and Peggy Cuciti, eds., *The Great Society and It's Legacy;* Sar A. Levitan, *The Great Society's Poor Law;* Charles Murray, *Losing Ground;* Frances Fox Piven and Richard Cloward, *Regulating the Poor;* and John E. Schwarz, *America's Hidden Success.* There are numerous studies of individual Great Society programs. An especially useful compilation is *The Food Stamp Program: History, Description, Issues, and Options,* prepared by the Staff of the Committee on Agriculture, Nutrition, and Forestry, U.S. Senate in April 1985. To gauge southern opinion on the Great Society and other developments of the 1960s, the Atlanta *Journal and Constitution,* edited by the famous southern liberal Ralph McGill, and the conservative New Orleans *Times-Picayune* offer a range of views.

On black migration, consult Rex R. Campbell et al., "Return Migration of Black People to the South"; Richard Raymond, "Determinants of Non-White Migration During the 1950s"; and Tommy Rogers, "Migration Attractiveness of Southern Metropolitan Areas." James L. Walker investigates *Economic Development* and *Black Employment in the Nonmetropolitan South. Employment of Blacks in the South* is the title of a collection edited by Ray Marshall and Virgil L. Christian, Jr.

William Chafe's masterly, *Civilities and Civil Rights* elucidates the black struggle for freedom in a "moderate" southern community. Elizabeth

Jacoway and David Colburn, eds., *Southern Businessmen and Desegregation,* collects case studies of the behavior of the business communities in various southern cities during the civil rights movement. Steven Lawson's *In Pursuit of Power* considers black politics in the South after 1965, while Earl Black investigates state governors' responses to black activism in *Southern Governors and Civil Rights.*

❂ Notes

Preface

1. John Brewer, *The Sinews of Power* (N.Y.: Knopf, 1989), xv–xvi. Robert Higgs, *Crisis and Leviathan* (N.Y.: Oxford Univ. Press, 1987), 237. See also William N. Parker, "The South in the National Economy, 1865–1970," *Southern Economic Journal* 46 (April 1980): 1045.
2. The idea of *Herrenvolk* modernization is introduced in Peter Coclanis and Lacy K. Ford, "The South Carolina Economy Reconstructed and Reconsidered," in Winfred B. Moore, Jr., et al., eds., *Developing Dixie: Modernization in a Traditional Society* (Westport, Conn.: Greenwood, 1988), 102–3.
3. William E. Leuchtenburg, "The Pertinence of Political History: Reflections on the Significance of the State in America," *Journal of American History* 73 (Dec. 1986): 585–600, quotation on p. 592. For one comparison of the size of the post-World War II American state in relation to those of Western Europe and Japan, see Ian Gough, "State Expenditure in Advanced Capitalism," *New Left Review* 92 (1975): 59; and Gough, *The Political Economy of the Welfare State* (London: Macmillan, 1979), 79. Recent scholarship on these issues is reviewed in Leuchtenburg, "The Pertinence of Political History"; and Theda Skocpol, "Bringing the State Back In: Strategies of Analysis in Current Research," in Peter Evans et al., eds., *Bringing the State Back In* (N.Y.: Cambridge Univ. Press, 1985), 3–43.
4. For an overview of this interpretation, see Richard Polenberg, "The Decline of the New Deal," in John Braeman, ed., *The New Deal: The National Level* (Columbus: Ohio State Univ. Press, 1975).
5. C. Vann Woodward, *Origins of the New South 1877–1913* (Baton Rouge: Louisiana State Univ. Press, 1951). Dwight B. Billings, Jr., *Planters and the Making of the "New South"* (Chapel Hill: Univ. of N.C. Press, 1979). Jonathan M. Wiener, *Social Origins of the New South: Alabama, 1860–1885* (Baton Rouge: Louisiana State Univ. Press, 1978).
6. V. O. Key, Jr., *Southern Politics: In State and Nation* (N.Y.: Vintage Books, 1949), ix.
7. Rupert Vance, "The Profile of Southern Culture," in John Shelton Reed and Daniel Singal, eds., *Regionalism and the South: The Selected Papers of Rupert Vance* (Chapel Hill: Univ. of N.C. Press, 1982), 43. The boundaries of the South and

the Sunbelt have generated considerable controversy. See James C. Cobb, *Industrialization and Southern Society 1877–1984* (Lexington: Univ. Press of Kentucky, 1984), 3; Michael O'Brien, *The Idea of the American South 1920–1941* (Baltimore: Johns Hopkins Univ. Press, 1979), 61–62, *passim;* Timothy G. O'Rourke, "The Demographic and Economic Setting of Southern Politics," in James F. Lea, ed., *Contemporary Southern Politics* (Baton Rouge: Louisiana State Univ. Press, 1988), 10–11; and Carl Abbott, "The American Sunbelt: Idea and Reality," *Journal of the West* 18 (July 1979): 5–18. The designation Sunbelt was first applied by Kevin Phillips in his *The Emerging Republican Majority* (N.Y.: Anchor Books, 1970); and Kirkpatrick Sale in *Power Shift: The Rise of the Southern Rim and Its Challenge to the Eastern Establishment* (N.Y.: Vintage Books, 1975). The two eminent historians of the 20th-century South lead the protests against applying the term "Sunbelt" to the South. Consult George B. Tindall, "1986: The South's Double Centennial," in Moore et al., eds., *Developing Dixie,* 330–31, 334; and C. Vann Woodward, *Thinking Back: The Perils of Writing History* (Baton Rouge: Louisiana State Univ. Press, 1986), 140.

Chapter 1

1. Roosevelt, letter to Conference on Economic Conditions of the South, 5 July 1938, in U.S. National Emergency Council, *Report on Economic Conditions of the South* (Washington, 1938), 1–2.
2. *Ibid.,* 21, 29.
3. Henry B. Kline, "Economic Motives for the Diffusion of National Defense Production in Depressed Agricultural Areas," Records of the TVA, Records of the TVA Commerce Department, RG 142, Federal Archives and Record Center, Region IV, East Point, Georgia (Hereafter FARC). U.S. Department of Commerce, Bureau of the Census, *Statistical Abstract of the United States, 1936,* p. 7.
4. On the loss of wealth, see Roger L. Ransom and Richard Sutch, *One Kind of Freedom* (N.Y.: Cambridge Univ. Press, 1977), 52–53, *passim.* On the credit structure of the postbellum South, consult Gerald D. Jaynes, *Branches Without Roots* (N.Y.: Oxford Univ. Press, 1986), 30–32, 45–48; C. Vann Woodward, *Origins of the New South 1877–1912* (Baton Rouge: Louisiana State Univ. Press, 1951), 183; and, Ransom and Sutch, *One Kind of Freedom,* 106–49, 181–87, For the effects of emancipation on the region's tax structure, see Peter Wallenstein, *From Slave South to New South* (Chapel Hill: Univ. of N.C. Press, 1987), 187–88; J. Mills Thornton, "Fiscal Policy and the Failure of Radical Reconstruction in the Lower South," in J. Morgan Kousser and James McPherson, eds., *Region, Race, and Reconstruction* (N.Y.: Oxford Univ. Press, 1982), 349–94; and Richard Sylla, "Long-Term Trends in State and Local Finances: Sources and Uses of Funds in North Carolina, 1800–1977," in Stanley Engerman and Robert Gallman, eds., *Long-Term Factors in American Economic Growth* (Chicago: Univ. of Chicago Press, 1986), 841. The landlords' new concern for local development is considered in Gavin Wright, *Old South, New South* (N.Y.: Basic Books, 1986), 17–123; and David L. Carlton, *Mill and Town in South Carolina, 1880–1920* (Baton Rouge: Louisiana State Univ. Press, 1982), 25–26, 59.

   Meanwhile, the newly slaveless landlords suddenly developed attachments to their land. Slaveholders, large and small, had migrated frequently, investing little in local community development. But postbellum "lords of acres, but not of men," as Henry Grady called them, held their wealth in land rather than in labor. Attempting to bolster land values, they introduced fertilizers, improved yields, and reclaimed for agriculture areas long considered exhausted. In fact, planters remained on the same land after the war and emancipation at nearly the same rate as they had been during the 1850s. This testified to the movability of antebellum planters and the relative stability of postwar "lords of acres." The essential matter is not that planters persisted on their land after the war, but that they had so little interest in land before the conflict that their propensity to abandon the land during the prosperous 1850s matched the rate of the disastrous 1860s. This acrimonious debate over "planter persistence" is joined by Jonathan Wiener, *Social Origins of the New South: Alabama 1860–1885* (Baton Rouge: Louisiana State Univ. Press, 1978); Dwight B. Billings, *Planters and the Making of the "New South"* (Chapel Hill: Univ. of N.C. Press, 1979); C. Vann Woodward, *Thinking Back: The Perils of Writing History* (Baton Rouge: Louisiana State Univ. Press, 1986), 72 74; and Wright, *Old South, New South,* 48–49. This dispute is analyzed in Harold Woodman, "Economic Reconstruction and the Rise of the New South, 1865–1900," in John B. Boles and Evelyn Thomas Nolen, eds., *Interpreting Southern History* (Baton Rouge: Louisiana State Univ. Press, 1987), 277.

5. James L. Roark, *Masters Without Slaves: Planters in the Civil War and Reconstruction* (N.Y.: W.W. Norton, 1977), 165. See also Jaynes, *Branches Without Roots;* Ransom and Sutch, *One Kind of Freedom;* and Wright, *Old South, New South.*
6. Jaynes, *Branches Without Roots,* 85–86, 128–30, 157, 216–44. For a different account of the emergence of sharecropping, consult Ransom and Sutch, *One Kind of Freedom.* The expansion of cotton culture and the decline of self-sufficiency in southern agriculture is described by many scholars. Consult Carlton, *Mill and Town in South Carolina,* 18; Paul D. Escott, *Many Excellent People: Power and Privilege in North Carolina, 1850–1900* (Chapel Hill: Univ. of N.C. Press, 1985), 81–82, 175–76; Steven Hahn, *The Roots of Southern Populism* (N.Y.: Oxford Univ. Press, 1983); Jacqueline Dowd Hall et al., *Like a Family: The Making of a Southern Cotton-Mill World* (Chapel Hill: Univ. of N.C. Press, 1987), 6–7; and Wright, *Old South, New South,* 34–36, 54. One historian of Alabama dates the decline of the yeomanry in that state to the 1850s. See J. Mills Thornton, *Power and Politics in a Slave Society: Alabama, 1800–1860* (Baton Rouge: Louisiana State Univ. Press, 1978), 299. This view of the decline of the southern yeomanry is questioned by Harold Woodman in his "Economic Reconstruction and the Rise of the New South," 279–80.
7. *Ibid.* The comparison of southern and northern rural earnings is based on the average agricultural income per worker. These figures (for average annual gross agricultural income per worker) are for the ten-state cotton-growing South and are drawn from Rupert Vance, "The Old Cotton Belt," in Reed and Singal, eds., *Regionalism and the South: Selected Papers of Rupert Vance,* 93. On the controlling role played by world cotton demand, see James C. Cobb, *Industrialization and Southern Society, 1877–1984* (Lexington: Univ. Press of Kentucky, 1984), 10–11; and Wright, *Old South, New South,* 56.

   Even the production of tobacco, the South's second cash crop, was tied to cotton demand. Tobacco culture spread south and west from its Virginia and

North Carolina homes during the cotton price collapse of the 1890s. See Pete Daniel, *Breaking the Land* (Urbana: Univ. of Illinois Press, 1985), 31.

8. A vivid example of tenant mobility is Ned Cobb's account of his career in Rosengarten, *All God's Dangers* (N.Y.: Avon Books, 1974), 112–15, 124, 151, 164, 186, 204, 240–43, 281–82. Cobb moved frequently from plantation to plantation, from sharecropping to cash tenancy, and even from farm labor to nonagricultural labor. Other anecdotal and quantitative evidence on mobility can be found in U.S. House, Hearings Before the Select Committee to Investigate the Interstate Migration of Destitute Citizens, *Interstate Migration,* 76th Cong., 3rd sess., Part 2, Montgomery Hearings, August 14–16 1940 (Washington, 1940), 436–43, 779–95. Peter Molyneaux, "Economic Nationalism and Problems of the South," *Arnold Foundation Studies in Public Affairs* 2 (Fall 1933): 2. Jaynes, *Branches Without Roots,* 109, 314–15. Wright, *Old South, New South,* 12–13, 65, 90. Daniel, *Breaking the Land,* 5–6. Jay Mandle, *The Roots of Black Poverty* (Durham: Duke Univ. Press, 1978), *passim.* Phillip J. Wood, *Southern Capitalism: The Political Economy of North Carolina, 1880–1980* (Durham: Duke Univ. Press, 1986), 26. For a counterargument, consult Jonathan M. Wiener, "Class Structure and Economic Development in the American South, 1865–1955," *American Historical Review* 84 (Oct. 1979): 973–75.
9. Howard W. Odum, *Southern Regions of the United States* (Chapel Hill: Univ. of N.C. Press, 1936), 42. Mandle, *Roots of Black Poverty,* 55, 63. The analysis of agricultural mechanization is drawn from Warren C. Whatley, "Institutional Change and Mechanization in the Cotton South: The Tractorization of Cotton Farming" (Ph.D. dissertation: Stanford University, 1982). See also Whatley, "Labor for the Picking: The New Deal in the South," *Journalof Economic History* 43 (Dec. 1983): 905–29.
10. National Emergency Council, *Report on Economic Conditions,* 37–39. John Moloney, "Some Effects of the Federal Fair Labor Standards Act Upon Southern Industry," *Southern Economic Journal* 9 (July 1942): 15–23. Calvin Hoover and B. U. Ratchford, *Economic Resources and Policies of the South* (N.Y.: Macmillan, 1951), 118.
11. Billings, *Planters and the Making of the "New South,"* 4–5. Wright, *Old South, New South,* 61–62. Carlton, *Mill and Town in South Carolina,* 8–23, 41, 61, 133–34. Hall et al., *Like a Family,* 24. Peter A. Coclanis and Lacy K. Ford, "The South Carolina Economy Reconstructed and Reconsidered: Structure, Output and Performance, 1670–1985," in Winfred Moore, Jr., et al., eds., *Developing Dixie* (Westport, Conn.: Greenwood, 1988), 93–110.
12. Escott, *Many Excellent People,* 210–11, 224. Wright, *Old South, New South,* 63, 162, 172–73. On the propensity of southern manufacturers to locate outside towns and cities, see Cobb, *Industrialization and Southern Society, passim;* and Chapter 6 below. On lumber, consult James Fickle, *The New South and the "New Competition": Trade Association Development in the Southern Pine Industry* (Urbana: Univ. of Illinois Press, 1980). For a provocative analysis of the relations between urban growth and industrial development, see Jane Jacobs, "Why TVA Failed," *New York Review of Books* 31 (10 May 1984): 41–47; and Jacobs, *Cities and the Wealth of Nations* (N.Y.: Random House, 1984).
13. Wood, *Southern Capitalism,* 68–69. Billings, *Planters and the Making of a "New South,"* 52–53. Hall et al., *Like a Family,* 33–36.
14. Escott, *Many Excellent People,* 224–25. Hall et al., *Like a Family,* 51–56, 77–

82, 105–9, 201. Wood, *Southern Capitalism,* 63. Wright, *Old South, New South,* 132–55.

15. U.S. House, *Interstate Migration,* Part 2, p. 738. David Carlton, "Unbalanced Growth and Industrialization: The Case of South Carolina," in Moore et al., eds., *Developing Dixie,* 114. Wright, *Old South, New South,* 9, 74–76, 147–55, 195–207. Hall et al., *Like a Family,* 183–201.
16. Maury Maverick, "Let's Join the United States," *Virginia Quarterly Review* 15 (1939): 64–70. "The South," New York *Post,* 8 April 1944. See also Molyneaux, "Economic Nationalism."
17. Roosevelt, informal extemporaneous remarks at Alabama Polytechnical Institution, Auburn, 30 March 1939, in Samuel I. Rosenman, ed., *The Public Papers and the Addresses of Franklin D. Roosevelt,* 1939 Volume, *War and Neutrality* (N.Y.: Macmillan, 1941), 182–83.
18. National Emergency Council, *Report on Economic Conditions,* 7–8, 37–39.
19. *Ibid.,* 29, 37–44. Rupert Vance, "Planning for the South" (1935), in Reed and Singal, eds., *Regionalism and the South: Selected Papers of Rupert Vance,* 84–85. See also Vance, "The Profile of Southern Culture" (1935), *ibid.,* 56.

    Already in 1933, the Southwest seemed to offer an alternative to low wage industrialization and a model for "advanced" industrialization. In his 1935 "Profile of Southern Culture," Rupert Vance referred to a new subregion of the South, "one possibly too young to chart accurately." There, in Oklahoma and the Gulf Southwest, Vance concluded, "has been erected largely on the foundations of gas and oil an industrial superstructure for the Southwest." After the Spindletop oil discovery in 1901 and the subsequent completion of the Houston Ship channel, the economy of the Southwest was reborn. A major refining center developed around Houston and the entire subregion realized rapid urban growth. Still, in 1935, Vance correctly noted the immaturity of this burgeoning area. See Vance, "Profile of Southern Culture," 52–53; and Joseph Pratt, *The Growth of a Refining Region* (Greenwich, Conn.: JAI Press, 1980), 3–4, 34–35, 49–54, 65.
20. On foundation activity in the South, consult George B. Tindall, *The Emergence of the New South 1913–1945* (Baton Rouge: Louisiana Univ. Press, 1967), *passim.* See also Wilma Dykeman and James Stokely, *Seeds of Southern Change* (N.Y.: W.W. Norton, 1962), 162–63; and Dewey W. Grantham, *The Regional Imagination* (Nashville: Vanderbilt Univ. Press, 1979), 163.
21. National Emergency Council, *Report on Economic Conditions.*
22. This view of the Solid South was prevalent in the 1930s, but it was not entirely accurate. The Redeemers and their successors endured a long, hard fight to consolidate their power. On the views of the southern past that prevailed in the 1930s, see Woodward, *Thinking Back,* 9–42. For an example of constitutional strictures against provision of public services, see U.S. House, *Interstate Migration,* Part 5, Oklahoma City Hearings, 19–20 Sept. 1940, p. 1795. Consult also Woodward, *Origins of the New South,* 65.
23. Woodward, *Origins of the New South,* 49–50, 58–62, 83. Escott, *Many Excellent People,* 194–95. Cobb, "Beyond Planters and Industrialists," 55.
24. Wallenstein, *From Slave South to New South,* 197ff. Woodward, *Origins of the New South,* 212–15. Jaynes, *Branches Without Roots,* 306, 312. Along with white supremacy, social control and low taxes defined the objectives of post-Reconstruction southern governments. The infamous convict lease system—the southern practice of meting out long prison sentences for petty offenses

and hiring out convicts for work—served all of those principles. Convict leasing transformed the penal system from a cost to the state to a supplier of revenue. It equipped local officials with a tool for harassing "uncooperative" blacks and aided a handful of planters, if only ineffectively, in their ongoing battle to secure a cheap, reliable labor force. Profits from the convict lease system built financial and political fortunes for state officials, among them the Bankhead dynasty in Alabama.

25. Wallenstein, *From Slave South to New South,* 24–25, 38–41, 56–61. Thornton, *Power and Politics,* 300–303. Thornton, "Fiscal Policy," 351, 534. In some states, like Georgia and Alabama, railroad and other promotional ventures supplemented the tax on slaves.
26. Thornton, "Fiscal Policy," 352, 378–86. Hall et al., *Like a Family,* 6–7. Michael R. Hyman, "Taxation, Public Policy, and Political Dissent: Yeoman Disaffection in the Post-Reconstruction Lower South," *Journal of Southern History* 55 (Feb. 1989): 49–76.
27. J. Morgan Kousser, *The Shaping of Southern Politics* (New Haven: Yale Univ. Press, 1974), 14–16, 236–37. Escott, *Many Excellent People, passim.* Woodward, *Origins of the New South,* 51, 76–86, 229–90. Hyman, "Taxation, Public Policy, and Political Dissent." In North Carolina, one index of the success of the Redeemer-Regular Democratic leadership was its ability to concentrate tax burdens on blacks and poor whites. See J. Morgan Kousser, "Progressivism—For Middle-Class Whites Only: North Carolina Education, 1880–1910," *Journal of Southern History* 46 (May 1980): 174–75.
28. Kousser, *Shaping of Southern Politics,* 80–123, 181, 195, 232–37, 262–62. Escott, *Many Excellent People,* 258. Rosengarten, *All God's Dangers,* 35–36. In Virginia, the early 20th-century electorate was so small that officeholders and state employees cast one-third of the votes. That fact paved the way for the emergence of the Byrd machine. In North Carolina, one of two southern states where the Republican party survived the turn of the century (Tennessee was the other), suffrage restriction turned both parties to the political right. See Kousser, *Shaping of Southern Politics,* 181, 195.
29. *Ibid.* V. O. Key, Jr., *Southern Politics: In State and Nation* (N.Y.: Vintage Books, 1949), 309–10. On the role of demagoguery in southern politics, see Kousser, *Shaping of Southern Politics,* 80–82. Consult also the alternative views advanced in Carlton, *Mill and Town in South Carolina,* 215–60; and Alan Brinkley, "The New Deal and Southern Politics," in James C. Cobb and Michael Namoroto, eds., *The New Deal and the South* (Jackson: Univ. Press of Mississippi, 1984). 109–12.
30. V. O. Key captured this symbiotic relationship between economic interest and government action in the early 20th-century South. Explaining the sway of North Carolina's economic "oligarchy," Key uncovered the relationship between politics and economic interests for the region: "The effectiveness of the oligarchy's control has been achieved through the elevation to office of persons fundamentally in harmony with its viewpoint. Its interests, which are often the interests of the state, are served without prompting." See Key, *Southern Politics,* 211. On Black Belt leadership of the disfranchisement movement, see Kousser, *Shaping of Southern Politics,* 83–123, 246–47. See also Wright, *Old South, New South,* 78–79, 123; and, Billings, *Planters and the Making of the "New South,"* 198–200.
31. George B. Tindall, *The Emergence of the New South 1913–1945* (Baton Rouge:

Louisiana State Univ. Press, 1967), 231–32, *passim.* Blaine A. Brownell, *The Urban Ethos in the South, 1920–1930* (Baton Rouge: Louisiana State Univ. Press, 1975), 159. Sylla, "Long-Term Trends in State and Local Finance," 830–31.

32. Cobb, "Beyond Planters and Industrialists," 56–58. Douglas L. Smith, *The New Deal in the Urban South* (Baton Rouge: Louisiana State Univ. Press, 1988), 4–5. Cobb, *Industrialization and Southern Society,* 31. Sylla, "Long-Term Trends in State and Local Finance," 831, 839–40. Tindall, *Emergence,* 223–24, 233. Since Tindall distinguished southern "business progressivism" from the standard national variety in 1967, historians have revised their interpretations of the Progressive movement in a manner that reduces the differences between the southern and northern versions of progressivism. For a review of this literature, consult Daniel Rogers, "In Search of Progressivism," *Reviews in American History* 10 (Dec. 1982): 113–32.
33. Kousser, "Progressivism—For Middle-Class Whites Only," 184–86. Cobb, *Industrialization and Southern Society,* 28–30. Brownell, *Urban Ethos in the South,* 26, 38.
34. On public health, see Tindall, *Emergence,* 275–82; and Woodward, *Origins of the New South,* 397, 425–26. On roads, consult Tindall, *Emergence,* 232–56. It is noteworthy that the federal government, through its Federal-Aid Highway Program supported this road-building zeal with matching funds. See below. The Mississippi saying is reported in Wilma Dykeman and James Stokely, *Neither Black Nor White* (N.Y.: Rinehart, 1957), 133.
35. Thornton, "Fiscal Policy," 378–82. Wallenstein, *From Slave South to New South,* 63–68, 154–55. Woodward, *Origins of the New South,* 399. After the initial cutbacks, the Redeemers slowly upgraded support for public education, but it remained well below national standards. Per pupil expenditures languished at less than half of the national average.
36. Edgar W. Knight, quoted in Tindall, *Emergence,* 262. Woodward, *Origins of the New South,* 402–6. Wallenstein, *From Slave South to New South,* 156–58. Tindall, *Emergence,* 259–62, 270–75. National Emergency Council, *Report on Economic Conditions,* 22, 26–32, *passim.* Odum, *Southern Regions,* 102–12, 122, 370. James C. Cobb, *The Selling of the South* (Baton Rouge: Louisiana State Univ. Press, 1982), 163. For a detailed study of educational development in Virginia, see William A. Link, *A Hard Country and a Lonely Place* (Chapel Hill: Univ. of N.C. Press, 1986).
37. On racial inequality, consult Kousser, "Progressivism—For Middle-Class Whites Only," 178–83, 191; and Escott, *Many Excellent People,* 184–85. On mill schools, see Hall et al., *Like a Family,* 127–29; and Carlton, *Mill and Town in South Carolina,* 96–104. If the children of an operative wanted to attend high school, they normally had to pay tuition because millhands fell outside the jurisdiction of the regular, free school systems. Few mill families could afford the expenditure.
38. Cobb, "Beyond Planters and Industrialists," 55–58.
39. Black politics between Reconstruction and disfranchisement is chronicled in Kousser, *Shaping of Southern Politics.* See also Woodward, *Origins of the New South,* 338.
40. For an evocative memoir of black life and black racial consciousness under Jim Crow in South Carolina, see Mamie Garvin Fields with Karen Fields,

Lemon Swamp and Other Places: A Carolina Memoir (N.Y.: Free Press, 1983), 24–31, 47–50; quotation on p. 30. On urban blacks, see also Brownell, *Urban Ethos in the South,* 37; and Dan T. Carter, "From Segregation to Integration," in Boles and Nolen, eds., *Interpreting Southern History,* 420–21. For an analysis of black leadership in the urban South, consult Raymond Gavins, "Urbanization and Segregation: Black Leadership Patterns in Richmond, Virginia, 1900–1920," *South Atlantic Quarterly* 79 (Summer 1980): 257–73. See also Raymond Gavins, *The Perils and Prospects of Southern Black Leadership: Gordon Blaine Hancock, 1884–1970* (Durham: Duke Univ. Press, 1977).

41. Arthur Raper, *A Preface to Peasantry* (Chapel Hill: Univ. of N.C. Press, 1936), 122. Daniel, *Breaking the Land,* 10, 160. Woodman, "Economic Reconstruction and the Rise of the New South," 293. Wright, *Old South, New South,* 65, 96–101. Rosengarten, *All God's Dangers,* 273.

42. Fields, *Lemon Swamp,* 160, 225–26, 234–38. Memories of the New Deal may have influenced Fields's recollections about the attitudes of black South Carolinians toward the federal government before 1933 because the New Deal rekindled and intensified those feelings (and tied them to the national Democratic party). Still, there is ample evidence that black southerners mistrusted state and local authorities.

43. *Ibid.,* 160. On southern opposition to the Keating-Owen child-labor legislation, see Hall et al., *Like a Family,* 58–59. On freight rates, see Woodward, *Origins of the New South,* 312–14. For a complete history of the freight-rate dispute, consult Robert A. Lively, *The South in Action: A Sectional Crusade Against Freight Rate Discrimination* (Chapel Hill: Univ. of N.C. Press, 1949). Kousser, *Shaping of Southern Politics,* 29–32, and Woodward, *Origins of the New South,* 255, chart the reaction to the Lodge Force Bill of 1890 and the Republican party's subsequent abandonment of black southerners. The exception to the rule of hostility to federal interference was the Federal-Aid Highway Program. Southerners enthusiastically supported that program, which, it is worth noting, provided matching funds for the South's favorite government program with a minimum of interference in the administration of those funds. See Tindall, *Emergence,* 240, 256.

This vigilance against federal action animated even Lister Hill, the Alabama congressman soon to become a champion of the New Deal in general and the Tennessee Valley Authority in particular. During the debate over federal development of the Muscle Shoals area before 1933, Hill insisted that the national government limit itself to aiding the development of private power and fertilizer manufacturing companies in the area. He opposed a TVA-style plan until the spring of 1933. See Virginia VanderVeer Hamilton, *Lister Hill: Statesman from the South* (Chapel Hill: Univ. of N.C. Press, 1987), 71.

44. *Cong. Record,* 72nd Cong., 1st Sess., 3 June 1932, p. 11931. For a biographical study of Bankhead's role in the New Deal, see Walter J. Heacock, "William B. Bankhead and the New Deal," *Journal of Southern History* 21 (Aug. 1955): 347–59. See also Pete Daniel, "The New Deal, Southern Agriculture, and Economic Change," in Cobb and Namoroto, eds., *The New Deal and the South* 37; and Frank Freidel, *F.D.R. and the South* (Baton Rouge: Louisiana State Univ. Press, 1965), 37.

State governments remained reluctant to pursue anti-Depression measures between 1929 and 1933. In North Carolina, for example, the administration of Gov. O. Max Gardner moved to balance the state budget by reducing teachers' salaries and replacing the property tax with a regressive general sales tax. Gardner pursued that course even though his advisers informed him that a sales tax would further depress consumer spending and deepen the Depression. See Wood, *Southern Capitalism,* 129–39; and Tindall, *Emergence,* 368–70.

45. Dr. Will Alexander, letter to Howard W. Odum, 13 Oct. 1934, Howard W. Odum Papers, Southern Historical Collection, Univ. of N.C. Library, Chapel Hill. To some extent, this local control of early New Deal programs in the South stemmed from the federal government's lack of bureaucratic capacity—it possessed neither sufficient personnel nor adequate organizational structure to administer its programs during Roosevelt's first term. But it also displayed an unwillingness to challenge the region's traditional power structure. Consult Alan Brinkley, "The New Deal and Southern Politics," in Cobb and Namoroto, eds., *The New Deal and the South,* 99–100; and Theda Skocpol, "Political Response to Capitalist Crisis," *Politics and Society* 10 (1980): 155–201.
46. Harold L. Ickes, *The Secret Diaries of Harold L. Ickes,* Vol. I, *The First Thousand Days, 1933–1936* (N.Y.: Simon and Schuster, 1954), 302. Brinkley, "The New Deal and Southern Politics," 98–102. Freidel, *F.D.R. and the South.* On Roosevelt's relationships with various southern leaders, see James F. Byrnes, *All in One Lifetime* (N.Y.: Harper & Bros., 1958); Jesse H. Jones, *Fifty Billion Dollars* (N.Y.: Macmillan, 1951); Martha H. Swain, *Pat Harrison: The New Deal Years* (Jackson: Univ. Press of Mississippi, 1978); and Bascom N. Timmons, *Garner of Texas* (N.Y.: Harper & Bros., 1948).
47. The region did, however, hold three cabinet posts—Tennessee senator Cordell Hull at State, Daniel Roper of South Carolina at Commerce, and Claude Swanson of Virginia at Navy, the South's traditional portfolio. See Freidel, *F.D.R. and the South,* 42.
48. Hoover and Ratchford, *Economic Resources,* 54. Census Bureau, *Statistical Abstract, 1934,* pp. 602–4. The price of a pound of tobacco plummeted from 18.3 cents in 1929 to 8.2 cents in 1931. Cotton prices crashed from 16.8 cents per pound in 1929 to 5.6 cents in 1931, rising to only 6.5 cents in 1932.
49. Franklin D. Roosevelt, Presidential Statement on the Merits of the Bankhead Bill, 21 April 1934, in Rosenman, ed., *Public Papers,* Vol. 3, *The Advance of Recovery and Reform, 1934* (N.Y.: Random House, 1935), 189.
50. David E. Conrad, *The Forgotten Farmers* (Urbana: Univ. of Illinois Press, 1965), 38–50. Edward L. Schapsmeier and Frederick H. Schapsmeier, "Farm Policy from FDR to Eisenhower: Southern Democrats and the Politics of Agriculture," *Agricultural History* 53 (Jan. 1979): 360. Arthur M. Schlesinger, Jr., *The Coming of the New Deal,* (Boston: Houghton Mifflin 1958), 40–54. Henry A. Wallace, *New Frontiers* (N.Y.: Reynal and Hitchcock, 1934).
51. Schapsmeier and Schapsmeier, "Farm Policy," 360. Tindall, *Emergence,* 392–96. U.S. Dept. of Agric., Agricultural Adjustment Administration, *Agricultural Conservation 1936: A Report of the Agricultural Adjustment Administration* (Washington, 1937), 22–24, 143. Oklahoma, Arkansas, and the eight coastal states from South Carolina to Texas formed the "Southern Division." Virginia, North Carolina, Tennessee, and Kentucky joined three other states in the "East Central Division." The divisions coincided somewhat with the prin-

cipal crops of the areas. All of the cotton-growing states, except Tennessee, for instance, were part of the Southern Division.

52. Lawrence J. Nelson, "Welfare Capitalism on a Mississippi Plantation in the Great Depression," *Journal of Southern History* 50 (May 1984): 226–29. Daniel, "New Deal and Southern Agriculture," 41. Conrad, *Forgotten Farmers,* 116. Paul E. Mertz, *The New Deal and Southern Rural Poverty* (Baton Rouge: Louisiana State Univ. Press, 1978), 41.
53. AAA, *Report of the Administration of the Agricultural Adjustment Act May 1933 to February 1934* (Washington, 1934), 19–26, and *A Report of the Administration of Agricultural Adjustment Act,* Feb. 15, 1934, to Dec. 31, 1934 (Washington, 1935), 45. For an official description of the entire New Deal agricultural program, see AAA, *Agricultural Adjustment 1937–38: Report of the Activities Carried on by the A.A.A.* (Washington, 1939).
54. *Ibid.* Beginning late in 1933, landowners could also secure cotton production loans at 10 cents per pound from the Commodities Credit Corporation. Landlords often extended that credit to their tenants at substantially higher interest rates. See Jones, *Fifty Billion Dollars,* 88–89; and Holmes, *New Deal in Georgia,* 263.
55. *Agricultural Adjustment in 1934,* pp. 10, 50ff, and *Report May 1933 to February 1934,* p. 37. *Administration, Agricultural Adjustment 1937–38,* p. 150. Census Bureau, *Statistical Abstract, 1936,* p. 634. *Statistical Abstract, 1937,* pp. 637–40. Gilbert Fite, "Southern Agriculture Since the Civil War," *Agricultural History* 53 (Jan. 1979): 16. Winter legumes gradually replaced cotton in some areas. In Alabama, production of winter legumes nearly tripled from 1933 to 1938. See U.S. House, *Interstate Migration,* Part 2, p. 734.
56. AAA, *Report May 1933 to February 1934,* p. 262.
57. Swain, *Pat Harrison,* 129–30. *Agricultural Adjustment 1934,* p. 54; and *Agricultural Adjustment 1937–38,* p. 111, for the results of cotton-control referenda. There were some southern opponents of the agricultural program. George Fort Milton of the *Chattanooga News* warned a Tennessee audience that "to this day the long-run impact upon our economy of this body blow to our great southern cash crop is insufficiently appreciated," George F. Milton Papers, Manuscript Division, Library of Congress, Box 31.
58. AAA, *Report May 1933 to February 1934,* p. 272. Whatley, "Labor for the Picking," 905, 908, 919. Wright, *Old South, New South,* 226–35. U.S. House, *Interstate Migration,* Part 5, p. 1795.
59. "Ain't Got No Screens," in Tom E. Terrill and Jerrold Hirsch, eds., *Such As Us: Southern Voices of the Thirties* (N.Y.: W.W. Norton, 1978), 56. Henry Richards, *Three Years Under AAA* (Washington: Brookings Institution, 1936), 112. U.S. House, *Interstate Migration,* Part 2, p. 444. For description of the benefit payments, consult the collected editions of AAA reports, *Reports 1933–1935* and *Reports 1936–1945* (Washington, 1934–46). The rice program proved an exception to the distribution mechanism. The AAA Rice Section distributed the entire allotment to whoever farmed, regardless of tenure. See Daniel, "New Deal and Southern Agriculture," 46.
60. Chester Davis, letter to David E. Conrad, quoted in Conrad, *Forgotten Farmers,* 54. Jack Temple Kirby, "The Transformation of Southern Plantations, c. 1920–1960," *Agricultural History* 57 (July 1983), 265. Harold Hoffsommer, "The AAA and the Cropper," *Social Forces* 13 (May 1935): 499. Conrad, *Forgotten Farmers,* 70–71.

61. Southern Committee for People's Rights, "The Status of Civil Rights in the South: Summary for 1934–1935," Jonathan W. Daniels Papers, Southern Historical Collection, Chapel Hill, N.C., Box 1. Bureau of the Census, *Historical Statistics of the United States from Colonial Times to 1970,* Part 1, p. 465. Tindall, *Emergence,* 409. Raper, *A Preface to Peasantry,* 34. T. J. Woofter et al., *Landlord and Tenant on the Cotton Plantation* (Washington, 1936). U.S. House, *Interstate Migration,* Part 2, p. 415.
62. Tindall discusses the spate of fictional works which included *Tobacco Road* (1932), H. H. Kroll's *I Was a Sharecropper* (1937), Charlie May Simon's *The Share-Cropper* (1937), and John Steinbeck's *The Grapes of Wrath* (1939) and culminated in James Agee and Walker Evans's documentary *Let Us Now Praise Famous Men* (1940). In addition to the Raper and Woofter volumes cited above, non-fiction accounts included: Charles S. Johnson, *The Shadow of the Plantation* (1934) and Johnson et al., *The Collapse of Cotton Tenancy* (1935). See Tindall, *Emergence,* 415–16. On the foundations, see Grantham, *The Regional Imagination,* 163; and Tindall, *Emergence,* 416. The troubles in Arkansas also focused attention on the issue. See Conrad, *Forgotten Farmers,* 83–104, 154–76; Mertz, *New Deal and Rural Poverty,* 32–38; and Howard Kester, *Revolt of the Sharecroppers* (N.Y.: Covici, Friede, 1936). For Interior Secretary Harold Ickes's interest in the tenant problem, see Harold L. Ickes, "The Crusade for Conservation," *Democratic Digest* 15 (April 1938): 23, in Secretary of Interior Papers, Harold L. Ickes Papers, Manuscript Division, Library of Congress, Box 105.
63. AAA, *Report May 1933 to February 1934,* p. 272. Henry A. Wallace, "Report of the Secretary of Agriculture," 8 May 1934, Records of the TVA, RG 142, TVA Board File Curtis-Morgan-Morgan 1933–57, Federal Archive and Record Center, Region IV, East Point, GA.
64. Richards, *Three Years Under AAA,* 12, reprints the text of paragraph 7. Conrad, *Forgotten Farmers,* 57, 136–53; Mertz, *New Deal and Rural Poverty,* 32–38; and Nelson, "The Art of the Possible: Another Look at the 'Purge' of the AAA Liberals in 1935," *Agricultural History* 57 (Oct. 1983): 416–35, describe the purge. On McCarl, see Conrad, *Forgotten Farmers,* 199–201.
65. Calvin B. Hoover, "Human Problems in Acreage Reduction in the South," Records of the Dept. of Agric., Records of the Agricultural Stabilization and Conservation Service, RG 145, National Archives, Washington, D.C. (hereafter cited as NA). Bureau of the Census; *Census of Agriculture,* 1935, Vol. II, *General Report: Statistics by Subject* (Washington, 1943), 465.
66. Oscar Goodbar Johnson, "Will the Machine Run the South?" *Saturday-Evening Post* 219 (31 May 1947). H. Clarence Nixon, *Forty Acres and Steel Mules* (Chapel Hill: Univ. of N.C. Press, 1938), 7. U.S. House, *Interstate-Migration,* Part 5, p. 1795. Daniel, "New Deal and Rural South," 55. From 1930 to 1938 the number of tractors in use in eight Old South cotton states (Alabama, Arkansas, Georgia, Louisiana, Mississippi, North Carolina, South Carolina, and Tennessee) increased by 50 percent, in Texas, by 62 percent, for the United States as a whole, only by 40 percent. Consult U.S. House, *Interstate Migration,* Part 5, p. 1952. Examples of the traditional interpretation include: Daniel, "New Deal and Rural South"; Daniel, "Transformation of the Rural South"; and Kirby, "Transformation of Southern Plantations." The rival view is drawn from: Whatley, "Labor for the Picking"; Whatley, "Institutional Change"; and Wright, *Old South, New South,* 226–35.

67. Pete Daniel, *Breaking the Land* (Baton Rouge: Louisiana State Univ. Press, 1986), 78. Alston and Ferrie, "Labor Cost, Paternalism and Loyalty in Southern Agriculture." Fite, "Southern Agriculture," 17. Numan V. Bartley, "Another New South," *Georgia Historical Quarterly* 65 (Summer 1981): 119–37. Bartley, "The Era of the New Deal as a Turning Point in Southern History," in Cobb and Namoroto, eds., *New Deal and the South,* 135–46.
68. *Ibid.*
69. Franklin D. Roosevelt, Presidential Statement on N.I.R.A., 16 June 1933, in Rosenman, ed., *Public Papers,* Vol. 2, *The Year of Crisis, 1933* (N.Y.: Random House, 1935), 251. For the origins and general history of the NRA, see: William E. Leuchtenburg, *Franklin D. Roosevelt and the New Deal, 1932–1940* (N.Y.: Harper & Row, 1963); Schleslinger, *The Coming of the New Deal;* and Michael M. Weinstein, *Recovery and Redistribution Under the N.I.R.A.* (N.Y.: North-Holland Publishing, 1980).
70. Roosevelt, Presidential Statement on N.I.R.A., 251–52. ". . . and by living wages I mean more than a bare subsistence level—I mean the wages of decent living," the president added (p. 252).
71. David R. Coker, letter to Clay Williams, 22 June 1934, Frank P. Graham Papers, Southern Historical Collection, Univ. of N.C., Chapel Hill. The reliability of Coker's recollection of General Johnson's remarks can not be confirmed, but accuracy aside, Coker's account reveals the hostility he and many other southerners felt toward the NRA.
72. U.S. National Recovery Administration, *A Handbook of NRA: Laws, Regulations, Codes* (Washington, 1933), 97–103. U.S. Dept. of Labor, Bureau of Labor Statistics, *Bulletin No. 663: Wages in Cotton-Goods Manufacturing* (Washington, 1938), 72. NRA, Research and Planning Division, *Hours, Wages and Employment Under the Codes* (Washington, 1935), 80–83. Leverett S. Lyon et al., *The National Recovery Administration: An Analysis and Appraisal* (Washington: Brookings Institution, 1935). 305–6. See also Tindall, *Emergence,* 436.
73. Average hourly earnings in southern plants rose from 20.5 cents in July 1933 to 34.3 cents in Dec. 1933. Bureau of Labor Statistics, *Bulletin 663,* p. 72. Charles F. Roos, *NRA Economic Planning* (Bloomington: The Principia Press, 1937), 159. On the benefits which the southern textile industry derived from the code, see Exhibits and Supplementary Briefs presented at the Meeting of the Textiles Committee, 14–17 Dec. 1938, Records of the Wages and Hours and Public Contracts Division of the Dept. of Labor, RG 155, Series 16, NA. Southern complaints appeared most frequently and forecfully in *Manufacturers' Record:* See H. C. Berckes, "Wage Differential Essential to Progress of South," *Manufacturers' Record* 103 (Sept. 1934): 18–19; John E. Edgerton, "To Protect the South Against Discrimination," 103 (July 1934); 19, 62; "Wage Discrimination," 103 (Jan. 1934): 13, 53; and "Southern Manufacturers Urge Changes in NRA," 104 (Jan. 1935): 23.
74. Edgerton, "To Protect the South," 19, 62. See also Robin Hood, "Some Basic Factors Affecting Southern Labor Standards," *Southern Economic Journal* 2 (1936): 57. Southern States Industrial Council. Wage Differential Committee, "Summary of Outstanding Factors Justifying a Wage Differential Between the South and Other Sections of the Country," 1935, Records of the NRA, RG9, Entry 39, NA.
75. Edgerton, "To Protect the South," 19, 62. See also the citations above from

Manufacturers' Record and Pat Harrison, "The Wages of Dixie," *Collier's,* 22 Jan. 1938.

76. *Ibid.*
77. Raleigh *News and Observer,* 19 Feb. 1934. Among the journalists were Jonathan Daniels of the Raleigh *News and Observer,* Lambert Davis of the *Virginia Quarterly Review,* George Fort Milton of the *Chattanooga News,* and Mark Ethridge of the *Louisville Courier-Journal.* See Lambert Davis, letter to Daniels, 18 March 1935, Daniels Papers. Hood, "Southern Labor Standards," 58. George Fort Milton, "The Duties of Educated Southerners," 10 May 1932, Milton papers, Box 31. Milton warned the employers of the South that "we must not be unheedful of the rights of Labor. A low wage scale will not prove an unmixed boon to the South. . . . If we do not protect our working men and women, who can prophesy the future of the South?"
78. Roos, *NRA Planning,* 154, 164. Dept. of Labor report, quoted on p. 164. NRA, *Hours, Wages and Employment,* 3. Citing his own study of rental costs, Roos believed that differences in the cost of living between the regions legitimated lower southern wages. Similar studies by University of Chicago Professor William F. Ogburn and the Works Progress Administration (WPA), however, reached the opposite conclusion. Those investigations discovered no significant differences in living costs between North and South. Most economists have sided with Roos in this dispute, arguing that whatever the relative costs of living, the eradication of differentials that had persisted for fifty years lacked economic sense. See William F. Ogburn, "Does It Cost Less to Live in the South?," *Social Forces* 14 (Dec. 1935): 211–14.
79. Roos, *NRA Planning,* 166. Southern States Industrial Council, "Summary of Outstanding Factors." The company was the Gulf States Paper Corporation of Tuscaloosa, Ala. See also Edgerton, "To Protect the South," 19, 62.
80. Southern States Industrial Council, "Summary of Outstanding Factors." Businesses like the Marietta Marble Company claimed that they could not pay blacks white wages for two reasons: blacks were inefficient so that it was unfeasible to hire one at a higher wage; the employer had an obligation to give preference to whites for high-paying jobs. See also Michael Holmes, "The Blue Eagle as 'Jim Crow Bird': The NRA and Georgia Black Workers," *Journal of Negro History* 57 (July 1972): 276.
81. Lorena Hickok, letter to Harry Hopkins, 18 Feb. 1934, in Lowitt and Beasley, eds., *One Third of a Nation,* 195. Gavins, *The Perils and Prospects of Black Leadership,* 54–62.
82. Roos, *NRA Planning,* 172–73. Raymond Walters, *Negroes and the Great Depression: The Problem of Economic Recovery* (Westport, Conn.: Greenwood, 1970), 103. Holmes, "'Jim Crow Bird.'"
83. NRA, Division of Economic Research and Planning, "North-South Wage Differentials," 30 March 1934. "Summary of the Wage Differential Reports with Considerations and Recommendations," n.d. Both in the Records of the NRA, RG 9, Entry 31, NA.
84. Leon C. Marshall, *Hours and Wages Provisions in NRA Codes* (Washington: Brookings Institution, 1935). Lyon et al., *National Recovery Administration,* 326–27. H. M. Douty, "Recovery and the Southern Wage Differential," *Southern Economic Journal* 4 (1937–38): 315–17.
85. Under the codes, hourly minimum wages ranged from 62 percent of the northern level in the fertilizer and steel industries to 92 percent in the cotton

textile and hosiery industries. Most codes established differentials of greater than 15 percent. Most of the region's nonagricultural workers, however, worked under codes with a small North-South differential. Lyon et al., *National Recovery Administration,* 173–80, 327.

Population and "time period" differentials also applied to approximately one-fifth of the codes and more than half of the employees under NRA jurisdiction. Since the South contained few large cities and southern industry typically operated in small, isolated villages, population differentials often served as regional wage scales. Time period differentials tied code wages to "the traditional wage structure" and acted indirectly as pouplation, racial, and geographic distinctions. The provisions also favored southern employers.

86. *Ibid.* NRA, Division of Economic Research and Planning, *Geographic and Population Differentials in Minimum Wages* (Washington, 1936). Wolters, *Negroes and the Great Depression,* 128–29. Douglas L. Smith, *The New Deal in the Urban South* (Baton Rouge: Louisiana State Univ. Press, 1988), 51–52.

The organization of the NRA further demonstrated its inability and unwillingness to mount a systematic assault on southern regional problems. There was no southern regional office. The NRA divided supervision of the states of the Old Confederacy among its Dallas, Atlanta, and Washington, D.C., branches. At the same time, the economic boundaries of the southern region varied from code to code. The NRA demarcated more than a dozen different "Souths." In only the fertilizer code, for example, was Delaware included in the low wage South. In that case, Delaware's fertilizer manufacturers employed large numbers of unskilled blacks, possessed little machinery, and maintained strong ties to the southern trade association. Those fluid boundaries served the interests of southern industry.

87. NRA, *Hours, Wages and Employment,* 3.
88. *Ibid.,* 45, 80–83. NRA, *Tabulation of Labor Provisions in Codes Approved by August 8, 1934* (Washington, 1935), Table X. Employment in southern establishments increased 15 percent during this period. The wages and hours regulations encouraged employers to spread employment among more workers at higher pay rates.
89. For the textiles industry, see Exhibits and Briefs Presented at the Meeting of the Textile Committee, 14–17 Dec. 1938. For the hosiery industry, see *Journal of Committee Proceedings,* n.d. For the lumber industry, see Digest of Record, Committee No. 30. All in Records of the Wage and Hour and Public Contracts Division of the Dept. of Labor, RG 155, Entry 16, NA. Consult also "Earnings and Hours of Labor in the Baking Industry," *Monthly Labor Review* 41 (Dec. 1935): 1589; "Earnings and Hours in the Iron and Steel Industry, 1933 and 1935," *Monthly Labor Review* 43 (Sept. 1936): 659–62; "Entrance Rates Paid to Common Labor, 1933," *Monthly Labor Review* 37 (Oct. 1933): 935; and *Monthly Labor Review* 53 (July 1941): 197–203.
90. NRA, *Tabulation,* Tables XVI–XVII. Marshall, *Wages and Hours Provisions.* "Entrance Rates Paid Common Labor, 1933," *Monthly Labor Review* 37 (Oct. 1933). "Entrance Rates Paid to Common Labor," *Monthly Labor Review* 42 (March 1936): 700–701. "Reports and Recommendations of Industry Committee No. 3 for the Establishment of Minimum Wage Rates in the Hosiery Industry, 15 May 1939," RG 155, Entry 16, NA. NRA, *Hours, Wages and Employment,* 49, 55, 63, 79–83. Weinstein, *Recovery and Redistribution,* 92. The growth rate of weekly earnings in low wage industries was more modest than

the change in hourly earnings. Average weekly earnings in those industries advanced 25 percent during the NIRA period, outpacing the 19.7 percent growth rate for high wage industries. The difference between the gains in weekly and hourly earnings reflected the 12.1 percent decline in average weekly hours in low wage industries. Hours per week in high wage industries increased 3.5 percent over the same period. See Roos, *NRA Planning,* 157.

In fact, raising the wages of unskilled labor to the legal minimum accounted for most alterations in the southern wage structure under the NRA. Many codes preserved "customary" or "equitable" differentials between the wages for various occupations within an industry. But southern employers resisted this "tapering above the minima." Fear of intraplant tensions arising from the failure to preserve occupational pay differentials, however, led a few southern industrialists to raise wages proportionately throughout their factories and motivated many to complain to the NRA. Joseph Winkers, the NRA's field adjustor in Savannah, reported that Georgia's employers' "biggest objection" to the minimum wage for unskilled labor was "the wishful hoping" that skilled labor would receive proportional pay increases. Joseph Winkers, memorandum to Atlanta Office, 24 Oct. 1935, Records of the NRA, RG 9, Series 525, FARC.

91. NRA, *Hours, Wages and Employment,* 112. "Employment, Hours, Earnings and Production," *Monthly Labor Review* 40 (March 1935): 544–45. Roos, *NRA Planning,* 166. Reports of Industry Committee No. 3.
92. Berckes, "Wage Differential," 19. *Monthly Labor Review* 44 (May 1937): 1143. "Hourly Earnings in the Lumber and Timber Products Industry," *Monthly Labor Review* 53 (July 1941): 197. On unenforceability, see Tindall, *Emergence,* 442. On post-NRA changes, see Roos, *NRA Planning,* 159; and "Entrance Rates of Common Laborers in Twenty Industries," *Monthly Labor Review* 45 (Dec. 1937): 1505.
93. Hall et al., *Like a Family,* 195–213. On the relative success of the southern textile industry during the 1930s, see U.S. House, *Interstate Migration,* Part 9, Washington Hearings, 5–10 Dec. 1940, p. 3699. U.S. House, Hearings Before the Select Committee Investigating National Defense Migration, *National Defense Migration,* 77th Cong., 1st Sess., Part 11, Washington Hearings, 24–26 March 1941, pp. 4422, 4429–30; and Michael A. Bernstein, *The Great Depression* (N.Y.: Cambridge Univ. Press, 1988), 130.
94. NRA, *Hours, Wages and Employment,* 112–13. On plant modernization under NRA in general, see "Modernization," *Manufacturers' Record* 103 (Feb. 1934): 20.
95. Roos, *NRA Planning,* 168. NRA, *Hours, Wages and Employment,* 80–83. "Employment, Hours, Earnings and Production," 544–45. Hall et al., *Like a Family,* 295–96.
96. Roos, *NRA Planning,* 172–73. Southern States Industrial Council, "Summary of Outstanding Factors." NRA, "North-South Wage Differentials." On skilled blacks, see Roos, *NRA Planning,* 173n. On the impact of the NRA on southern blacks in general, see J. Wayne Flynt, "The New Deal and Southern Labor," in Cobb and Namoroto, eds., *New Deal and the South,* 85; and Harvard Sitkoff, "Impact of the New Deal on Black Southerners," in Cobb and Namoroto, *New Deal and the South,* 117–34. The quotation is from Holmes, "'Jim Crow Bird,'" 27.
97. *New York Times,* 9 Jan. 1934, p. 13. Holmes, *New Deal in Georgia,* 196. Wol-

ters, *Negroes and the Great Depression,* 125. The contrast between the cotton textile and fertilizer codes best evinced White's charge. Relatively few blacks labored in the southern textile industry for which the NRA set the minimum wage at 92 percent of the northern rate. In the southern fertilizer industry, however, blacks made up more than three-quarters of the work force. The southern minimum wage represented less than 70 percent of the rates applicable to the overwhelmingly white workers in the northern and midwestern branches of the industry. See Marshall, *Hours and Wages Provisions;* NRA, *Geographic and Population Differentials;* and, Wolters, *Negroes and the Great Depression,* 128–29.

The tactic of exploiting racial animosities to inhibit other conflict has been a central theme of southern and indeed American history. See Flynt, "New Deal and Southern Labor"; George M. Fredrickson, *White Supremacy* (N.Y.: Oxford Univ. Press, 1981); Edmund S. Morgan, *American Slavery American Freedom* (N.Y.: W. W. Norton, 1975); and, Woodward, *Origins of the New South.*

98. William L. Mitchell, memorandum to Martin, 15 Aug. 1935, RG9, Series 525. Questionnaire submitted to Regional Directors Together with Their Answers, RG9, Series 525. Mitchell, memorandum to William Galvin, 31 Oct. 1935. RG9, Series 525. For example, the lumber and cotton garments industries opposed; the fertilizer and fabricated metals industries approved. See also C. K. Ramond, memorandum to A. B. Hammond, 23 Oct. 1935, RG9, Series 525. All in FARC.
99. Robert H. Gamble, letter to Donald Richberg, 31 May 1935, Records of the NRA, RG9, Series 525, FARC.
100. *Ibid.* Fred C. Rogers, letter to Ernest L. Tutt, 13 Aug. 1935, Records of the NRA, RG9, Entry 57, NA. *Joint Hearings before the Committee on Education and Labor, U.S. Senate and the Committee on Labor, House of Representatives on the Fair Labor Standards Act of 1937,* 75th Cong., 1st sess. (Washington, 1937), 174. Bureau of Labor Statistics, *Bulletin No. 663,* p. 72. "Entrance Rates of Common Laborers in 20 Industries," 1501. A. F. Hinrichs, "Average Hourly Earnings in Manufacturing, 1933 to 1936," *Monthly Labor Review* 44 (April 1937): 828–58, details post-NRA earnings declines in the major southern industries.
101. Vance, "Planning the Southern Economy," 85.
102. Schlesinger, *Coming of the New Deal,* 274.
103. Lorena Hickok, letters to Harry Hopkins, 14 Jan. 1934, and, 10 Feb. 1934, in Lowitt and Beasley, eds., *One Third of a Nation,* 147, 185. Howard B. Myers, "Relief in the Rural South," *Southern Economic Journal* 3 (Jan. 1937): 282. On policy, see the *Monthly Report* of the U.S. Federal Emergency Relief Administration, published monthly from May 1933 through June 1936. Donald S. Howard, *The WPA and Federal Relief Policy* (N.Y.: Russell Sage Foundation, 1943), describes the practice under the WPA.
104. Odum, *Southern Regions,* 86. Howard, *WPA,* 74–76. Texas, Arkansas, Alabama, and Louisiana followed the "surplus commodities only" policy. Virginia received less than three-quarters of its relief funds from Washington.
105. Federal Emergency Relief Administration (FERA), *Monthly Report,* 1 Dec. to 31 Dec. 1933 (Washington, 1934), 64; *Monthly Report,* 1 Nov. to 30 Nov. 1933, p. 64; *Monthly Report,* 1 May to 31 May 1934, p. 42; *Monthly Report,* 1 Dec. to 31 Dec. 1934, p. 21; *Monthly Report,* 1 Feb. to 28 Feb. 1935, p. 31.

Monthly Report, 1 June to 30 June 1936, p. 188. For example, by July 1935 the southern rates had increased to a range between $8.79 in Oklahoma and $26.17 for Louisiana. The national average, however, had reached $129.64. *Monthly Report,* 1 Sept. to 30 Sept. 1935, p. 53. Myers, "Relief in the Rural South," 283. Myers was the WPA official.

106. Tindall, *Emergence,* 474–75; Holmes, *New Deal in Georgia,* 75–96, *passim.* Census Bureau, *Statistical Abstract, 1936,* p. 322.
107. *Ibid.* Howard, *WPA,* 158–61, 178. In May 1935 the minimum monthly WPA wage in every state outside the South exceeded $32; in only nine of those non-South states did it fail to reach $40. The southern standards ranged between $19 and $21. In addition to low benefits, Deep South states certified a smaller percentage of their population as eligible for WPA and employed a smaller percentage of those certified than other regions. See U.S. House, *Interstate Migration,* Part 9, p. 3630.
108. Myers, "Relief in the Rural South," 283–88.
109. "Differences in Living Costs in Northern and Southern Cities," *Monthly Labor Review* 49 (July 1939): 23. Howard, *WPA,* 78–79, 172. Holmes, *New Deal in Georgia,* 77.
110. F. C. Harrington, quoted in Howard, *WPA,* 162, See also *ibid.,* 160–61, 178.
111. Howard, *WPA,* 285–86. Holmes, *New Deal in Georgia,* 77–78. Sitkoff, "Impact of New Deal on Black Southerners," 123–24. Fields, *Lemon Swamp,* 224–31. As a percentage of the national average, the rate of incidence of WPA employment grew between 1936 and 1938 from 49 to 55 in North Carolina, 75 to 97 in South Carolina, 66 to 78 in Georgia, 97 to 108 in Florida, 65 to 81 in Alabama, 72 to 82 in Mississippi, 91 to 100 in Arkansas, 101 to 103 in Kentucky. The incidence rates dropped over that period in Texas, Tennessee, Oklahoma, and Virginia, but shot up again in 1939 in Texas and Tennessee. During the same period, there was little change in the rates for the New England states and a drop in the Middle Atlantic states. Howard, *WPA,* 540–41.
112. Lorena Hickok, letter to Harry Hopkins, 18 April 1934, in Lowitt and Beasley, eds., *One Third of a Nation,* 215. Slim Jackson, "Them That Needs," WPA, Federal Writers Project, *These Are Our Lives* (N.Y.: W.W. Norton, 1975), 366. Brooks Hays, *A Southern Moderate Speaks* (Chapel Hill: Univ. of N.C. Press, 1959), 19. For a discussion of the broader implications of "New Deal principles" among the southern population, see James C. Cobb, *The Selling of the South* (Baton Rouge: Louisiana State Univ. Press, 1982), 32–33; Daniel, *Breaking the Land,* 73–78; and Holmes, *New Deal in Georgia,* 319–20.
113. Tindall, *Emergence,* 535. Robert Caro, *Lyndon Johnson: The Path to Power* (N.Y.: Knopf, 1982).
114. Eugene Talmadge, "Return to Democratic Principles," *Manufacturers' Record* 103 (Dec. 1934): 18–19. Tindall, *Emergence,* 475. During the 1930s most southern states created state planning boards and state public welfare departments so as to receive and control New Deal-generated public works and social welfare funds. Consult Smith, *New Deal in the Urban South,* 90, 146; and, Albert Lepawsky, *State Planning and Economic Development in the South: NPA Committee of the South Report No. 4* (Kingsport, Tenn.: NPA, 1949), 14.
115. John Easton, "You're Gonna Have Lace Curtains," in WPA, Federal Writers Project, *These Are Our Lives,* 15–16.
116. George F. Milton, speech at Norris Dam, Tennessee, 4 March 1936, Milton

Papers, Box 30. Milton made clear his opposition to other aspects of the New Deal: "TVA has no flavor of ploughing under rows of cotton, of stifling puny pigs or mighty kilowatts." For a TVA official's views, see David E. Lilienthal, *The Journals of David E. Lilienthal,* Vol. I, *The TVA Years, 1939–1945* (N.Y.: Harper and Row, 1964), 52–53.

117. Harcourt A. Morgan, "Decentralization of Industry: Address Before Southeastern Division of National Electric Light Association," n.d., Records of the TVA, RG 142, TVA Board File Curtis-Morgan-Morgan, FARC. Gilbert Banner, "Toward More Realistic Assumptions in Regional Economic Development," in John R. Moore, ed., *The Economic Impact of TVA* (Knoxville: Univ. of Tennessee Press, 1967), 127. John V. Krutilla, "Economic Development: An Appraisal," in Roscoe Martin, ed., *TVA: The First Twenty Years: A Staff Report* (University and Knoxville: Univ. of Alabama Press and Univ. of Tennessee Press, 1956), 220.

118. Morgan, "Decentralization of Industry." John P. Ferris, memorandum to H. A. Morgan, 5 Sept. 1933, Records of the TVA, RG 142, TVA Board File Curtis-Morgan-Morgan, FARC. Eventually, the TVA directors engaged in a bitter feud over the nature of the TVA, a feud pitting A. E. Morgan against David Lilienthal. H. A. Morgan came around to Lilienthal's views. On the conflict, consult Thomas McCraw, *Morgan versus Lilienthal* (Chicago: Loyola Univ. Press, 1970). See also Chapter 4 below.

119. Ferris, memo to Morgan, 5 Sept. 1933. Earle S. Draper, "What Is Regional Planning," Lecture at Harvard Univ., Cambridge, Mass., 14 Jan. 1935, p. 14, attached to Draper, letter to Harold L. Ickes, 13 Feb. 1935, Secretary of Interior Papers, Ickes Papers, Box 364.

120. David E. Lilienthal, *TVA: Democracy of the March* (Westport, Conn.: Greenwood, 1953), 125. Gordon R. Clapp, "The Meaning of TVA," in Martin, ed., *TVA: The First Twenty Years,* 9.

121. TVA Commerce Dept., "Manufacturing in the Tennessee Valley Compared with the Southeast and the United States," 7 May 1940, Records of the TVA, RG 142, TVA Board File Curtis-Morgan-Morgan, FARC. Lilienthal, *TVA,* ix. James P. Pope, "Regional Development as a Dynamic Force," *Journal of Politics* 5 (Feb. 1943). Between 1933 and 1937, for example, wages soared 82.6 percent in the Tennessee Valley. They grew 68.5 percent in the Southeast as a whole, 56 percent across the nation.

122. Roger D. Raymo and Robert M. LaForge, "New Industries in the Tennessee Valley," 20 Oct. 1938, Records of the TVA, RG 142, TVA Board File Curtis-Morgan-Morgan, FARC. TVA Commerce Department, interoffice memorandum, 2 June 1941, Records of the TVA, RG 142, Records of TVA Commerce Department, FARC. A former TVA chief economist, Stefan Robock, identified this problem in the 1960s. See Robock, "An Unfinished Task: A Socio-Economic Evaluation of the TVA Experiment," in Moore, ed., *Economic Impact of TVA,* 115–16. A contemporary historian has also written on this subject. See Jane Jacobs, "Why TVA Failed," *New York Review of Books* 31 (10 May 1984): 41–47; and Jacobs, *Cities and the Wealth of Nations* (N.Y.: Random House, 1984).

123. National Emergency Council, *Report on Economic Conditions of South, passim.* TVA Commerce Dept., "Industrial Production in Southeast Showing Deficient and Surplus Production by Industries," July 1940, Records of the TVA, RG 142, TVA Commerce Department, FARC. William Miernyk, "The

Changing Structure of the Southern Economy," in E. Blaine Liner and Lawrence Lynch, eds., *The Economics of Southern Growth* (Durham: SGPB, 1977), 45. A ten-state southern area (excluding Texas, Oklahoma, and Kentucky) contributed a whopping 22.5 percent of the total U.S. textile manufactures and 22.7 percent of the forest products, but only a paltry 2 percent of the machinery and 3.1 percent of the transportation equipment.

124. David E. Lilienthal, "An Increased Income for the South, Our Central Problem," address before the Institute of Public Affairs of the Univ. of Georgia, Athens, 29 Oct. 1936, Hugo L. Black Papers, Manuscript Division, Library of Congress, Box 217.
125. Roosevelt, letter to Conference on Economic Conditions in the South, 5 July 1938, in National Emergency Council, *Report on Economic Conditions of the South,* 1. Franklin D. Roosevelt, *Complete Presidential Press Conference of Franklin D. Roosevelt* (N.Y.: DaCapo, 1972), IX, 436–48.

Chapter 2

1. Jesse H. Jones, *Fifty Billion Dollars,* 263.
2. "South Building Factories," *Business Week,* 5 Dec. 1936, p. 37. For another report on the resurgence of southern industry, see O. C. Huffman, "The South Resurgent," *Manufacturers' Record* 103 (June 1934): 22, 56.
3. Richard Lowitt, "The TVA 1933–45," in Erwin C. Hargrove and Paul K. Conkin, eds., *TVA: Fifty Years of Grass-Roots Bureacracy* (Urbana: Univ. of Illinois Press, 1983), 49–53. Lorena Hickok, letter to Harry Hopkins, 6 June 1934, in Richard Lowitt and Maurine Beasely, eds., *One Third of a Nation: Lorena Hickok Reports on the Great Depression* (Urbana: Univ. of Illinois Press, 1981), 269.
4. Stuart Chase, "TVA: The New Deal's Greatest Asset," *The Nation* 142 (3 June 1936): 703. WPA, *These Are Our Lives,* 240. NEC, *Report on Economic Conditions,* 30–39, 62–63. Census Bureau, *Historical Statistics,* 225. Hoover and Ratchford, *Economic Resources,* 48. Odum, *Southern Regions,* 139, 472.
5. WPA, *These Are Our Lives,* 3, 11, 210, 240.
6. Memorandum adopted at the Meeting of the Southern Policy Conference in Atlanta, 25–28 April 1935, Graham Papers. See also the contents of the folder marked "Southern Policy Committee, 1935." Morton B. Sosna, *In Search of the Silent South* (N.Y.: Columbia Univ. Press, 1977), 89, 121. Tindall, *Emergence,* 606. Thomas A. Kreuger, *And Promises To Keep* (Nashville: Vanderbilt Univ. Press, 1967), 54, 195.
7. Daniel J. Singal, *The War Within* (Chapel Hill: Univ. of N.C. Press, 1982), 115–24, 149, 303. Sosna, *Silent South,* 56–57, Rupert P. Vance, *The South's Place in the Nation* (Washington: Public Affairs Committee Pamphlets, No. 6, 1936), 11. Woodward, *Thinking Back,* 17–20. Michael O'Brien, *The Idea of the American South* (Baltimore: Johns Hopkins Univ. Press, 1979), 48–49.
8. Odum, *Southern Regions,* 1, 245–90; the quotations are from pp. 1, 249. Tindall, *Emergence,* 584–87. Tindall, *The Ethnic Southerners* (Baton Rouge: Louisiana State Univ. Press, 1976), pp. 100–13. Singal, *The War Within,* 115, 124, 149, 303. Sosna, *Silent South,* 56–57.
9. Odum, *Southern Regions,* 15, 337–38, 353. Walter Prescott Webb, *Divided We Stand: The Crisis of Frontierless Democracy* (N.Y.: Farrar & Rinehart, 1937), dramatized the colonial thesis. In Webb's hands, the metaphor described not merely the objective condition of the South, but the North's systematic plun-

dering of the other sections. *Divided We Stand* indicted the Republican party as the agent of northern imperialism and proposed a "sectional tariff" allowing an "import duty" on all northern goods shipped to the South and West. Webb's polemic was the first of a series of adaptions of the colonial concept to political debate that culminated in the *Report on Economic Conditions of the South.* For the evolution of the concept of the South as a colonial economy, see Tindall, *Ethnic Southerners,* 209–23.

10. Brooks Hays, *Politics Is My Parish: An Autobiography* (Baton Rouge: Louisiana State Univ. Press, 1981), 109–23. Hays, *A Southern Moderate Speaks* (Chapel Hill: Univ. of N.C. Press, 1959), 17–20. Jonathan Daniels, *A Southerner Discovers the South* (N.Y.: Macmillan, 1938). Steve Davis, "The South as 'The Nation's No. 1 Economic Problem': The NEC Report of 1938," *Georgia Historical Quarterly, 62* (Spring 1978): 120.
11. Daniels, *Southerner Discovers the South,* 12. Clark H. Foreman, "Decade of Hope," *Phylon 12* (1951): 139. Sosna, *Silent South,* 66. Davis, "The South as 'The Nation's No. 1 Economic Problem,'" 120. Leon H. Keyserling, Oral History Interview, conducted by Jerry N. Hess, 3–19 May 1971 (Independence: Harry S Truman Presidential Library, 1975). Erich W. Zimmerman, letter to Frank P. Graham, 11 Jan. 1937, Graham Papers. Nancy Weiss, *Farewell to the Party of Lincoln* (Princeton: Princeton Univ. Press, 1983), 71, 73. Hamilton, *Lister Hill,* 74, 87.
12. John Salmond, *A Southern Rebel: The Life and Times of Aubrey Willis Williams 1890–1965* (Chapel Hill: Univ. of N.C. Press, 1983), 42, 53–54. The notions of relatively autonomus state managers pursuing interests vis-à-vis a dominant class, and even against an entire socio-economic system, has been developed by a number of social scientists. For a summary of recent research in this vein, see Skocpol, "Bringing the State Back In," 3–43. On the application of these ideas to United States history, see Leuchtenburg, "The Pertinence of Political History: Reflections on the Significance of the State in America," 585–600. See also Theda Skocpol, *States and Social Revolutions* (N.Y.: Cambridge Univ. Press, 1979), 25–27. It is worth noting that southern liberals did not believe expanded state action in the South could (and, in many cases, even that it should) end segregation. See Chapter 5 below.
13. Hugo L. Black, speech accepting the Jefferson Medal of the Southern Conference for Human Welfare, 23 Nov. 1938, Speeches, Writings and Related Materials, Black Papers. Zimmerman, letter to Graham, 11 Jan. 1937, Graham Papers. Lucy R. Mason, letter to Frank P. Graham, 28 Nov. 1939, Graham Papers. Clark H. Foreman, "Decade of Hope," 137. Kreuger, *And Promises To Keep,* 25–29. Daniels, *Southerner Discovers the South,* 335. See also C. Vann Woodward's reminiscences of southern liberalism in the 1930s, *Thinking Back,* 14.
14. David E. Lilienthal, Notes on a conversation with Franklin D. Roosevelt, 12 Sept. 1936, in *The Journals of David E. Lilienthal,* Vol. 1, *The TVA Years 1939–1945* (N.Y.: Harper and Row, 1964), 64.
15. Lowitt, "The TVA, 1933–45," 58–59. Charles H. Houston and John P. Davis, "TVA: Lily-White Reconstruction," *The Crisis* 41 (Oct. 1934): 290–311. John P. Davis, "Plight of the Negro in the Tennessee Valley," *The Crisis* 42 (Oct. 1935): 294–305. John P. Davis, "A Black Inventory of the New Deal," *ibid.,* (May 1935): 141–54. Walter White, "U.S. Department of (White) Justice," in Howard Zinn, ed., *New Deal Thought* (N.Y.: Bobbs-Merrill, 1966), 331–38.

Franklin Roosevelt, letter to Walter White, quoted in Walter White, *A Man Called White* (N.Y.: Viking Press, 1948), 148. Walter White, letter to Jonathan Daniels, 27 Oct. 1936, Daniels Papers, Box 7. James T. Patterson, *Congressional Conservatism and the New Deal* (Lexington: Univ. of Kentucky Press, 1967), 156–57. Arthur M. Schlesinger, Jr., *The Age of Roosevelt:* Vol. 3, *The Politics of Upheaval* (Boston: Houghton Mifflin, 1960), 437–38. Harvard Sitkoff, "The Impact of the New Deal on Black Southerners," in Cobb and Namoroto, eds., *The New Deal and the South,* 117–34. Gavins, *The Perils and Prospects of Black Leadership,* 74.

16. Patterson, *Congressional Conservatism,* 1–31, 67. Robert Garson, *The Democratic Party and the Politics of Sectionalism, 1941–1948* (Baton Rouge: Louisiana State Univ. Press, 1974), x, 7. Schlesinger, *Politics of Upheaval,* 580–81. Tindall, *Emergence,* 607–13.
17. Schlesinger, *Politics of Upheaval,* 421–43, 581. Samuel Lubell, *The Future of American Politics* (N.Y.: Harper & Row, 1951), 57, *passim.* Sitkoff, "Impact of New Deal on Black Southerners," 130. Harold F. Bass, Jr., "Presidential Party Leadership and Party Reform: Franklin D. Roosevelt and the Abrogation of the Two-thirds Rule," *Presidential Studies Quarterly* 18 (Spring 1988): 303–18.
18. "Dealer's Choice," *Business Week,* 7 Nov. 1936, p. 13. Ickes, *Secret Diary,* Vol. II, p. 131. Carl N. Degler, "American Political Parties and the Rise of the City: An Interpretation," *Journal of American History* 51 (June 1964): 51–60. Schlesinger, *Politics of Upheaval,* 421–43. Lubell, *Future of American Politics,* 43–68. Garson, *Democratic Party,* 7–8. Alan Brinkley, "The New Deal and Southern Politics," 113. Harvard Sitkoff, *A New Deal for Blacks,* (N.Y.: Oxford Univ. Press, 1978), 89–90, 100–01. Weiss, *Farewell to the Party of Lincoln,* 21, 29, 180–87, 205, 228–29.
19. A. Cash Koeniger, "The New Deal and the States: Roosevelt Versus the Byrd Organization in Virginia," *Journal of American History* 68 (March 1982): 876–77, 880. Koeniger's study of Virginia takes issue with James MacGregor Burns's assertion that FDR refused to exploit his administration's patronage powers to attempt to recast the Democratic party in the South. Consult James MacGregor Burns, *Roosevelt: The Lion and the Fox, 1882–1940* (N.Y.: Harcourt Brace Jovonavich, 1956), 378–80. Patterson, *Congressional Conservatism,* 45, 65–67, 79, 97, 112, 136–41ff. Swain, *Pat Harrison,* 154. On this same fear of an enlarged Court, see Harold L. Ickes, *The Secret Diary of Harold L. Ickes,* Vol. II, *The Inside Struggle 1936–1939* (N.Y.: Simon and Schuster, 1954), 153. On relief, consult Alston and Ferrie, "Labor Cost, Paternalism and Loyalty in Southern Agriculture," 95. See also Timmons, *Garner of Texas,* 240–41, *passim;* Brinkley, "New Deal and Southern Politics"; and Garson, *Democratic Party,* 9–11.
20. Winfred B. Moore, Jr., "The Unrewarding Stone: James F. Byrnes and the Burden of Race, 1908–1944," in Bruce Clayton and John A. Salmond, eds., *The South Is Another Land* (Westport, Conn.: Greenwood, 1987), 14–16; Byrnes quoted on p. 14.
21. Smith, *The New Deal in the Urban South,* 259. Sitkoff, *A New Deal for Blacks,* 99.
22. Rosengarten, *All God's Dangers,* 456. Sitkoff, *New Deal for Blacks,* 89. Weiss, *Farewell to the Party of Lincoln,* 227, 234. Ickes, *Secret Diary,* Vol. II, p. 153.
23. On the principle of *Herrenvolk* development in the South, see Coclanis and Ford, "The South Carolina Economy Reconstructed and Reconsidered," 102–3.

24. Carter Glass, letter to W. Gordon McCabe, 14 Feb. 1938, quoted in Patterson, *Congressional Conservatism,* 257. Tindall, *Emergence,* 618. The term "small town big man" was popularized by Atlanta *Constitution* editor Ralph McGill. See his *The South and the Southerner* (Boston: Little, Brown, 1963). The term "county-seat elites" is Tindall's. A compelling example of the federal government's replacement of the local elite as the place of petition and identification for ordinary southerners is the testimony of one North Carolina farm laborer. Expressing his devotion to the New Deal to a WPA interviewer, John Easton remembered that the "government shore give us enough when it paid for Amy's leg operation," WPA, *These Are Our Lives,* 11.
25. Daniels, *Southerner Discovers South,* 48. Claude Pepper, "A New Deal in Reconstruction," *Virginia Quarterly Review* 15 (Autumn 1939): 551–60.
26. Roosevelt, "The United States Is Rising and Rebuilding on Sounder Lines," address at Gainesville, Ga., 23 March 1938, in Rosenman, ed., *The Public Papers,* 1938 Vol., *The Continuing Struggle for Liberalism* (N.Y.: Macmillan, 1941), 167.
27. *Ibid.* Roosevelt, informal extemporaneous remarks at Dallas, Texas, 12 June 1936, in Rosenman, ed., *Public Papers,* Vol. 5, *The People Approve* (N.Y.: Random House, 1938), 215–16. Roosevelt, address at Thomas Jefferson Day Dinner, New York, New York, 25 April 1936, in *Public Papers,* Vol. 5, p. 181. Roosevelt, informal extemporaneous remarks at Forth Worth, Texas, 10 July 1938, in *Public Papers,* 1938 Vol. Roosevelt, address at the Univ. of Georgia, Athens, 11 Aug. 1938, *Public Papers,* 1938 Vol., p. 472. Roosevelt, informal extemporaneous remarks at Greenville, S.C., 11 Aug. 1938, in *Public Papers,* 1938 Vol., p. 474. Roosevelt, informal extemporaneous remarks at Tuskegee Institute, Tuskegee, Ala., 30 March 1939, in Rosenman, ed., *Public Papers,* 1939 Vol., *War—and Neutrality* (N.Y.: Macmillan, 1941), 178–81.
28. Press conference with members of the American Society of Newspaper Editors, in Rosenman, ed., *Public Papers,* 1938 Vol., p. 264.
29. Lee Collier, "The Solid South Cracks," *The New Republic* 94 (23 March 1938): 185–86. Tindall, *Emergence,* 622. Patterson, *Congressional Conservatism,* 238. Kreuger, *And Promises To Keep,* 42–43. The Roosevelt administration first surmised that southern opponents of the New Deal might be vulnerable to a "purge" in the 1938 elections in the spring of 1935. Secretary of Interior Harold Ickes traveled to Raleigh, N.C., where his speech, a bitter denunciation of conservative senators, was enthusiastically received. Ickes, at least, returned from this experience believing that New Dealers could unseat such conservatives in the next election. See Ickes, *Secret Diary,* Vol. II, pp. 95–96.
30. James C. Cobb, *The Selling of the South* (Baton Rouge: Louisiana State Univ. Press, 1982), 5–15. WPA, *These Are Our Lives,* 210.
31. Foreman, "Decade of Hope," 137–40. Davis, "The South as 'The Nation's No. 1 Economic Problem,'" 121. Kreuger, *And Promises To Keep,* 13–14. Roosevelt added his own explanatory notes to the volumes of his public papers which appeared before his death. His 1941 note on the background to the *Report on Economic Conditions of the South* reiterates his intention that the report stick to the facts. The *Report* was, in Roosevelt's words, "the first concise, adequate statement of the problems of the South." Roosevelt, Note to Document 88, in Rosenman, ed., *Public Papers,* 1938 Vol., p. 422. According to Claude Pepper, FDR would have remained cautious about the South (he had, in fact, stayed away from Pepper's campaign) if not for Pepper's surprising victory. See

Claude D. Pepper with Hays Gorey, *Pepper: Eyewitness to a Century* (N.Y.: Harcourt Brace Jovanovich, 1987), 71.

32. *Ibid.* Roosevelt, letter to Lowell Mellett, 22 June 1938, in *Public Papers,* 1938 Vol., p. 384. Roosevelt, letter to the Conference on Economic Conditions of the South, 4 July 1938, *ibid.,* p. 421. NEC, *Report on Economic Conditions,* pp. 1–3. On Mellet, see Lilienthal, *Journals,* Vol. I, p. 248.
33. NEC, *Report on Economic Conditions,* 4–8, 12, 17, 21–34, 61–63; quotations from pp. 20, 61.
34. *Ibid.,* 1–3. Numan V. Bartley, "The Era of the New Deal as a Turning Point in Southern History," in Cobb and Namoroto, eds., *The New Deal and the South,* 143. That commitment to federal sponsorship of regional development was not lost on the *Report*'s audience. Representatives of Pennsylvania demanded similar concern for their state. The Keystone State, they insisted, and not the South, represented economic problem number one. See "National Emergency Council, 1938," in the Graham Papers.
35. NEC, *Report on Economic Conditions, passim.* For descriptions of the actual activities of the Farm Security Administration and National Youth Administration in the region between 1938 and 1941, see U.S. House, *Interstate Migration,* Part 2, pp. 708, 721–23; and, U.S. House, *National Defense Migration,* Part 11, Washington, D.C., Hearings, 15–17 July 1941, pp. 6374–75, respectively. On the poll tax, consult Weiss, *Farewell to the Party of Lincoln,* 250.
36. Daniels, "Democracy Is Bread," *Virginia Quarterly Review* 14 (Autumn 1938): 482–83. "Economic Problem No. 1," *Commonweal* 28 (19 Aug. 1938): 417. "The Plight of the South," *The New Republic* 96 (24 Aug. 1938): 61. Maverick, "Let's Join the United States," 64–75. Coordinating Committee of the Citizens' Fact Finding Movement, "Georgia Annotations to the *Report on Economic Conditions of the South,*" March 1939, Records of the TVA, RG 142, TVA Board File Curtis-Morgan-Morgan 1933–57, FARC. Davis, "The South as 'The Nation's No. 1 Economic Problem,'" 122. One northern magazine, however, ridiculed the NEC Report. See Garret Garrett, "The Problem South," *Saturday Evening Post* 211 (8 Oct. 1938): 86–91.
37. Fitzgerald Hall, letter to Frank Graham, 26 Aug. 1938, Frank P. Graham Papers. Fitzgerald Hall, "Comments on the *Report of the Economic Conditions of the South,*" 7 Sept. 1938, Frank P. Graham Papers. David R. Coker, letter to Frank P. Graham, 7 July 1938, Frank P. Graham Papers. "Is the South the Nation's Economic Problem?," *Manufacturers' Record* 107 (Aug. 1938): 13–15. WPA, *These Are Our Lives,* 15.

    Frank Graham summarized the responses to the *Report.* "The reaction," he concluded, "was almost as varied as the people in the South themselves." Requests for copies of the *Report* were enormous all over the South; Graham noted that schools, civic clubs, women's clubs were making wide use of it. See Frank Graham, memorandum on the letter of Mr. Fitzgerald Hall, president of the Southern States Industrial Conference, concerning the *Report on Economic Conditions of the South,* Frank P. Graham Papers. Graham, "Summary of Reaction to N.E.C. *Report,*" Frank P. Graham Papers.
38. Roosevelt, address at Barnesville, Ga., 11 Aug. 1938, in Rosenman, ed., *Public Papers,* 1938 Vol., pp. 466–70. David, "The South as 'The Nation's No. 1 Economic Problem,'" 124–25. Frank Freidel, "The South and the New Deal," in Cobb and Namoroto, eds., *The New Deal and the South,* 33. Foreman, "Decade of Hope," 137–40. Foreman, letter to Jonathan Daniels, 8 June 1948, Jona-

than W. Daniels Papers. "The Plight of the South," 61. *New York Times,* 21 Aug. 1938.

39. Roosevelt, address at Barnesville, Ga., 11 Aug. 1938. Roosevelt, informal extemporaneous remarks at Greenville, S.C., 11 Aug. 1938, in Rosenman, ed., *Public Papers,* 1938 Vol., pp. 471–72. Koeniger, "New Deal and the States," 876–80. James A. Farley, *Jim Farley's Story* (N.Y.: McGraw-Hill, 1948), 95–96, 120–22, 134–44. Byrnes, *All in One Lifetime,* 102–3. James C. Cobb, "Not Gone, but Forgotten: Eugene Talmadge and the 1938 Purge Campaign," *Georgia Historical Quarterly* 59 (Summer 1975): 197–209.
40. Byrnes, *All in One Lifetime,* 102–3. Farley, *Jim Farley's Story,* 120–22. Cobb, "Not Gone, but Forgotten," 204. Ellison Smith, quoted in Brinkley, "New Deal and Southern Politics," 108–9. Patterson, *Congressional Conservatism,* 283. Reflecting back on the purge, Vice President Garner remembered Roosevelt's confession that he had voted for a Republican, distant cousin Theodore Roosevelt, against Democrat Alton Parker in the 1904 presidential election. To Garner, that confession disclosed FDR's tendency to value "ideological compatibility" over party loyalty, a tendency Garner disdained and held responsible for the purge. See Timmons, *Garner of Texas,* 231, 148.

    Political punishment was not the only motivation for the cut-off of PWA aid to Georgia, but it played an important role in the decision. See Smith, *New Deal in the Urban South,* 109.
41. Farley, *Jim Farley's Story,* 144, *passim.* Tindall, *Emergence,* 649. Brinkley, "New Deal and Southern Politics," 111. Tindall, *Ethnic Southerners,* 96.
42. NEC, *Report on Economic Conditions,* 22–23, 29, 38–44.
43. *Roosevelt Press Conferences,* Vol. IX, pp. 436–48. Roosevelt, Annual Message to Congress, 3 Jan. 1938, in Rosenman, ed., *Public Papers,* 1938 Vol., p. 13. Roosevelt, address at Gainesville, Ga., p. 167.
44. New York *Herald Tribune,* 21 May 1938, p. 13. Orme Phelps, *The Legislative Background of the Fair Labor Standards Act* (Chicago, 1939), 5–8, 43–64. The constitutional issues surrounding the replacement of the NRA by a bill such as the FLSA were considered by Robert Jackson of the Dept. of Justice in his testimony before a Joint Congressional Committee. Consult *Joint Hearings Before Committee on Education and Labor, U.S. Senate and the Committee on Labor, House of Representatives on the Fair Labor Standard Act of 1937* (Washington, 1937), 1–90.

    In the immediate aftermath of the 1935 Supreme Court decision terminating the NRA, Congress passed the Walsh-Healy Act to restore federal control over wages and hours by restricting the disposition of U.S. Government contracts to firms that met specified national labor standards. The faultily drafted law, however, applied only to contracts in excess of $10,000, a restriction which exempted 99 percent by number and 75 percent by total value of federal contracts from the act's coverage. Furthermore, since firms could evade the mandated labor standards simply by refusing to accept government contracts, the act could only have been effective in cases where government purchases formed a significant component of total demand. Only in the shipbuilding industry was such the case. In the major southern industries, like textiles (0.48%), tobacco (0.04%), chemicals (0.43%), and forest products (0.25%), government purchases accounted for less than 1 percent of total production in fiscal years 1937, 1938, 1939, and 1940. See E. F. Denison, "The Influence of the Walsh-Healy Public Contracts Act Upon Labor Conditions," *Journal of Political Economy* 49 (April 1941): 228–42.

45. The legislative history of the FLSA is charted in Leuchtenburg, *Franklin D. Roosevelt and the New Deal,* 261–64; Patterson, *Congressional Conservatism,* 153, 242–46; Phelps, *Legislative Background,* 1–8, 43–64; and Tindall, *Emergence,* 533–36. Contemporary analyses of the congressional debate can be found in the *New York Times,* 25 May, 5 June, and 12 June 1938. Consult also "Differential Differences," *Time* 31 (9 May 1938): 10–11; and "Congress Balks," *The New Republic,* 18 Aug. 1937. For the actual debate in Congress, consult House of Representatives, 75th Cong., 3rd sess., Report 2182, and, U.S. Senate, 75th Cong., 1st sess., Report 884 (prepared by Senator Black). The text of the final version of the act is printed in the *New York Times,* 13 June 1938, p. 8. For the Roosevelt quotation, see Roosevelt, memorandum for Mac [Marvin MacIntyre], 13 Jan. 1938, quoted in Irving Bernstein, *A Caring Society* (Boston: Houghton Mifflin, 1985), 317.
46. *Ibid.*
47. *Ibid.*
48. *Joint Hearings on the FSLA,* 271–72, 427, 943.
49. *Joint Hearings on the FLSA,* 571–75.
50. Gavins, *Perils and Prospects of Southern Black Leadership,* 54–62. Wilma Dykeman and James Stokely, *Seeds of Southern Change* (N.Y.: W.W. Norton, 1962), 27. On civil-rights groups and minimum wages after 1938, see below.
51. See, for example, the testimony in *Joint Hearings on Fair Labor Standards Act.*
52. Pat Harrison, "The Wages of Dixie," *Collier's,* 22 Jan. 1938, p. 11. Swain, *Pat Harrison,* 163–65, 195–97. "Congress Balks," 32.
53. J. H. Eddy, letter to Hugo L. Black, 1 July 1937, Black Papers, Box 159. Troy Peanut Company, telegram to Hugo L. Black, 14 July 1937, Black Papers, Box 160. Paul B. Fuller, letter to Hugo L. Black, 8 July 1937, Black Papers, Box 160. A. B. Carroll, letter to Hugo L. Black, 9 June 1937, Black Papers, Box 159.
54. *Joint Hearings on the FLSA,* 356, 477, 591, 761. R. F. Darrah, letter to Hugo L. Black, 13 July 1937, Black Papers, Box 159. William L. Lloyd, letter to Hugo L. Black, 21 June 1937, Black Papers, Box 159. Robert B. Anderson, letter to Hugo L. Black, 21 June 1937, Black papers, Box 159. Harrison, "Wages of Dixie," 11, 46.
55. *Joint Hearings on the FLSA,* 763, 767. W. R. Holley, letter to Hugo L. Black, 16 June 1937, Black Papers, Box 159. Presidents of the First National Bank of Mobile, Merchants' National Bank of Mobile, and American National Bank and Trust Company, letter to the Bankers of Alabama, 19 June 1937, Black Papers, Box 160. Harrison, "Wages of Dixie," 11, 46. Fitzgerald Hall, letter to Frank P. Graham, 26 Aug. 1938, Graham Papers.

Walter Prescott Webb offered the most sophisticated of these "carpetbagger arguments." Webb contended that by encouraging mechanization, the North by virtue of its monopoly on the manufacture of heavy machinery only extended its economic domination of the South and West. See Webb, *Divided We Stand,* 123, 153.

Even Jonathan Daniels, a supporter of the legislation, evinced some sympathy for this view. "So far as literate America is concerned," Daniels wrote in

1938, "the South is in some sense a foreign country now almost as menacing industrially as Japan and in some degree regarded in a similar light." See Daniels, letter to Carl Ferguson, 7 May 1938, Daniels Papers.

56. *Joint Hearings on the FLSA,* 761–63. For other examples of this reasoning, see Harrison, "Wages of Dixie," 11, 46; and C. W. Mizell, letter to Hugo L. Black, 21 June 1937, Black Papers, Box 160.

57. *Joint Hearings on the FLSA,* 768. Southern Pine Industry Committee, "A Brief Outline of Basic Considerations with Respect to Minimum Wages and Maximum Hours in the Southern Pine Lumber Industry," n.d., Black Papers, Box 164. Harrison, "Wages of Dixie," 11, 46. The Gallup poll appeared in the *New York Times,* 1 June 1938. It is noteworthy that low-income southerners opposed geographic differentials. The role of the regional wage differential in the "growth mentality" of southern industrialists is elucidated in Tindall, *Ethnic Southerners,* 211.

58. Roosevelt, "President Recommends Minimum Wages and Maximum Hours Legislation," letter to Congress, 24 May 1937, in Rosenman, ed. *Public Papers,* 1937 Vol., pp. 212–13. Roosevelt, Fireside Chat on Recommended Legislation, 12 Oct. 1937, in Rosenman, ed., *Public Papers,* 1937 Vol., p. 435. House of Representatives, 75th Cong., 3rd sess., Report 2182, pp. 6–7. *Joint Hearings on the FLSA,* 193, 309–17.

59. *Joint Hearings on the FLSA,* 94–95, 126, 147–49, 371–72. Patterson, *Congressional Conservatism,* 153–54. Another Massachusetts Republican, Rep. Joseph W. Martin, behaved similarly in the House. Patterson, *Congressional Conservatism,* 243–45. See also Thomas Ferguson and Joel Rogers, *Right Turn* (N.Y.: Hill and Wang, 1986), 46–47; and Thomas Ferguson, "From Normalcy to New Deal: Industrial Structure, Party Competition, and American Public Policy in the Great Depression," *International Organization* 38 (Winter 1984): 41–94.

60. *Joint Hearings on the FLSA,* 866, 937ff., 973. Hugo L. Black, letter to William E. Yerby, 3 July 1937, Black Papers, Box 164. Michael Bernstein has noted that the political platforms of cooperative associations such as the Chamber of Commerce and the National Association of Manufacturers obscure the position of various industries and sectors represented by such national bodies. Still, while many northern industries had little reason to oppose the law and kept silent, with the exception of New England textiles, no northern industry testified in favor of it. See Michael A. Bernstein, *The Great Depression* (N.Y.: Cambridge Univ. Press, 1988), 198–99.

61. *Ibid.* "For the Wage and Hour Bill," *The Democratic Digest* 15 (April 1938): 3, in Secretary of Interior File, Ickes Papers, Box 105. I. Bernstein, *A Caring Society,* 143. *Joint Hearings on the FLSA,* 94–95, 126, 147–49, 371–72. Patterson, *Congressional Conservatism,* 153–54. Ferguson, "From Normalcy to New Deal," 41–94.

62. Pepper, "A New Deal in Reconstruction," p. 556. *Joint Hearings on the FLSA,* 125, 309, 405–6, 792–96, 806–7. House of Representatives, 75th Cong., 1st Sess., Report 1452, p. 8. Hugo L. Black, letter to Donald Comer, Black Papers, Box 159. Black, letter to J. A. Adams, 15 July 1937, Black Papers, Box 159. Roosevelt, Press Conference with Editors of Trade Papers, in Rosenman, ed., *Public Papers,* 1938 Vol., p. 198. Bureau of Labor Statistics economist Isador Lubin, one of the architects of minimum wage regulation, explained the "single Maxim" behind the proposed legislation in similar terms: "namely, that the

welfare and profits of no private business shall interfere with the welfare of the Nation as a whole. . . . Minimum wages, maximum hours are expressions of this maxim." *Joint Hearings on the FLSA,* 309. For a detailed analysis of the rationale behind minimum-wage regulation as a vehicle for regional development, see the introduction to Chapter 3, below.

63. George Fort Milton, Draft of "The South Do Move," 1939, Speeches, Articles and Book File, Milton Papers, Box 30. Roosevelt, letter to Louise Charlton, 19 Nov. 1938, in "Report of the Proceedings of the Southern Conference for Human Welfare," 20–23 Nov. 1938, Graham Papers.
64. Lilienthal, letter to Newton Arvin, Feb. 1939, in Lilienthal, *Journals,* Vol. I, pp. 79–82.
65. *Ibid. Congressional Record,* 76th Cong., 1st sess., 11165–66. John Temple Graves, "The South Still Loves Roosevelt," *The Nation* 149 (1 July 1939): 11–13.

Chapter 3

1. Roosevelt, informal extemporaneous remarks at Fort Worth, 447. Maverick, "Let's Join the United States," 72. See also the testimony of Wage and Hour Administrator Philip Fleming before the Tolan Committee, in U.S. House, *Interstate Migration,* 76th Cong., 3rd sess., Part 8, Washington, D.C., Hearings, 29 Nov.–3 Dec. 1940.
2. Bureau of Labor Statistics, "Labor Conditions in the South," *Monthly Labor Review* 47 (Oct. 1938): 748. J. Ronnie Davis, *The New Economics and the Old Economists* (Ames: Iowa State Univ. Press, 1971), x–xi, 94–99. Robert Lekachman, *The Age of Keynes* (N.Y.: Random House, 1966), 82–89. Lawrence R. Klein, *The Keynesian Revolution* (N.Y.: Macmillan, 1947), 107–9. David E. Kaun, "Economics of the Minimum Wage: The Effects of the Fair Labor Standards Act, 1945–1960" (doctoral dissertation, Stanford Univ., 1963), 41–45. Consult also Marriner S. Eccles, *Beckoning Frontiers: Public and Private Recollections* (N.Y.: Macmillan, 1951), 287–323; and Robert Collins, *The Business Response to Keynes, 1929–1964* (N.Y.: Columbia Univ. Press, 1981), 6–11.
3. At the Congressional hearings on the FLSA, one witness testified that "the depression had come about not because wages were too high, but because those who controlled large amounts of money . . . took the position publicly expressed by Mr. Albert Wiggin, then Chairman of the Board of the Chase National Bank, that securities had been liquidated, commodity prices had been liquidated, and therefore, it was time that wages must be liquidated. And that, sir, to my knowledge, increased the severity of our depression as nothing else did." The House report on the FLSA echoed that diagnosis, posing the minimum-wage law as a check against ruinous deflation, as did South Carolina's New Deal governor, Olin D. Johnston, who claimed that a minimum wage was "the only protection against a new depression." See U.S. House of Representatives, 75th Cong., 3rd sess., Report 2182, p. 6, *Joint Hearings* on the FLSA, 155–56, 605; and "For the Wage and Hour Bill," 3. Consult also Davis, *New Economics and Old Economists, passim;* and Lekachman, *Age of Keynes,* 91.
4. Roosevelt, informal remarks at Fort Worth, p. 447. Roosevelt, address at Gainesville, Ga., 167. See also NEC, *Report on Economic Conditions,* 61; and Pepper, "New Deal in Reconstruction," 556.
5. Bureau of Labor Statistics, *Bulletin No. 898,* p. 102. *Joint Hearings on the FLSA,*

309–10. M. Bronfenbrenner has noted that advocates of minimum wages reverse the marginal productivity theory of wage distribution, so that productivity is dependent on wages, at least at the lower end of the wage scale, rather than wages being determined by productivity. See M. Bronfenbrenner, "Minimum Wages, Unemployment and Relief: A Theoretical Note," *Southern Economic Journal* 10 (July 1943): 52.

6. Roosevelt, "Address at Thomas Jefferson Day Dinner," in Rosenman, ed., *Public Papers,* Vol. 5, p. 181. Odum, *Southern Regions,* 355. NEC, *Report on Economic Conditions of the South,* 39. *Joint Hearings on the FLSA,* 309, 764. For a discussion of how predictions of the impact of the minimum wage depended on assumptions about the management and operation of low wage firms, see Kaun, "Economics of the Minimum Wage," 15–31.
7. Lorena Hickok, letter to Harry Hopkins, 5 Feb. 1934, in Lowitt and Beasley, eds., *One Third of a Nation,* 173. NEC, *Report on Economic Conditions,* 41–44.
8. *Ibid. Joint Hearings on the FLSA,* 275, 595, 846–47. When Mississippi embarked on its first industrial promotion campaign in the 1930s, state officials discovered that many southern employers, and most firms interested in the state program, desired all-female work forces. The Industrial Commission therefore made a special effort to attract firms that would employ adult males. See Cobb, *Selling of the South,* 2–3, 23.
9. Richard Leche, quoted in the *New York Times,* 6 Oct. 1938. Some southern employers warned that the minimum wage would displace black laborers. Charles H. Robertson of the Ablemarle Paper Manufacturing Company, for example, asserted his "social obligation to these people most of whom are in the marginal class of laborers." See the "Official Report of Proceedings Before the Wage and Hour Division of the Department of Labor in the Matter of a Hearing on the Minimum Wage Recommendation of Industry Committee No. 11, for the Pulp and Primary Paper Industry," 20–21 May 1940, Records of the Wage and Hour and Public Contracts Division of the Department of Labor, RG 155, Entry 16, Industry Committee No. 11 for the Pulp and Primary Paper Industry, NA. See also Lorena Hickok, letter to Harry Hopkins, 18 Feb. 1934, in Lowitt and Beasley, eds., *One Third of a Nation,* 196. Consult also Wright, *Old South, New South,* 177–97, 223–25.
10. John Temple Graves II, "Wage-Hour Law Outlook Different for the Country," *New York Times,* 19 June 1938. Harrison, quoted in *New York Times,* 16 Aug. 1938. *New York Times,* 22–23, 26 Oct. 1938.
11. *New York Times,* 27 Oct., 31 Oct. 1938. Bureau of Labor Statistics, *Bulletin No. 898,* 98–99. The figures on the 25-cent minimum include Arizona and New Mexico in the South. More accurate data are available for April 1939 that anticipate the impact of the 30-cent minimum. At that date, 18 percent of covered employees in the South earned less than 30 cents an hour; in the rest of the country, only 2.7 percent fell below that standard. The only trustworthy figures on average hours come from the same April 1939 Bureau of Labor Statistics survey: 24.5 percent of southern laborers worked in excess of 42 hours per week as opposed to 17.6 percent of workers outside the region.
12. Bureau of Labor Statistics, *Bulletin No. 898,* 98–104. Moloney, "Some Effects of the Fair Labor Standards Act," 17–23. H.M. Douty, "Minimum Wage Regulations in the Seamless Hosiery Industry," *Southern Economic Journal* 8 (Oct. 1941): 176–90. "Hourly Earnings in the Lumber and Timber Products Industry," *Monthly Labor Review* 53 (July 1941). David E. Kaun, "The Economics of

the Minimum Wage," 70, 145–46. John M. Peterson and Charles T. Stewart, Jr., *Employment Effects of Minimum Wage Rates* (Washington: American Enterprise Institute, 1969), 60–63. Elmer Andrews asserted that spreading employment was one of the FLSA's goals. See *New York Times,* 27 Oct. 1938.

13. Douty, "Minimum Wage Regulations in Seamless Hosiery." A.F. Hinrichs, "Effects of the 25-Cent Minimum Wage on Employment in the Seamless Hosiery Industry," *Journal of the American Statistical Association* (March 1940): 13–23. Kaun, "Economics of the Minimum Wage," 70, 145–46. Peterson and Stewart, Jr., *Employment Effects of Minimum Wage Rates,* 60–63. Bureau of Labor Statistics, *Bulletin No. 898,* 103–5.

    Critics of the Wages and Hours law noted that federal analysts made no effort to isolate the impact of the minimum wage from the general economic expansion occasioned by the war in Europe. Furthermore, the Labor Dept. based its assessments on the number of wage earners rather than the more slowly rising number of total man-hours. That accounting procedure obscured the FLSA's impact on the total demand for labor in southern manufacturing. Economists of differing political persuasions concur on the inadequacy of Department of Labor studies. See Kaun, "Economics of Minimum Wage," 70, and Peterson and Stewart, *Employment Effects,* 56–59.

14. Dept. of Labor, "Entrance Rates of Common Laborers, July 1938," *Monthly Labor Review* 48 (Jan. 1939): 170–73. "Working Conditions of Pecan Shellers in San Antonio," *ibid.* (March 1939): 549–51. Sheldon C. Menefee and Orin C. Cassimore, *The Pecan Shellers of San Antonio* (Washington: Federal Works Agency, 1940). Moloney, "Some Effects of the Fair Labor Standards Act."

    The cottonseed products industry suffered similar drastic changes. Between the 1937–38 and 1938–39 seasons, hourly wage rates jumped 30 percent. Employers moved to hold down labor costs by slashing their labor force 19 percent. They also abandoned the twelve-hour day, even though the industry was exempt from the hours provisions of the FLSA. The adoption of machinery—large cooking and crushing equipment that allowed the replacement of seven unskilled workers with one skilled operator—accounted for part of the savings. Like the pecan shellers, some cottonseed products manufacturers substituted already available technology for labor-intensive production processes after the FLSA shifted the relative costs of machinery and unskilled labor. See Moloney, "Some Effects of the Fair Labor Standards Act."

    Some economists have challenged the dramatic conclusions of the studies of these two industries, citing the studies, short-term frame of reference and failure to isolate minimum-wage effects from general economic trends. See Peterson and Stewart, *Employment Effects,* 58–59.

15. "Report and Recommendations of Industry Committee No. 3 for the Establishment of Minimum Wage Rates in the Hosiery Industry," 15 May 1939. "Official Report of Proceedings Before the Wage and Hour Division of the Dept. of Labor in the Matter of the Public Hearing on the Recommendations of the Hosiery Committee, Concerning Minimum Wage Rates for the Hosiery Industry," 12 June 1939. "A Brief Presented by the National Association of Hosiery Manufacturers, Inc. to the Hosiery Industry Committee created under the Provisions of the Fair Labor Standards Act of 1938," May 1939. All in RG 155, Entry 16, Industry Committee No. 3 for the Hosiery Industry, NA. Research and Statistics Branch, Wage and Hour Division, Department of Labor, "Minimum Wages in the Seamless Hosiery Industry," March 1941. RG

155, Entry 16, Industry Committee No. 21 for the Seamless Hosiery Industry. Douty, "Minimum Wage Regulation in Seamless Hosiery."

Along with the reduction in employment of knitters, the Labor Department presented an increase in the number of "machine fixers" on the payrolls of southern hosiery firms between 1939 and 1940 as evidence of the trend toward automation. See Department of Labor, "Minimum Wages in Seamless Hosiery," 42.

16. "Digest of Record, 1941," RG 155, Entry 16, Industry Committee No. 30 for the Lumber and Timber Products Industry, NA. Defense production allowed the switch to exclusively intrastate commerce. It also complicated the tracking of products that indirectly or eventually ended up in interstate commerce, making enforcement impossible.
17. Bureau of Labor Statistics, *Nineteenth Report on Average Hourly Earnings in the Cotton-Goods Industry,* 15 Feb. 1938. Bureau of Labor Statistics, *Bulletin No. 663: Wages in Cotton-Goods Manufacturing,* Nov. 1938. "Exhibits and Supplementary Briefs Presented at the Meeting of the Textiles Committee," 14–17 Dec. 1938. "Statement of the Minority of Industry Committee No.1 Submitted to the Administrator of the Fair Labor Standards Act of 1938," 22 May 1939. All in RG 155, Entry 16, Industry Committee No. 1 for the Textiles Industry, NA. Solomon Barkin, "Brief Presented on Behalf of the Textile Workers Union of America," 14 April 1941. Bureau of Labor Statistics, *Average Hourly Earnings in the Cotton-Goods Industry, September 1940 and April 1941,* 14 April 1941. Both in RG 155, Entry 16, Industry Committee No. 25 for the Textiles Industry, NA.
18. On the long-standing economic pressures faced by the principal southern industries, consult M. Bernstein, *The Great Depression,* 130. An example of the efforts to extend FLSA coverage is the testimony of Secretary of Labor Frances Perkins and Wage and Hour Administrator Philip Fleming before the Tolan Committee in 1940. See U.S. House, *Interstate Migration,* Part 8, pp. 3334ff.
19. F. Ray Marshall, *Labor in the South* (Cambridge: Harvard Univ. Press, 1967), 299.
20. John Temple Graves II, "Wage-Hour Law Outlook Different for the Country," *New York Times,* 19 June 1938.
21. "Exhibits and Supplementary Briefs Presented at the Textiles Committee." The southern textile industry's relative lack of organization before 1938 is discussed in Wright, *Old South, New South,* 223. Murchison's militance distressed southern liberals who had hoped for "enlightened leadership" when Murchison became C-TI president. Jonathan Daniels lamented the situation that Murchison, "since becoming head of C-TI, is steadily voicing the old-fashioned reactionary ideas of the cotton textile industry." Daniels, letter to Frank P. Graham, 13 Feb. 1936, Graham Papers.

On the squelching of dissent, see, for example, Hyman Battle, letter to Frank P. Graham, 20 June 1939; and, Kemp D. Battle, letter to Graham, 20 June 1939, Frank P. Graham Papers. The tenacity and discipline of the southern resistance frightened some witnesses and committee members. John Burns of the American Lace Operatives Union expressed relief that none of the plants in his industry lay below the Mason-Dixon line. "My heart was wringing when I heard them talking about the South," Burns sighed. "We have nothing like that at all. God bless us, we are free!" See "Transcript of Hearings, Industry Committee No. 1," 14–17 Dec. 1938, RG 155, Entry 16, Industry Committee No.1 for the Textile Industry, NA.

22. "Statement of the Minority of Industry Commitee No.1." Elmer F. Andrews, "Findings and Opinion of the Administrator in the Matter of the Recommendations of Industry Committee No. 1 for Minimum Wage Rates in the Textile Industry, 1939." Alabama Mills, Inc., letter to Elmer F. Andrews, n.d. J. W. Sanders Cotton Mill, Inc., letter to Andrews, n.d. These are two of more than 200 identical letters sent to the administrator. See RG 155, Entry 16, NA. See also "Official Report of Proceedings . . . of the Hosiery Committee."
23. "Statement of the Minority of Industry Committee No.1." "Transcript of Hearings, Industry Committee No. 1," 14–17 Dec. 1938, RG 155, Entry 16, Industry Committee No.1 for the Textile Industry, NA. For a sample of northern manufacturing opinion on this issue, see the above and also "Statement of Russell Fisher, President of the National Association of Cotton Manufacturers (Boston)"; Eugene Phelan, "Brief on Behalf of Greater New Bedford for the Establishment of a Single 32½ Cent Hourly Minimum in the Cotton Textile Industry"; and William E. G. Batty, "Brief Supporting Recommendation of Industry Committee No. 1 Presented in Behalf of the New Bedford Textile Council," RG 155, Entry 16, Industry Committee No. 1 for the Textile Industry, NA. Consult also U.S. House, *Interstate Migration,* Part 8, pp. 3368.
24. *Ibid.*
25. "Memorandum of Public Members on Testimony Before Industry Committee No. 9," Aug. 1940, Graham Papers. "Brief of the National Association of Hosiery Manufacturers." "Official Report of Proceedings . . . of the Hosiery Committee." "Transcript of Hearings, Industry Committee No. 1."
26. "Transcript of Hearings, Industry Committee No.1." "Official Report of Proceedings . . . of the Hosiery Committee." "Statement of Minority of Industry Committee No. 1." On paternalism, see "Official Report of Proceedings Before the Wage and Hour Division of the Department of Labor in the Matter of a Hearing on the Minimum Wage Recommendation of Industry Committee No. 11, for the Pulp and Primary Paper Industry," 20–21 May 1940, RG 155, Entry 16, Industry Committee No. 11 for the Pulp and Primary Paper Industry, NA, and, "The Industrial South," *Fortune* 18 (Nov. 1938): 123.
27. "Summary of the Carriers' Argument on Mechanization," Aug. 1940, Graham Papers. "Transcript of Hearings, Industry Committee No. 1." See also "Official Report of Proceedings . . . of the Hosiery Committee."
28. "Transcript of Hearings, Industry Committee No. 1." So passionate was the dispute over plant mechanization that it caused the only rift in the southern management position. A few genuinely liberal southern employers supported the Wages and Hours law on principle; one forward-thinking mill owner declared his support for the FLSA because he sought a Dept. of Labor regional office for his home town. Still, of the few southern manufacturers in favor of raising the minimum wage, nearly all operated automated plants. They did not oppose regional wage differentials, as northern employers did, but desired a higher base rate. In 1940, however, the voices of these southern businessmen remained faint. See "Official Report of Proceedings . . . of the Hosiery Committee," Transcript of Hearings, Industry Committee No. 1," and, R. W. Johnson, letter to Mark Ethridge, 17 April 1940, Frank P. Graham Papers.
29. "Transcript of Hearings, Industry Committee No.1." "Brief Presented in

Behalf of the Textile Workers Organizing Committee of the C.I.O. in Connection with the Establishment of a Minimum Wage in the Textile Industry," Dec. 1938, RG 155, Entry 16, Industry Committee No.1 for the Textile Industry.

30. "Brief of the American Federation of Hosiery Workers." Philip B. Fleming, "Findings and Opinion of the Administrator in the Matter of the Recommendation of Industry Committee No. 25 for Minimum Wage Rates in the Textile Industry," 13 June 1941, RG 155, Entry 16, Industry Committee No. 25, NA. Hodges, *New Deal Labor Policy and the Southern Cotton Textile Industry, 1933–1941,* 33, 178–79.
31. Bureau of Labor Statistics, *Bulletin No. 663,* 121. "Transcript of Hearings, Industry Committee No. 1." "Official Report of Proceedings . . . of the Hosiery Committee." "Digest of Record, Industry Committee No. 30, Lumber and Timber Products Industry Committee." Andrews, "Findings and Opinion of Administrator in the Matter of . . . the Textile Industry." Bureau of Labor Statistics, "Differences in Living Costs in 5 Northern and 5 Southern Cities," 28 May 1939. Andrews, "Findings and Opinion of the Administrator in the Matter of the Recommendation of Industry Committee No. 3 for Minimum Wage Rates in the Hosiery Industry," 18 Aug. 1939, RG 155, Entry 16, Industry Committee No. 3 for the Hosiery Industry, NA. Fleming, "Findings and Opinion of the Administrator in the Matter of the Recommendation of Industry Committee No. 11 for Minimum Wage Rate in the Pulp and Primary Paper Industry," 15 July 1940, RG 155, Entry 16, Industry Committee No. 11, NA.
32. Roosevelt, "Press Conference with the Editors of Trade Papers," 8 April 1938, in Rosenman, ed., *Public Papers,* 1938 Vol., pp. 192–203. Roosevelt, "Fireside Chat on Recommended Legislation," 12 Oct. 1937, in Rosenman, ed., *Public Papers,* 1937 Vol., pp. 434–36.
33. Department of Labor, Wage and Hour and Public Contracts Division, *Annual Report for the Fiscal Year Ended June 30, 1942* (Washington, 1942), 3–4. Bureau of Labor Statistics, *Bulletin No. 898,* 44. Hoover and Ratchford, *Economic Resources,* 126. The employment figure is for production workers in manufacturing 1939–47. Typical Industry Committee wage orders are "Report and Recommendation of Industry Committee No. 21 for the Establishment of a Minimum Wage Rate in the Seamless Hosiery Industry," April 1941, RG 155, Entry 16, Industry Committee No. 21 for the Seamless Hosiery Industry, NA; and Philip B. Fleming, "Findings and Opinion of the Administrator in the Matter of the Recommendation of Industry Committee" 17 Oct. 1941, RG 155, Entry 16, Industry Committee No. 30 for the Lumber and Timber Products Industry, NA.
34. U.S. House, *Interstate Migration,* Part 8, pp. 3370. On broader economic conditions, consult Bernstein, *The Great Depression,* 28–29, 146–52. I developed this counterfactual speculation during a session at the 1988 Meetings of the Organization of American Historians. I am indebted to Professor Otis Graham of the Univ. of North Carolina for his insightful commentary.
35. Alan Richards, *War Labor Boards in the Field* (Chapel Hill: Univ. of N.C. Press, 1953), 229. U.S. Dept. of Labor, Wage and Hour and Public Contracts Division, *Annual Report for the Fiscal Year Ended June 30, 1942,* quoted in Bureau of Labor Statistics, *Bulletin No. 898,* 107. NWLB, *The Termination Report of the National War Labor Board* (Washington, 1947). These lists of NWLB and RWLB

members should be compared with the lists of Industry Committee members in RG 155, NA.

36. Richards, *War Labor Boards,* 29–31, 66–69. NWLB, *Termination Report,* 784, *passim.* As with the New Deal agencies, the NWLB's first decentralization plan divided the southern states among many jurisdictions. Arkansas fell under the supervision of the Kansas City office, while the Cleveland board set wages for Kentucky. Most alarming to business interests, the original plan placed Virginia with the mid-Atlantic states. Southern politicians and business interests lobbied to transfer the Old Dominion to the Southeastern region; they believed a separate southern board would protect the regional wage differential and the section's anti-union environment. Organized labor, on the other hand, feared that southern trade associations would dominate a regional board. The issue soon came to a head. In the official opinion on the matter, Frank Graham reiterated his commitment to raising wages in the South. He assigned Virginia to the Atlanta office, reasoning that the inclusion of Virginia would tend to lower wages in the Mid-Atlantic region and to raise them in the South. The particulars of this decentralization plan occasioned such debate because most observers believed that the new regional boards would act autonomously. "What do you think we got you for," NWLB Chairman William H. Davis asked a meeting of regional chairmen, "except to exercise your judgement?" Theodore Kheel, memorandum to Regional Chairmen, 3 March 1944, Records of NWLB, RG 202, Entry 142, NA. Lloyd K. Garrison, memorandum to Regional Chairmen, 2 Oct. 1942, Records of NWLB, RG 202, Entry 142, NA. Eighth RWLB, Resolution, 3 Aug. 1944, Records of the NWLB, RG 202, Entry 412, NA. David B. Harris, letter to George H. Mead, 8 Aug. 1944, Records of the NWLB, RG 202, Entry 412, NA. NWLB, *Termination Report,* 715. William H. Davis, quoted in Richards, *War Labor Boards,* 122. See also *ibid.,* 266–67.
37. David B. Harris, letter to William H. Davis, 11 July 1944, Records of the NWLB, RG 202, Entry 412, NA. Harris, letter to Floyd McGown, 20 April 1944, Records of the NWLB, RG 202, Entry 412, NA. Robert Brumby, letter to the members of the Industry Advisory Council, 12 July 1944, Records of the NWLB, RG 202, Entry 412, NA. Wayne Morse, "Opinion in the Matter of the Cotton Garment Industry and the Amalgamated Clothing Workers of America, Case Numbers 111-1641D, 111-1862D, and 111-1546D," 9 Dec. 1943, Records of the NWLB, RG 202, Entry 412, NA. Louisiana Manufacturers Association, Texas State Manufacturers Association, and Associated Industries of Oklahoma, "Plea to the NWLB for Continuation of Its Established Policies of Maintaining Geographical Wage Differentials and Disposing of Voluntary and Dispute Cases on the Regional Level," 14 Aug. 1944, Records of the NWLB, RG 202, Entry 67, NA. Marshall, *Labor in the South,* 228.
38. M. T. Van Hecke, Report to Executive Director, 11 Dec. 1943, Records of NWLB, RG 202, Entry 67, NA. L. Metcalfe Walling, speech before the Meeting of AFL Representative from 11 Southern States, Atlanta, Ga., 17 Jan. 1943, Records of the NWLB, RG 202, Records of the Fourth RWLB, Entry 201, Federal Archive and Record Center, Region IV, East Point, GA. Walling (of the Labor Dept.) added: "Before I came down here I talked to a number of Southern Senators and Congressmen. They were of one mind that I should carry heartfelt appreciation." See also statement by Mr. Philip Murray, president of the Congress of Industrial Organizations, in Support of the Pepper 65

Cents Minimum Resolution, n.d., Records of the NWLB, RG 202, Entry 31, NA. For the stridency of national labor representatives, see George Meany et al., Petition of AFL Members of Board, 16 March 1943, Records of the NWLB, RG 202, Entry 27, NA; and NWLB, *Termination Report.*

39. Frank P. Graham, address delivered over Station KRLD, Dallas, 8 March 1943, Records of the NWLB, RG 202, Entry 31, NA. Frank P. Graham, telegram to M.T. Van Hecke, 17 March 1943, Records of the NWLB, Records of Fourth Regional War Labor Board, Entry 201, FARC. Graham, telegram to Van Hecke, 17 March 1943. NWLB, *Termination Report,* 197–98. For an example of business concern with General Order 30, see Robert E. Brumby, letter to Russell J. Cooper, 8 Dec. 1944, Records of the NWLB, RG 202, Entry 412, NA.
40. NWLB, *Termination Report, passim.*
41. *Ibid.,* 665.
42. The Atlanta Board decided 14 percent of its voluntary cases, involving 50 percent of affected employees, on the basis of substandards. In the Dallas board's jurisdiction, 27 percent of the cases and 28 percent of covered employees received adjustments under the substandards policy. In every other region, this policy accounted for fewer than 14 percent of the workers and cases under review. The statistics for dispute cases reflected the same pattern. NWLB, *Termination Report,* 202–23, 810, 821. Eighth Regional War Labor Board, Minutes, 6–8 July 1943, Records of the NWLB, RG 202, Entry 239, NA.
43. Ben B. Gossett and Harold McDermott, memorandum to Van Hecke, 18 Feb. 1944, Records of the NWLB, RG 202, Entry 31, NA. Brumby, letter to George H. Mead, 18 March 1944, Records of the NWLB, RG 202, Records of the National Wage Stabilization Board, Entry 412, NA.
44. M. T. Van Hecke, memorandum to NWLB, 8 Dec. 1944. Robert Segrest, memorandum to Fourth Regional War Labor Board, 4 Dec. 1944. Both in Records of the NWLB. RG 202, Entry 67, NA. Frank P. Graham, address delivered over Station KRLD, Dallas, 8 March 1943, Records of the NWLB, RG 202, Entry 31, NA. NWLB, *Termination Report,* 195. NWLB vice chairman George Taylor declared that the Board retained a "separate duty" to upgrade low paying jobs. See George W. Taylor, draft message to Congress, attached to letter to William H. Davis, 23 April 1945, Records of the NWLB, RG 202, Entry 32, NA, and, opinion of George W. Taylor in the matter of thirty Michigan and Northern Wisconsin lumber companies and the International Woodworkers of America, 8 July 1945, Records of the NWLB, RG 202, Entry 31, NA.
45. NWLB, Resolution of 29 July 1943, Records of the NWLB, RG 202, Entry 27, NA. "Weekly Summary of Region VIII Activities," 3 Nov. 1943, Records of the NWLB, RG 202, Entry 239, NA. David B. Harris, letter to Floyd McGown, 20 April 1944, Records of the NWLB, RG 202, Records of the National Wage Stabilization Board, Entry 412, NA. Fourth Regional War Labor Board, Summary, Records of the NWLB, RG 202, Entry 77, NA. Garrison, memorandum to Regional Chairmen, 2 Oct. 1943, Records of NWLB, RG 202, Entry 142, NA. McGown, letter to Theodore Kheel, 24 Jan. 1944, Records of the NWLB, RG 202, Entry 67, NA. NWLB, *Termination Report,* 715.
46. David B. Harris, letter to George H. Mead, 8 Aug. 1944, Records of the NWLB, RG 202, Entry 412, NA.

47. William H. Davis, letter to His Excellency, the ambassador of Mexico, 22 June 1943, Records of the NWLB, RG 202, Series 31, NA. William H. Davis, memorandum to Board, 9 Sept. 1944, Records of the NWLB, RG 202, Series 31, NA. Frank P. Graham, opinion in the matter of the Southport Petroleum Company and the Oil Workers' International Union, 5 June 1943, Graham Papers.
48. NWLB, *Termination Report.* See also Directive Order of NWLB, 24 Feb. 1945, in Case #111-7360-D, Records of the NWLB, RG 202, Series 32, NA.
49. Jonathan Daniels, letter to Frank P. Graham, 7 June 1943, Frank P. Graham Papers.
50. Frank P. Graham, NWLB opinion in the matter of the Aluminum Company of America, American Magnesium Company, and the Aluminum Workers of America, 18 Aug. 1942, Graham Papers. NWLB, *Termination Report,* 197. The assault on the regional differential proceeded in several steps. For a detailed description with complete documentation, see Bruce J. Schulman, "From Cotton Belt to Sunbelt" (doctoral dissertation, Stanford Univ., 1987), 127–29.
51. L. Metcalfe Walling, address before Meeting of the American Federation of Labor Representatives from Eleven Southern States, Atlanta, 17 Jan. 1943, Records of the NWLB, RG 202, Series 201, FARC. Nelson Lichtenstein, *Labor's War at Home* (N.Y.: Cambridge Univ. Press, 1982), 177.
52. Theodore Kheel, memorandum to Regional Chairmen, 2 March 1944, Records of the NWLB, RG 202, Series 27, NA. Frank P. Graham, "The Struggle for Freedom To Organize," *Journal of the National Education Association* 31 (Oct. 1942): 213, in Graham Papers. Graham, "Address over Station KRLD."
53. Richards, *War Labor Board in the Field,* 196. Marshall, *Labor in the South,* 229. NWLB, *Termination Report,* 649, 663. Eighth Regional War Labor Board, "Region VIII Policy on Maintenance of Membership," n.d., Records of the NWLB, RG 202, Records of the National Wage Stabilization Board, Series 412, NA. Eldridge Haynes, letter to William H. Davis, 6 Nov. 1944, Records of the NWLB, RG 202, Series 31, NA. On the southern union environment, consult Flynt, "The New Deal and Southern Labor," 69–82.
54. Marshall, *Labor in the South,* 230. Telephone message for George Taylor, 13 Dec. 1944, Records of the NWLB, RG 202, Series 32, NA. Weekly Summary of Regional Board Activities, 5 Aug. 1944, Records of the NWLB, RG 202, Series 239, NA. Weekly Summary of Regional Board Activities, 28 March 1944, Records of the NWLB, RG 202, Series 239, NA. Union leaders in Region VIII, telegram to William H. Davis, 24 Aug. 1944, Records of the NWLB, RG 202, Series 67, NA.
55. "Questions and Problems," Records of the NWLB, RG 202, Records of the Historical Section, Summary of the Reports of Regional Chairmen 1943–45, Series 77, NA. Marshall, *Labor in the South,* 242–43. NWLB general counsel Jesse Frieden informed Florida officials: "We believe that as a matter of law the Board is no more bound by a limiting provision of a state constitution than of a state statute." Jesse Freiden, memorandum to William H. Davis, 27 Dec. 1944, Records of the NWLB, RG 202, Series 31, NA.
56. Lichtenstein, *Labor's War at Home,* 24–25. Marshall, *Labor in the South,* 225–37. Pratt, *Growth of a Refining Region,* 162–63. Tindall, *Emergence,* 514–16.
57. The Textile Workers Union of America (CIO) reported a 100 percent annual turnover rate in Alabama textile mills. Since the wage-stabilization program limited the wage increases one could expect in any given job and the tight war labor market offered plentiful better-paid employment in other lines of work,

textile workers had little reason to support unions. Textile Workers Union of America, CIO, statement before the War Labor Board in the Matter of the Cotton-Rayon Industry and the Textile Workers Union of America, Atlanta, 27 March 1944, Records of the NWLB, RG 202, Series 201, FARC. See also Marshall, *Labor in the South,* 236–46: Lichtenstein. *Labor's War at Home,* 209–11: and Hall et al., *Like a Family,* 353–54.

58. Marshall, *Labor in the South,* 247–54.
59. *Ibid.,* 226–27, 266–68. See also Flynt, "New Deal and Southern Labor," 67–70.
60. *Ibid.*
61. Garson, *Democratic Party and the Politics of Sectionalism,* 217. Wood, *Southern Capitalism,* 151. Southern concerns figured prominently in President Truman's decision to veto the bill. According to Leon Keyserling, a native South Carolinian and a member of Truman's Council of Economic Advisors, administration officials feared that Taft-Hartley would stymie labor organization in the South. At the same time, they concluded that to conciliate southerners in the Congress was fruitless. See Leon H. Keyserling, Oral History Interview, Washington, D.C., 3–19 May 1971, Harry S Truman Presidential Library, pp. 65–69.
62. Lichtenstein, *Labor's War at Home,* 238–239. Anti-labor sentiment was a potent force in southern elections during the 1940s and 1950s, particularly in Texas, where Pappy O'Daniel rose to prominence on an anti-labor, anti-overtime pay platform. Texas supporters of the New Deal such as Lyndon Johnson and Sam Rayburn were wary of and catered to these sentiments. See C. Dwight Dorough, *Mr. Sam* (N.Y.: Random House, 1962), 334, 539.
63. Franklin D. Roosevelt, message on the State of the Union, 11 Jan. 1944, in Samuel I. Roseman, ed., *Public Papers,* 1944–45 Vol., *Victory and the Threshold of Peace, 1944–45* (N.Y: Harper and Bros., 1950), 32–44. Sosna, "More Important Than the Civil War? The Social Impact of World War II on the South," a paper presented at the Meetings of the Southern Historical Association, Memphis, 5 Nov. 1982, p. 7. Wright, *Old South, New South,* 254. Marshall, *Labor in the South,* 298–300, 350.
64. The journalist was Pare Lorentz of McCall's magazine. See U.S. House, *National Defense Migration,* Part 11, p. 4295.
65. Bureau of Labor Statistics, *Bulletin No. 898,* 16–26. Donald B. Dodd and Wynelle S. Dodd, *Historical Statistics of the South 1790–1970* (University: Univ. of Alabama Press, 1973), 74–76. Jack Temple Kirby, "The Southern Exodus, 1910–1960," *Journal of Southern History* 49 (Nov. 1983): 585–600. Tindall, *Emergence,* 703, 710.
66. Horace C. Hamilton, "Educational Selectivity of Net Migration from the South," *Social Forces* 38 (Oct. 1959): 33–41.
67. U.S. Fair Employment Practices Committee (FEPC), *First Report, July 1943–December 1944* (Washington, 1945), 90; and *Final Report, July 1943–December 1944,* (Washington, 1945), 34–35, 89. "The Negro's War," Fortune 25 (July 1942). U.S.House, *National Defense Migration,* Part 16, Washington, D.C., Hearings, 15–17 July 1941, p. 6382; and Part 17, Washington, D.C., Hearings, 18–21 July 1941, p. 6729. Coclanis and Ford, "The South Carolina Economy Reconstructed and Reconsidered," 102–3. Because so few blacks could find defense employment in the early period of defense production, blacks accounted for a relatively small component of migration to northern and west-

ern defense areas before 1942. In this early period, the destination of black southerners tended to be southern cities or Washington, D.C. See U.S. House, *National Defense Migration,* Part 27, p. 10327; and Part 28, p. 10854.

68. U.S. FEPC, *Final Report,* 12–13, 34–35; and *First Report,* 2–10. "The Negro's War." John M. Blum, *V Was For Victory* (N.Y.: Harcourt Brace Jovanovich, 1976), 185–99. Richard Polenberg, *One Nation Divisible* (N.Y.: Pelican Books, 1980), 34, 75–76. Charles Kesselman, *The Social Politics of FEPC* (Chapel Hill: Univ. of N.C. Press, 1948), 15, *passim.*

69. Joseph J. Persky and John F. Kain, "Migration, Employment and Race in the Deep South," *Southern Economic Journal* 36 (Jan. 1970), 268–76. Other studies of black employment in the South reiterated this conclusion. The FEPC discovered that nonwhites accounted for only 27 percent of the increase in manufacturing employment in a six-state Deep South region (Alabama, Florida, Georgia, Mississippi, South Carolina, and Tennessee), even though blacks accounted for a larger proportion of the population and a larger proportion of those who had been unemployed before the defense boom. See FEPC, *Final Report,* 33.

    In 1955 the Committee of the South of the National Planning Association published several case studies of black employment in the South. A study of Durham, N.C., concluded that the stresses of war and reconversion had little altered the racial employment patterns in the Durham area. Similar studies of the upper South and the textile industry revealed the persistent exclusion of blacks from skilled positions, managerial positions, and the textile industry in general. Consult National Planning Association, *Selected Studies of Negro Employment in the South* (Washington: NPA, 1955), 167, 179, 205–7.

70. For the Census South (the South plus West Virginia, Maryland, Delaware, and the District of Columbia), the number of employed women grew 35 percent from 1940 to 1950, as opposed to 45 percent in the North Central states, 83 percent in the West and 39 percent for the nation as a whole (the Northeast registered a 25 percent gain). See *Statistical Abstract, 1952,* 187. See also D'Ann Campbell, *Women at War with America* (Cambridge: Harvard Univ. Press, 1984); and, Carl N. Degler, *At Odds* (N.Y.: Oxford Univ. Press, 1980), 362–435.

71. Kaun, "Economics of the Minimum Wage," *passim.* Southern Conference for Human Welfare, Report of the Committee on Foreign Trade of the Washington Committee, n.d. [1945], Frank P. Graham Papers.

72. Franklin D. Roosevelt, message on the State of the Union, 6 Jan. 1945, in Rosenman, ed., *The Public Papers,* 1944–45 Vol. Franklin D. Roosevelt, address at Thomas Jefferson Day Dinner, New York, 25 April 1936, in Rosenman, ed., *The Public Papers,* Vol. 5, p. 181. The Federal Reserve Board agreed. It proclaimed in 1945 that "Employment by itself is not sufficient—it must be employment at useful pursuits and at adequate wages." See Board of Governors of the Federal Reserve System, *Postwar Economic Studies, No. 1, Jobs, Production, and Living Standards* (Washington, 1945), 8.

73. "Entrance Rates of Common Laborers in Twenty Industries," *Monthly Labor Review* 45 (Dec. 1937): 1501–5. Bureau of Labor Statistics, *Bulletin No. 898,* 57–95; and "Hourly Entrace Rates Paid to Common Laborers," July 1943.

74. Hearings Before a Subcommittee of the Committee on Labor and Public Welfare, U.S. Senate, *Fair Labor Standards Act Amendments,* 80th Cong., 2nd sess., 19 April-4 May 1948 (Washington, 1948), 559–614. John V. Van Sickle, *Plan-*

ning for the South (Nashville: Vanderbilt Univ. Press, 1947), viii–ix, 97, 189, 193. John V. Van Sickle, "Regional Aspects of the Problem of Full Employment at Fair Wages," *Southern Economic Journal* 13 (July 1946): 36–45.

75. Chester Bowles, quoted in Kaun, "Economics of the Minimum Wage," 9. George Taylor, letter to William H. Davis, 23 April 1945, with accompanying draft message to Congress on amending the FLSA, Records of the NWLB, RG 202, Entry 32, NA. Statement of Secretary of Labor Lewis B. Schwellenbach Before the House Committee on Labor on Proposed Amendments to the FLSA, 24 Oct. 1945, Records of Secretary of Labor Lewis B. Schwellenbach, General Subject File. Schwellenbach, letter to Robert Taft, 6 April 1948, Records of Secretary of Labor Lewis B. Schwellenbach, General Subject File. Testimony of Secretary of Labor Lewis B. Schwellenbach before Subcommittee on Wages and Hours, House of Representatives, 17 Nov. 1947, Records of Secretary of Labor Lewis B. Schwellenbach, General Subject File of David A. Morse, Undersecretary of Labor. All in Records of the Dept. of Labor, RG 174, NA. The administration's position was adumbrated by Richard Lester, a Princeton economist associated with the NWLB. See Richard Lester, "Effectiveness of Factory Labor: North-South Comparisons," *Journal of Political Economy* 54 (Feb. 1946): 60–75; and, Lester, "Southern Wage Differentials," *Southern Economic Journal* 13 (April 1947): 386–94.
76. Robert R. Nathan, memorandum for Fred Vinson, 18 June 1945, Records of the Office of War Mobilization and Reconversion, RG 250, NA. Memorandum for the Board, 28 Aug. 1946, Records of the NWLB, RG 202, Records of the National Wage Stabilization Board, Series 400, NA.
77. Claude Pepper, Draft Report of Senate Committee on Education and Labor, sent to Lewis B. Schwellenbach, 25 Feb. 1946, Records of the Department of Labor, RG 174, Records of Secretary of Labor Lewis B. Schwellenbach, General Subejct File, NA.
78. Theodore Winslow, memorandum to Secretary of Labor Maurice J. Tobin, 29 Sept. 1949, Records of the Dept. of Labor, RG 174, Records of Secretary of Labor Maurice J. Tobin, General Subject File, NA.
79. Subcommittee on Labor of the Committee on Labor and Public Welfare, U.S. Senate, *Amending the Fair Labor Standards Act of 1938,* 84th Cong., 1st sess. (Washington, 1955), 349, 684–713, 755–813, 948–49, 176, 454, 1451. Department of Labor, Notes on Hearings, Records of the Dept. of Labor, RG 174, Records of Secretary James P. Mitchell, No. 13, 23 June 1955; No. 14, 23 June 1955; No. 20, 30 June 1955, NA. Kaun, "Economics of the Minimum Wage," 218.
80. U.S. Senate, *Amending the FLSA,* 113–14, 481. For the Southern Garment Manufacturers Association endorsement, see William McComb, memorandum to James P. Mitchell, 23 July 1954, Records of the Dept. of Labor, RG 174, Records of Secretary James P. Mitchell, NA. See also Dept. of Labor, Notes on Hearings, No. 9, 15 June 1955; and Kaun, "Economics of the Minimum Wage," 217–20. The state of North Carolina even ran an advertisement in the late 1940s declaring "ONLY FAIR DEALERS WANTED" (emphasis in original). North Carolina "wants no one to seek location within its borders expecting long hours of work at low pay. Sweat shop operators are unwelcome." Admittedly, few southern boosters were willing to stop using cheap labor as a selling point in 1950, and this advertisement was a highly unusual one. Still, it points to the waning of the ferocious opposition to higher wages of the 1930s

and early 1940s. The advertisement is reprinted in Albert Lepawsky, *State Planning and Economic Development in the South* (Kingsport: National Planning Association, 1949).

81. *Ibid.* In March 1955, 43 interests owned 507 plants employing 345,500 persons out of a total 656,000 production workers in the textile industry. See U.S. Senate, *Amending the FSLA,* 481. Before the FLSA, observers referred to the South as sprouting "mills like Mushrooms." See Hodges, *New Deal Labor Policy and the Southern Cotton Textile Industry,* 9. Consult also Kaun, "Economics of the Minimum Wage," 156–57, *passim.*
82. See Chapters 4, 6, and 7 below. On skill upgrading in the four states of the Southwest (Texas, Arkansas, Oklahoma, and Louisiana) between 1940 and 1960, see Robert F. Smith, "Employment and Economic Growth: Southwest," *Monthly Labor Review* 91 (March 1968): 27. On the lesser regional differentials for skilled labor, see H.M. Douty, "Wage Differentials: Forces and Counterforces," *Monthly Labor Review* 91 (March 1968): 75–77. On the effect of minimum-wage legislation on returns to skill, see Ronald J. Krumm, *The Impact of the Minimum Wage on Regional Labor Markets* (Washington: American Enterprise Institute, 1981), 3, 28–33, 58.
83. Alto B. Cervin, letter to A. Colman Barrett, 3 Oct. 1945, Records of the NWLB, RG 202, Records of the National Wage Stabilization Board, Series 412, NA. W. Willard Wirtz, letter to Chairman, Region IV, 13 Dec. 1946, Records of the NWLB, RG 202, Records of the National Wage Stabilization Board, Series 397, NA.
84. Donald M. Nelson, "The South's Economic Opportunity," *American Mercury* 59 (Oct. 1944): 422–27.

Chapter 4

1. Thomas Wallner, quoted in Bruce Catton, *The War Lords of Washington* (N.Y.: Harcourt, Brace, 1948), 180. John Dos Passos, *State of the Nation* (Boston: Houghton Mifflin, 1944), 67.
2. For reports on the relative success of southern manufacturing during the Great Depression, see "Southern Industrial Mobilization," *Manufacturers' Record* 109 (July 1940): 25; "The Industrial South," *Fortune* 18 (Nov. 1938): 48, 118; and Melvin H. Baker, "Buying Power in the South Has Increased Rapidly," *Manufacturers' Record* 109 (Dec. 1940): 24.
3. Roosevelt, address at Little Rock, Ark. 10 June 1936, in Rosenman, ed., *Public Papers,* Vol. 5, pp. 199–200. "An Urgent Call to the People of the South," Draft Statement, Minutes of the Southern Conference for Human Welfare, 31 Oct. 1941, Graham Papers. Odum, *Southern Regions,* 22–23, 55. For Rupert B. Vance, Odum's most devoted student, the future of the South rested on the acquisition of the "Balance Wheel"—industry. See Vance, *Wanted: The South's Future for the Nation* (Atlanta: Southern Regional Council, 1946), 6–9, 16. Examples of contemporary works stressing a theory of economic growth stages tied to the expansion of manufacturing are: Colon Clark, *The Conditions of Economic Progress* (London: Macmillan, 1940); August Losch, "The Nature of Economic Regions," *Southern Economic Journal* 5 (July 1938): 71–80.
4. Odum, *Southern Regions,* 57. Joint Committee on the Economic Report, *The Impact of Federal Policies on the Economy of the South,* 81st Cong., 1st sess. (Washington, 1949), 6, 14. U.S. Dept. of Commerce, Bureau of the Census, *Sixteenth*

Census of the United States, 1940: Manufactures 1939, Vol. 1 (Washington, 1942), 44. Frederick L. Deming and Weldon Stein, *Disposal of Southern War Plants;* National Planning Association, Committee of the South, Report No. 2 (Washington, NPA, 1949), 7–8.

5. National Resource Planning Board, Southeastern Regional Planning Council, "Regional Resource Development Plan: Report for 1942, Region 3, Atlanta, Georgia," Dec. 1942, p. 2, in Graham Papers. Southern Governors Conference, "Made in the South Carries a Powerful Appeal to Southern Consumers" (advertisement), *Manufacturers' Record* 108 (Sept. 1939): 76. Deming and Stein, *Disposal of Southern War Plants*, 6. NEC, *Report on Economic Conditions*, 8. Further data on the composition of prewar southern industry can be found in: Glenn McLaughlin and Stefan Robock, *Why Industry Moves South*, National Planning Association, Committee of the South, Report No. 3 (Washington: NPA, 1949), 15; Bureau of Labor Statistics, *Bulletin No. 898*, 29–31; and Hoover and Ratchford, *Economic Resources*, 126–27.
6. Transcript of Proceedings of Southern Governors Conference Held at Monteleone Hotel, New Orleans, 15–17 March 1941, pp. 70–74, Graham Papers. James C. Worthy, *Shaping an American Institution: Robert E. Wood and Sears, Roebuck* (Urbana: Univ. of Illinois Press, 1984), 179–80, 195–96. "The Industrial South," 120. Rupert B. Vance, *All These People* (Chapel Hill: Univ. of N.C. Press, 1945), 251. For a discussion of the merchandising department at Sears, see Gordon L. Weil, *Sears, Roebuck, U.S.A.* (N.Y.: Stein and Day, 1977), 131, *passim*. Donald Nelson, the chairman of the Textile Industry Committeee and later chairman of the War Production Board entered government service after a career in the Sears merchandising department. See Donald Nelson, *Arsenal of Democracy* (N.Y.: Harcourt, Brace, 1946), 62.
7. *The Report on Economic Conditions* asserted the necessity of a federal role in southern economic reconstruction (p. 4). For the opinions of liberal southerners affiliated with the federal government, consult Southern Conference on Interstate Problems, Résumé and Resolutions, 25–27 Jan. 1940, p. 6, Graham Papers. See also Frank P. Graham, remarks before the Southern Governors Conference, Mobile Bay off Dauphin Island, Ala., 15–17 Sept. 1941, p. 12, Graham Papers. On philanthropic efforts in the South and their ties to southern liberals, see Dewey W. Grantham, *The Regional Imagination*, 163; and, Wilma Dykeman and James Stokely, *Seeds of Southern Change* (N.Y.: W. W. Norton, 1962), 162–63. On the political implications of federal intervention in the southern economy, see Chapter 5 below.
8. Gordon R. Clapp, letter to John P. Ferris, 30 July 1942, Records of the TVA, RG 142, Records of the TVA Commerce Dept. TVA Commerce Dept., "Relation of the TVA to Industrial Development," 17 Feb. 1941, Records of the TVA, RG 142, Records of the TVA Commerce Dept. TVA Commerce Department, "Prospective Sites in the Tennessee Valley Area for an Army War Production Plant," July 1942, Records of the TVA, RG 142, Records of the TVA Commerce Dept. John B. Blandford, "Adaptation of Regional Activities to the National Emergency," 21 Sept. 1939, Records of the TVA, RG 142, Records of the TVA Board Curtis-Morgan-Morgan 1933–57. TVA Commerce Department, "Electricity as a Catalyst to Stimulate Economic Development," 6 Jan. 1943, Records of the TVA, RG 142, Records of the TVA Commerce Dept. TVA, "Opportunities for Postwar Development of Tennessee Valley Resources," May 1944, Records of the TVA, RG 142, Records of the TVA

Board Curtis-Morgan-Morgan 1933–57. All in FARC. Lilienthal, *Journals,* Vol. 1, p. 104. Erwin C. Hargrove, "The Task of Leadership: The Board Chairmen," in Hargrove and Conkin, *TVA,* 98. For Morgan's version of the ouster, see Arthur E. Morgan, *The Making of the TVA* (Buffalo: Prometheus Books, 1974).

9. Lilienthal, *Journals,* Vol. 1, pp. 166–68, 359–61, 460. U.S. House, *National Defense Migration,* Part 32, Huntsville Hearings, 7–8 May 1942, pp. 12017–27.
10. Clapp, letter to Ferris, 30 July 1942. John Ferris, "Industrial Development in the Southeast," address before the Spring Meeting of the American Society of Engineers, Chattanooga, 3 April 1946, Records of the TVA, RG 142, Records of the TVA Commerce Dept., FARC, p. 13. U.S. House, *National Defense Migration,* Part 32, pp. 12014–15. William Bruce Wheeler and Michael J. McDonald, "The 'New Mission' and the Tellico Project, 1945–70," in Hargrove and Conkin, *TVA,* 169. Vernon Ruttan, "The TVA and Regional Development," in Hargrove and Conkin, *TVA,* 151–56.
11. Lilienthal, *Journals,* Vol. 1, pp. 290–92, 302–4. U.S. House, *National Defense Migration,* Part 32, p. 12023. *Our Own Worst Enemy: The Impact of Military Production on the Upper South* (New Market, Tenn.: Highlander Research an Education Center, 1983), 102. Wright, *Old South, New South,* 172–76.
12. Ferris, "Industrial Development in the Southeast," 13. Vance, *Wanted: The South's Future for the Nation,* 6–9, 16. During the 1930s, Colin Clark and Allan G.B. Fisher developed a theory of economic growth stages tied to the expansion of manufacturing. See Clark, *The Conditions of Economic Progress;* and Harvey S. Perloff et al., *Regions, Resources, and Economic Growth* (Baltimore: Johns Hopkins Univ. Press, 1960), 58–60. August Losch applied that model to southern economic development in "The Nature of Economic Regions," 71–80.
13. Ferris, "Industrial Development in the Southeast," 13. David E. Lilienthal, "How One Southern Region Has Developed Its Natural Resources: A Report on TVA," address before the Rotary Club of New Orleans, 18 Sept. 1946, Records of the TVA, RG 142, Records of the TVA Board Curtis-Morgan-Morgan 1933–57. W.K. McPherson, memorandum to Howard K. Menhinick, 20 Dec. 1948, *Ibid.* Gordon R. Clapp, letter to Morris Cooke, 20 Oct. 1950, *ibid.* TVA Commerce Dept., "War Industries in the Tennessee Valley Region," 30 June 1943, Records of the TVA, RG 142, Records of the TVA Commerce Dept. All in FARC. Lilienthal, TVA, 36. For example, the 122-county TVA region experienced greater increases than the Southeast and the United States as a whole in total bank deposits 1935–49, retail sales 1939–48, employment in manufacturing establishments 1939–48, and per capita income payments 1929–48. However, the Southeast as a whole exceeded the TVA region in percentage increase in value added by manufacturing 1939–47. See the charts accompanying Clapp, letter to Cooke, 20 Oct. 1950.
14. H.C. Nixon, "The South After the War," *Virginia Quarterly Review* 20 (Summer 1944): 323. David E. Lilienthal, *TVA: Democracy on the March* (Westport, Conn.: Greenwood, 1944, 1953), 41.
15. Joint Committee on the Economic Report, *Impact of Federal Policies on the South,* 15. Army and Navy Munitions Board, *Geographical Directory of Industrial Allocations 1940* (Washington, 1940), Records of the WPB, RG 179, No. 221.41X, NA. Walter J. Matherly, "The Postwar Economy of the South," address, Atlanta, June 1943, attached to letter to Frank P. Graham, 19 July

1943, Frank P. Graham Papers. Gov. Prentice Cooper, "Defense Industries in Tennessee," address over WSM, Nashville, 3 Nov. 1941, Records of Governor Prentice Cooper, 1939–45, Governor's Office Records, Tennessee State Library and Archives, Nashville, Box 16. U.S. House, *National Defense Migration,* Part 27, pp. 10258–66.

The early neglect of the South, however exaggerated, drew the ire of many of the region's champions. Resolution of the Southern Governors Conference, 15 March 1941, in Frank P. Graham Papers, and in Records of the WPB, RG 179, No. 261.3, NA. Chester C. Davis, testimony before the Southern Governors Conference, March 1941, p. 30, Graham Papers. Emerson Ross, memorandum to Stacy May, 10 April 1941, Records of the WPB, RG 179, No. 261.3, NA. Stacy May, letter to General Edwin Watson, 8 April 1941, Records of the WPB, RG 179, No. 261.3, NA. May, letter to Watson, 9 April 1941, *ibid.*

Examples of behind-the-scenes efforts to bring war plants to the South include: Burnet Maybank, letter to Donald Nelson, 17 Dec. 1942; telephone memorandum, 17 Dec. 1942; Nelson, letter to Maybank, 22 Dec. 1942; and Houlder Hudgins, letter to Maybank, 26 Dec. 1942, all in Records of the WPB, RG 179, No. 241C, NA. See also Estes Kefauver, letter to Philip D. Reed, 12 March, Records of the WPB, RG 179, No. 574.2C, NA., and the heated controversy over the cancellation of the proposed Rusk, Texas, pig-iron plant proposal, Nov. 1942–Jan. 1943, Records of the WPB, RG 179, No. 511.424, NA.

By September 1941 the South had made up the stagger on its slow start. By that date the region had received 21.4 percent of the dollar value of defense-facilities contracts. That figure, while less than the region's share of population, exceeded its share of value added by manufactures or wage earners in manufacturing in 1939. See U.S. House, *National Defense Migration,* Part 27, pp. 10258–66.

16. W. N. Mitchell, "Field Organization of the War Production Board," 16 Nov. 1942, Records of the WPB, RG 179, No. 077.1, NA., pp. 1, 46, On the procurement agencies, see John M. Blum, *V Was for Victory* (N.Y.: Harcourt Brace Jovanovich, 1976), 120–22; Nelson, *Arsenal,* 368–90; Herman M. Somers, *Presidential Agency* (Cambridge: Harvard Univ. Press, 1950), 24–29; Byrnes, *All in One Lifetime,* 171–73, *passim;* and R. Elberton Smith, *The Army and Economic Mobilization,* in U.S. Dept. of the Army, *The United States Army in World War II* (Washington, 1959), 235–38, 415. Military control of the WPB particularly worried David Lilienthal, who blamed WPB chairman Donald Nelson for the state of affairs. See Lilienthal, *Journals,* Vol. 1, p. 504.
17. Deming and Stein, *Disposal of Southern War Plants,* 14, 17. See also *Manufacturers' Record,* Vols. 110–14, especially the monthly tabulations by Samuel Lauver. The figure on construction expenditures is computed from Lauver, "Southern Construction Increases," *Manufacturers' Record* 114 (Oct. 1945); 48. See also WPB, Programs and Statistics Bureau, Facilities Branch, "War Manufacturing Facilities Authorized Through December 1944," Records of the WPB, RG 179, No. 221.3R, NA.
18. Deming and Stein, *Disposal of Southern War Plants,* 17. Glenn McLaughlin, "Notes on the Effects of the War on the South," 30 Oct. 1943, attached to Stacy May, letter to Jonathan Daniels, 30 Oct. 1943, Records of the WPB, RG 179, No. 260.21, NA. Serviceman quoted in John Temple Graves, *The Fighting South* (N.Y.: G. P. Putnam's Sons, 1943), 104. Of the fifty largest military installations in the continental United States during World War II, in terms of

costs in dollars, twenty-five were located in the South. See Smith, *The Army and Economic Mobilization.* The term "military-payroll complex" was coined, to the best of the author's knowledge, by Morton Sosna in a personal conversation with the author during the summer of 1988.

19. U.S. House, *Interstate Migration,* 76th Cong., 3rd sess., Part 10, Washington, D.C., Hearings, 11 Dec. 1940-26 Feb. 1941, p. 3877. U.S. House, *National Defense Migration,* Part 11, pp. 4574, 4737–44; Part 17, pp. 6853, 6861–62; Part 32, pp. 12118–19.
20. *Ibid.,* Part 32, p. 12072. For the relocation activities of the FSA in general, consult U.S. House, *National Defense Migration,* Part 28, p. 10871, and Part 32, pp. 12059, 12071–72.
21. Southern Conference for Human Welfare, Report of the Committee on Foreign Trade of the Washington Committee, p. 3, Graham Papers. Matherly, "Postwar Economy of the South," 8. W. Harry Vaughan, "Southern Industry at War: Weathervaned for Peace?," address before the Chattanooga Engineers Club, 24 May 1943, Records of the TVA, RG 142, Records of the TVA Board Curtis-Morgan-Morgan 1933–57, FARC. Frederic C. Lane, *Ships for Victory* (Baltimore: Johns Hopkins Univ. Press, 1951), 34–35, 142.
22. Holt Ross, "Higgins Ships," reprinted in Lane, *Ships for Victory,* 187–88. Blum, *V Was for Victory,* 110–11. Sosna, "More Important Than the Civil War?," 9–11.
23. Donald Nelson, letter to Franklin D. Roosevelt, 21 Oct. 1942, Records of the WPB, RG 179, No. 314.4342, NA. Roosevelt, Memorandum to Nelson, 22 Oct. 1942, Records of WPB, RG 179, No. 314.4342, NA. Blum, *V Was for Victory,* 110–11. Sosna, "More Important Than the Civil War?," 9–11. Lane, *Ships for Victory,* 184–94. Eighth Regional War Labor Board, "Summary of Regional Board Activities," 24 June 1944, Records of the NWLB, RG 202, Records of Region VIII, Series 239, p. 6.
24. WPB, Programs and Statistics Bureau, Facilities Branch, Report, 1 June 1945, pp. 5–15, Records of the WPB, RG 179, No. 221.3R, NA. W.A. Hauck, *Steel Expansion for War,* 14 June 1945, Records of the WPB, RG 179, No. 512.422R, NA. Eighth Regional War Labor Board, "Weekly Summary of Eighth Regional Board Activities," 7 Aug. 1943, Records of the NWLB, Records of Region VIII, Series 239, NA. Jones, *Fifty Billion Dollars,* 402. Nelson, *Arsenal,* 297. The war years witnessed the expansion and diversification of Texas steel-making, especially the move from primary to fabricated metal products, although the war also expanded production of primary materials. The American Iron and Bridge Division of U.S. Steel (1940), Chicago Bridge and Iron (1941), American Smelting and Refining (1943), and Bethlehem Steel (1947) all arrived on the Gulf Coast after 1940. In primary metals, Armco Steel, Alcoa, and Dow Chemical's Magnesium Recovery Plant were products of the war years. See Pratt, *Growth of a Refining Region,* 107, 141–42.
25. Wood, quoted in Marvin Hurley, *Decisive Years for Houston* (Houston Chamber of Commerce, 1966), 126–27. Deming and Stein, *Disposal of Southern War Plants,* 13. WPB, Programs and Statistics Bureau, Facilities Branch, Report, 1 June 1945, pp. 5–10. Jonathan W. Daniels, letter to William D. Carmichael, 18 March 1944, Jonathan W. Daniels Papers. Pratt, *Growth of a Refining Region,* 107, 141–42.

    The construction of the first pipelines from the Gulf oil fields to the East Coast illustrates one of the ways federal investment enhanced the Southwest's locational advantages and accelerated development on the Gulf Coast. See Pratt, *Growth of a Refining Region,* 94–95.

26. *Ibid.* On the control of North Carolina politics by textile and tobacco interests, see V. O. Key, *Southern Politics* (N.Y.: Vintage Books, 1949), 211–15. The textile industry's continued opposition to the reorganization of its labor force by federal policy is discussed in Chapter 3 above. North Carolina led the South toward modern, high technology industry in the 1950s and 1960s. But unlike other southern states, which used defense research to finance their high technology facilities, North Carolina's Research Triangle Park received little direct stimulus from the national defense establishment. See Chapter 6 below.
27. Forty-five southern cities, twenty-one of them in the Southwest, were among the 102 metropolitan areas in which the ratio of wartime expansion to prewar employment exceeded the national average of 2.6 to 1; that included phenomenal expansions like Baton Rouge's 31:1, and Houston's 23:1. WPB, Programs and Statistics Bureau, Facilities Branch, Report, 1 June 1945, pp. 28–30, 51–53. "Defense Boom in Dixie," *Time* 37 (17 Feb. 1941): 75.

    The Southeastern Regional Planning Commission of the National Resource Planning Board (NRPB) commented on the greater marginal impact of federal dollars in the South during the course of the conflict. Noting the small absolute increases in southern industrial capacity in comparison with the Northeast and North Central regions, the commission asserted that "in light of the distinctive long-range needs of the Southeast, their importance is considerably magnified. No previous wave of industrial growth and promotional effort in the region was brought an increase in industrial capacity as large as that which has occurred with the defense program." See NRPB, "Regional Resources Development Plan," 7.
28. U.S. House, *Insterstate Migration of Destitute Citizens,* Part 2, p. 756. Lilienthal, *Journals,* Vol. 1, p. 461. For efforts to dispel this prejudice against southern labor, see U.S. House, *National Defense Migration,* Part 32, p. 12019. Organized labor also opposed the building of war industry in the South, preferring the rehabilitation of old factories in unionized areas to the construction of new ones in the anti-labor South. For the CIO view, consult U.S. House, *Insterstate Migration of Destitute Citizens,* Part 8, p. 3419; for the AFL, Part 9, p. 3693.
29. *The Southern Patriot* 3 (April 1945). John Kenneth Galbraith, *A View from the Stands* (Boston: Houghton Mifflin, 1986), 389–90. WPB chairman Nelson worked for Sears, Roebuck and served as chairman of the Textile Industry Committee. Taylor, vice chairman and eventually chairman of the NWLB, served on the Hosiery Industry Committee. On FDR's fondness for the South, see Freidel, *FDR and the South.*
30. Lane, *Ships for Victory,* 46–50, 142, 184. Bureau of Programs and Statistics, Facilities and Contracts Branch, WPB, "Characteristics of Manufacturing Facilities Authorized July 1940–May 1944," 22 Jan. 1945, p. 19, Records of the WPB, RG 179, No. 220.1R, NA. WPB, "Proposed Statement of Policy on the Location of New Defense Facilities," n.d. (placed in chronological file at 18 July 1945, but appears to have been prepared in July 1940), Records of the WPB, RG 179, No. 223R, NA. Chester C. Davis, "Statement on Policy on Selection of Sites for New Defense Industry," 5 Aug. 1940, attached to Davis, letter to Donald Nelson, 8 Aug. 1940, Records of the WPB, RG 179, No. 223.1, NA.
31. *Ibid.* See also U.S. House, *Interstate Migration of Destitute Citizens,* Part 8, pp.

3213–14; Part 10, p. 3859; and, U.S. House, *National Defense Migration,* Part 16, pp. 6546–66. For a summary of priorities in procurement, see Smith, *Army and Economic Mobilization,* 261–63. The Housing Committee of the Twentieth Century Fund, an influential private policy research and philanthropic organization echoed this call for regional dispersion of defense plants. This committee counted Frank P. Graham among its members. See Twentieth Century Fund, *Housing for Defense* (N.Y.: Twentieth Century Fund, 1940), 127–28.

32. Glenn McLaughlin, "Notes on the Effects of the War on the South," 30 Oct. 1943, attached to Stacy May, letter to Jonathan Daniels, 30 Oct. 1943, Records of the WPB, RG 179, No. 260.21, NA. Deming and Stein, *Disposal of Southern War Plants,* 16. "Defense Boom in Dixie," 75. Joint Committee on the Economic Report, *Impact of Federal Policies on the South,* 15. Wright, *Old South, New South,* 241. Even though employment in war plants fell more precipitously in the South after the cessation of hostilities than in other regions, non-war plant payrolls stood up better in the South. See Joint Committee on the Economic Report, *Impact of Federal Policies on the South,* 15.
33. Dos Passos, *State of the Nation,* 67, 73.
34. On the replacement of agriculture by manufacturing as the leading component of income payments, see Hoover and Ratchford, *Economic Resources,* 60.
35. Dos Passos, *State of the Nation,* 69, 89, 93. Wright, *Old South, New South,* 241–49. Hoover and Ratchford, *Economic Resources,* 90, 108. Bureau of Planning and Statistics, Planning Division, WPB, "Regional and Industry Impacts of War Production," 7 Feb. 1944, p. 61., Records of the WPB, RG 179, No. 223R, NA. Pete Daniel has investigated this "Southern Enclosure," noting both the ameliorating effect of the war in siphoning displaced tenants into war jobs and the failure of federal policy to keep tenants on the land. See Daniel, *Breaking the Land,* 168; and Daniel, "Rural South," 243. On the refusal of rural southerners to accept deferments, see U.S. House, *National Defense Migration,* Part 32, p. 12009.
36. Richard Day, "The Economics of Technological Change and the Demise of the Sharecropper," *American Economic Review* 57 (June 1967): 427–49. Gilbert Fite, "Recent Progress in the Mechanization of Cotton Production in the United States," *Agricultural History* 24 (1950): 19–28. Fite, "Mechanization of Cotton Production Since World War II," *Agricultural History* 54 (1980): 190–207. For a discussion of the various technologies involved in the gradual mechanization of cotton cultivation, consult Charles R. Sayre, "Cotton Mechanization Since World War II," *Agricultural History* 53 (Jan. 1979): 105–24. On prisoner labor in the rural South, see Morton Sosna, "Stalag Dixie," unpublished paper in the author's possession.
37. Hoover and Ratchford, *Economic Resources,* 95–96, 103. Daniel, *Breaking the Land,* 244. Harry B. Fornari, "The Big Change: Cotton to Soybeans," *Agricultural History* 53 (Jan. 1979): 245. The eight cotton states were Georgia, Alabama, South Carolina, North Carolina, Tennessee, Mississippi, Arkansas, and Louisiana. Joint Committee on the Economic Report, *Impact of Federal Policies on the South,* 44. Tindall, *Emergence,* 704. Pete Daniel, *Standing at the Crossroads* (N.Y.: Hill and Wang, 1986), 139. Fite, "Southern Agriculture Since the Civil War," 19–20. Wright, *Old South, New South,* 241–49.

    Even though southern farm wages tripled during World War II, John Dos Passos detected an unwillingness on the part of landlords to pay farm workers on a par with industrial workers, even when they could afford to do so. "What worries the farmer," one southern landowner told Dos Passos, "is if his hired labor gets used to making industrial wages they won't want to work for less when this is over." Dos Passos, *State of the Nation,* 78. Unlike cotton, tobacco production increased during the war, despite the diminution of the rural labor force and the nonessential character of the crop, a testimony to the importance of cigarettes on the battlefields. See Daniel, *Breaking the Land,* 256–57. For a general discussion of wartime agricultural change, consult Board of Governors of the Federal Reserve System, *Postwar Economic Studies, No. 2, Agricultural Adjustment and Income* (Washington, 1945). For accounts of agricultural diversification in the early years of the war, consult U.S. House, *National Defense Migration,* Part 28, pp. 10791–92, 10865–66, and Part 32, p. 12002.

38. Board of Governors of the Federal Reserve System, *Agricultural Adjustment,* 51, 66.
39. "This is the Story of Tennessee" (advertisement), *Manufacturers' Record* 114 (Jan. 1945): 81. Arthur [Handley?], letter to Jonathan Daniels, 25 Jan. 1946, Daniels Papers.
40. Hoover and Ratchford, *Economic Resources,* 57, 129, 138. Joint Committee on the Economic Report, *Impact of Federal Policies on the South,* 16–19. Deming and Stein, *Disposal of Southern War Plants,* 19, 22–23. Victor R. Fuchs, *Changes in the Location of Manufacturing in the United States Since 1929* (New Haven: Yale Univ. Press, 1962), 5–6, 25. Thomas H. Naylor and James Clotfelter, *Strategies for Change in the South* (Chapel Hill: Univ. of N.C. Press, 1975), 10.
41. Dos Passos, *State of the Nation,* 124. Southern Conference for Human Welfare, Report of the Committee on Foreign Trade of the Washington Committee, p. 4. Arthur [Handley?], letter to Jonathan Daniels, 25 Jan. 1946, Jonathan W. Daniels Papers. "Southern Militarism," *Southern Exposure* 1 (Spring 1973): 61.
42. Arthur [Handley?], letter to Jonathan Daniels, 25 Jan. 1946, Daniels Papers. Tindall, *Emergence,* 699–700. Perloff et al., *Regions, Resources, and Economic Growth,* 455. "Economic Center of Gravity Shifts South and West," *Manufacturers' Record* 113 (Jan. 1944): 27. Bureau of Planning and Statistics, Planning Division, WPB, "Regional and Industry Impacts of War Production," 21. The expanding steel center in Texas keyed expansion in metal-products employment throughout the Southwest 1939–54. See Perloff et al., *Regions, Resources, and Economic Growth,* 455. The rising economic and political prominence of Houston was trumpeted by Secretary of the Treasury John W. Snyder in 1947. See Snyder, address before the Houston Chamber of Commerce, 6 Dec. 1947, Jesse H. Jones Papers, Manuscript Division, Library of Congress, Washington, D.C., Box 24.
43. Nixon, "South After the War," 321.
44. Southern Conference for Human Welfare, Report of the Committee on Foreign Trade of the Washington Committee, p. 3. McLaughlin and Robock, *Why Industry Moves South,* xiii. Bureau of Planning and Statistics, Planning Division, War WPB, "Regional and Industry Impacts of War Production," ii. Deming and Stein, *Disposal of Southern Plants,* 41.
45. Atlanta, of course, had been a leading regional industrial center since the 1880s. George W. Healy, Jr., letter to Jonathan Daniels, 31 Jan. 1946, Daniels Papers. McLaughlin and Robock, *Why Industry Moves South.* Jerry Jeran, "South Offers Greatest Economic Future of Any Section of the United States," *Manufacturers' Record* 112 (Nov. 1943): 32. James C. Cobb, *Industrialization and Southern Society 1877–1984* (Lexington: Univ. Press of Kentucky,

1984), 52, 65. Robert A. Sigafoos describes the rapid process of reconversion in Memphis. See Sigafoos, *Cotton Row to Beale Street* (Memphis: Memphis State Univ. Press, 1979), 210–11.

46. McLaughlin and Robock, *Why Industry Moves South,* 26–46, 60, 79.
47. *Ibid.* Cobb, *Industrialization and Southern Society,* 52–43.
48. At the same time, Westinghouse built a heavy electrical equipment plant in Houston to serve growing Southwest industry. Market-oriented lines were prominent even among the low-paying entrants into the southern economy; the shoe industry, for instance, fanned into the South from St. Louis. Arkansas, which housed no shoe plants before 1945, had eleven by the end of 1948. *Ibid.* See also National Planning Association, *Selected Studies of Negro Employment,* 3, 125–26.
49. McLaughlin and Robock, *Why Industry Moves South,* 55–75, 119. Between 1945 and 1960, employment in the low wage, slow growth group declined nationally while rising 15 percent in the South. The rapid growth industries, however, expanded twice as fast in the South as in the United States as a whole. Thus, while the absolute size of the low wage, labor-oriented sector increased, the relative dominance of those lines over the southern economy was declining. See Fuchs, *Changes in Location* 154–59; Alfred J. Watkins and David C. Perry, "Regional Change and the Impact of Uneven Urban Growth," in Watkins and Perry, eds., *The Rise of the Sunbelt Cities* (Beverly Hills: Sage Publications, 1977), 42; and, Kaun, *Economics of the Minimum Wage,* 126–35.
50. McLaughlin and Robock, *Why Industry Moves South,* 7–9. Joint Committee on the Economic Report, *Impact of Federal Policies on the South,* 28–29. Hoover and Ratchford, *Economic Resources,* 73, 77. John Temple Graves, letter (and proposed column) to Jonathan Daniels, 5 Sept. 1945, Daniels Papers.
51. The defenders of southern labor believed that uplifting southern labor—raising wages and skill levels—and uplifting southern laborers amounted to the same thing. They did not realize that different demographic groups would suffer the dislocations and reap the benefits of federally sponsored economic development after 1944. See Chapters 6 and 7 below. Organized labor, for example, demanded "that the South be completely integrated into the national economy and participate fully in the nation's post-war efforts to expand our economy." See Amalgamated Local 453, UAW-CIO, Resolution on the South, n.d. [1944], Daniels Papers.
52. Ellis G. Arnall, *The Shore Dimly Seen* (N.Y.: J.B. Lippincott, 1946), 20. In the 1950s, constituent letters to southern members of Congress, like Texas senator Tom Connally, reveal an implicit assumption that the federal government has an obligation to locate defense plants in areas with little industry. See, for example, Thomas G. Pollard, letter to Connally, 15 Feb. 1952, Thomas T. Connally Papers, Manuscript Division, Library of Congress, Washington, D.C. On the Chaunce-Vought move, see Barry Bluestone and Bennett Harrison, *The Deindustrialization of America* (N.Y.: Basic Books, 1982), 25.
53. Deming and Stein, *Disposal of Southern War Plants,* 1. Harry S Truman, "Synthetic Rubber: Recommendations of the President," transmitted to Congress, Jan. 1950, Records of the Dept. of Labor, RG 174, Records of Secretary of Labor Maurice J. Tobin, General Subject File, NA.
54. Leon Keyserling, Oral History Interview, Truman Library, pp. 46, 87–92, 118, 152–57. Hamilton Q. Dearborn and Leon Keyserling, memorandum to the Executive Secretary of the National Security Council, 8 May 1950, *Foreign*

Relations of the United States, 1950, Vol. 1 (Washington, 1972), 307–11. Alonzo Hamby, *Beyond the New Deal* (N.Y.: Columbia Univ. Press, 1973), 298–300, 415, 453. For the early adumbration of military Keynesianism, see Council of Economic Advisors, *Fifth Annual Report to the President* (Washington, 1950).

55. Howard W. Odum, letter to Jonathan W. Daniels, 25 Jan. 1946, Daniels Papers. R.E. Bolton, *Defense Purchasers and Regional Growth* (Washington: Brookings Institution, 1966), 166. Watkins and Perry, "Regional Change," 40–50. For a discussion of defense procurement policies after 1952, see Chapter 6 below.
56. Marguerite Owen, letter to Jonathan W. Daniels, 11 March 1946, Jonathan W. Daniels Papers. *Blue Book of Southern Progress, 1951,* 28. Fuchs, *Changes in Location,* 61–62. In the immediate postwar years, the region also registered comparative gains in value added by manufactures.
57. Joint Committee on the Economic Report, *Impact of Federal Policies on the South,* 2, 18, 21–22, 29–30, 37, 46, 80. In 1933 southern per capita income made up 48 percent of U.S. per capita income.
58. Graves, *The Fighting South,* 226. Cobb, *Industrialization and Southern Society,* 1, 118.

Chapter 5

1. Franklin D. Roosevelt, "The Nine Hundred and Twenty-ninth Press Conference," 28 Dec. 1943, in Rosenman, ed., Public Papers, 1943 Vol., *The Tide Turns* (N.Y.: Harper and Bros., 1950), 569–71. For an overview of the argument that the New Deal ceased as a force for reform by 1938, see Richard Polenberg, "The Decline of the New Deal," in John Braeman, ed., *The New Deal: The National Level* (Columbus: Ohio State Univ. Press, 1975). Robert Garson is one of few historians who perceives the continuation of New Deal into the 1940s, particularly with respect to the South. Consult Garson, *Democratic Party and Sectionalism,* 53. For the cancer image, I am indebted to Ralph McGill's famous remark that southerners "seeking new enterprises never saw themselves as the carriers of a virus which was to destroy the status quo in their towns." McGill, quoted in William Nicholls, *Southern Tradition and Regional Progress* (Chapel Hill: Univ. of N.C. Press, 1960), 156–57.
2. NEC, *Report on Economic Conditions,* 4. Frank Graham, remarks before the Southern Governors Conference, Minutes of the Meeting of the Southern Governors Conference, Mobile Bay off Dauphin Island, Ala., 15–17 Sept. 1941, p. 12, Graham Papers. Arnall, *Shore Dimly Seen,* 24–25. Southern Conference on Interstate Problems, Résumé and Resolutions, 25–27 Jan. 1940, p. 6, Graham Papers. Lyndon B. Johnson, letter to J.A. Krug, 23 Feb. 1949, Records of the Department of the Interior, RG 48, File 8-1, Part 20, Reclamation Bureau, NA (on microfilm of "Documents Relating to Lyndon Baines Johnson"). President Roosevelt indicated his own support of variable federal grants to equalize the regional economic unbalance. He recommended that the government "make proportionately larger Federal grants-in aid to the States with limited fiscal capacity." Roosevelt, Message of 16 Jan. 1939, quoted in *Congressional Record,* 76th Cong., 1st sess., p. 11135.
3. B. F. Farrar, letter to Tom Connally, 29 April 1940, Connally Papers, Box 126. By 1940 some southerners already believed that federal funds and public

works employment had transformed the political economy of the South. See Graves, "The South Still Loves Roosevelt," 11.

4. Vance, *All These People,* 367–68, 378, 432. Odum, *Southern Regions,* 362–63, 365. Census Bureau, *Statistical Abstract, 1942,* 137. Key, *Southern Politics,* 161. Vance, *Wanted: The South's Future for the Nation,* 21. The stress on expenditures for development-oriented services like roads and schools, and the relative lack of commitment to welfare programs which offer little aid to economic growth, was a prominent feature of southern state fiscal policy in the 1950s and 1960s. See Alan M. Huber, "A Fiscal History of the South Since 1940," a paper presented before the Social Science History Workshop, Stanford Univ., 5 Dec. 1985, pp. 2–3, 26.
5. Advisory Commission on Intergovernmental Relations, *The Role of Equalization in Federal Grants* (Washington, Jan. 1964), 5–6. Sam Rayburn, quoted in Dorough, *Mr. Sam,*
6. Advisory Commission on Intergovernmental Relations, *Role of Equalization,* 63. The negative correlation between grants and income in 1963 was very small, but it did represent a departure from the positive correlation which operated before World War II. Advisory Commission on Intergovernmental Relations, *Periodic Congressional Reassessment of Federal Grants-in Aid to State and Local Governments,* (Washington, June 1961), 8. The southern official was Mrs. Walter Humphrey, director of the Madison County Department of Public Welfare, Huntsville, Ala. Consult U.S. House, *National Defense Migration,* Part 32, pp. 11990–91. On the demand for reform of the matching system, see Arnall, *Shore Dimly Seen,* 24, 142–43: and Van Sickle, *Planning for the South,* 216–17.
7. Discrimination against urban areas was only one of the common objections to equalization provisions in the 1940s and 1950s. For a summary of those objections, see Advisory Commission on Intergovernmental Relations, *Role of Equalization,* 70.
8. *Congressional Record,* 76th Cong., 1st sess., pp. 8830, 8848–51, 8901–12, 11138, 11146. The tally on the Connally Amendment includes those senators who paired on the roll call as well as those who recorded a vote. The three southern opponents of the Connally Amendment were die-hard opponents of the New Deal and fiscal conservatives opposed to almost all federal expenditures: Harry Byrd and Carter Glass of Virginia, and Josiah Bailey of North Carolina. See *Congressional Record,* 76th Cong., 1st sess., pp. 8904–08.
9. *Ibid.* The final vote for the adoption of the Conference Report on Social Security was 59 to 4, with Connally, Georgia senator Richard Russell, and Louisiana senator Allen Ellender joining the minority because of the omission of the Connally Amendment. Harrison did not vote. The House, where the wealthier states of the urban Northeast were most strongly represented, refused to consider such an amendment. See *Congressional Record,* 76th Cong., 1st sess., p. 11146. On Connally's continued efforts, see S. 1946, "A Bill To Amend the Social Security Act by Providing for Special Federal Aid to Certain States, and for other purposes," 77th Cong., 1st sess., 6 Oct. 1941, Tom Connally, press release: "Connally Bill Would Increase Old Age Pay," 23 July 1941; and Paul V. McNutt, letter to Walter F. George, 3 Sept. 1941. All in Tom Connally Papers, Library of Congress. The variable aid formula that was ultimately adopted is spelled out in Law and Legislative Reference Section, Social Security Administration, "Legislation Relating to Social Security," 1946, pp. 50–52. The 79th Congress approved a scheme very much like the Connally

Amendment in the "Social Security Act Amendments of 1946." The bill was proposed by North Carolina congressman Robert Doughton. For subsequent revisions of the aid formula, see the 1950s revisions described in Advisory Commission on Intergovernmental Relations, *Role of Equalization,* 148–50.

10. U.S. Dept. of Transportation, Federal Highway Administration, *America's Highways 1776–1976* (Washington, 1977), 249–60. *Congressional Record,* 84th Cong., 2nd sess., p. 12644. Luther H. Hodges, address by Secretary of Commerce Luther H. Hodges at dedication, Interstate Route 40, Hickory, N.C., 26 May 1961, Luther H. Hodges Papers, Southern Historical Collection, Univ. of N.C. Library, Chapel Hill. Cobb, *Selling of the South,* 78, 99. Advisory Commission on Intergovernmental Relatins, *Role of Equalization,* 133. Randall M. Miller, "The Development of the Modern Urban South: An Overview," in Randall M. Miller and George Pozetta, eds., *Shades of the Sunbelt* (Westport, Conn.: Greenwood, 1988), 7. The 1944 Federal-Aid Highway Act actually designated the National System of Interstate Highways for the first time, but Congress appropriated no funds for them until 1956. See Mark H. Rose, *Interstate: Express Highway Politics 1941–1956* (Lawrence: Regents Press of Kansas, 1979), x, 26, 37–39, 76–89. The economic impact of the highway system is further analyzed in Chapter 6 below.
11. U.S. Dept. of Commerce, Civil Aeronautics Administration, press release: "CAA Announces the 1952 Federal Airport Program," 1 Nov. 1951, Tom Connally Papers, Library of Congress. *Congressional Record,* 79th Cong., 1st sess., p. 8519. *Ibid.,* 2nd Sess., pp. 4222, 4234. U.S. Dept. of Transportation, Federal Aviation Administration, *FAA Historical Fact Book: A Chronology 1926–1971* (Washington, 1974), 277. Advisory Commission on Intergovernmental Relations, *Role of Equalization,* 146. Lepawsky, *State Planning,* 87. On the comity between southern congressmen and the Pentagon, see Chapter 6 below.
12. *Ibid.*
13. Advisory Commission on Intergovernmental Relations, *Role of Equilization,* 5–11, 31–37, 135, 168–75. Another program of benefit to the South in the immediate postwar years was the Advance Planning Program, a public works appropriation under the aegis of the Federal Works Agency. See Lepawsky, *State Planning,* 142–44.
14. Census Bureau, *Statistical Abstract,* 1957, p. 414; *Statistical Abstract,* 1942, p. 248. In his detailed study of North Carolina, Richard Sylla noted that the federal contribution to state and local revenues in North Carolina grew from nearly zero in 1930 to 7 percent in 1942, and shot up to 27 percent by 1977. See Sylla, "Long-Term Trends in State and Local Finance," 847. It was 17.1 percent in 1959.
15. U.S. House, *Interstate Migration of Destitute Citizens,* Part 2, p. 642; and *Interstate Migration of Destitute Citizens* Part 5, p. 1919. Abe Fortas, letter to Henry Stimson, 21 May 1945, Records of the Dept. of Interior, RG 48, Administrative-General-Power Development, File 1-310, Part 5, NA (on microfilm of "Documents Relating to Lyndon Baines Johnson").
16. Advisory Commission on Intergovernmental Relations, *Role of Equilization,* 20. Hugo L. Black, letter to Jesse Hearin, 1 Oct. 1946, Black Papers, Box 34. Robert D. Reischauer, discussion of Rudolph Penner, "Reforming the Grants System," in Peter Mieskowski and William Oakland, eds., *Fiscal Federalism and Grants-in-Aid* (Washington: Urban Institute, 1979), 138–39. The inverse relationship between a state's personal income and its share of grant dollars

became more pronounced after 1962. In 1967, grants accounted for 4 percent of personal income in the poorest states (Mississippi, Arkansas, Alabama, South Carolina, and the non-southern state of West Virginia). The national average was 2.5 percent. See Rudolph G. Penner, "Reforming the Grants System," in Mieskowski and Oakland, eds., *Fiscal Federalism and Grants-in-Aid,* 116.

17. Paul Courant et al., "The Stimulative Effects of Intergovernmental Grants: Or Why Money Sticks Where It Hits," in Mieskowski and Oakland, *Fiscal Federalism and Grants-in-Aid,* 5–21. Similarly, matching grants, which alter relative prices, exerted stronger price effects in the South because of the sliding scales under which poorer regions received larger matching contributions.
18. Key, *Southern Politics,* 310–11. Nicholls, *Southern Tradition and Regional Progress,* 1–5, *passim.* Numan V. Bartley, *The Rise of Massive Resistance* (Baton Rouge: Louisiana State Univ. Press, 1969), 17. Rupert Vance described this apparent hypocrisy as the South seeming to say to the federal powers: "Give us our money and don't ask us what we intend to do with it." Vance, "The South Considered as an Achieving Society" (1967), in Reed and Singal, eds., *Regionalism and the South,* 293.
19. Cash, *Mind of the South,* 434–35. Key, *Southern Politics,* 5–10, 493, 578–81. Bartley, *Rise of Massive Resistance,* 18–19. Nicholls, *Southern Tradition and Regional Progress,* 59–60. For general background on state politics in the South, see Key, *Southern Politics,* 16–20, 46–47, 82, 130–31, 142.
20. *Ibid.* On the Georgia county unit system, consult Key, *Southern Politics,* 119; William Anderson, *The Wild Man from Sugar Creek* (Baton Rouge: Louisiana State Univ. Press, 1975), 16; and Numan V. Bartley, *From Thurmond to Wallace: Political Tendencies in Georgia 1948–1968* (Baltimore: Johns Hopkins Univ. Press, 1970), 14–18.
21. *Ibid.* The morphology of southern politics has been a subject of debate since the publication of Key's seminal work. In "The New Deal and Southern Politics," Alan Brinkley suggests a two-way division between populists and conservatives. See Brinkley, "The New and Southern Politics," 111–15. Numan Bartley's threefold division into Neo-Bourbons, Neo-Populists, and Business Conservatives, and William Harvard's trinity of "Bourbons," "Whigs," and "Populists," both reflect the changes in southern politics described in Chapter 5 and Conclusion below. Consult Bartley, *Rise of Massive Resistance,* 20.; and William C. Havard, Jr., "Southern Politics: Old and New Style," in Louis D. Rubin, Jr., ed., *The American South* (Baton Rouge: Louisiana State Univ. Press, 1980), 49–53.
22. Key, *Southern Politics,* 3.
23. *Ibid.,* 256–57. Sam Jones, "Will Dixie Bolt the New Deal?," *Saturday Evening Post,* 6 March 1943. One southerner succinctly summarized the "declining influence of the Democratic party" in the South by pointing to union activity, the growth of federal bureaucracy, and interference with state's rights on the matter of the Negro. See Ralph P. Jones, letter to Tom Connally, 20 Jan. 1943, Connally Papers. Josephus Daniels criticized the talk of rebellion. "'About this time of year,' as old almanacs put it," Daniels wrote in a draft editorial which he never ran, but sent instead to his son, "look out for reactionaries." Josephus Daniels, letter to Jonathan Daniels, 17 Feb. 1942, Daniels Papers. See also Jonathan Daniels, letter to Carl Brandt, 8 March 1943; and Daniels, memorandum for Marvin McIntyre, 4 Feb. 1943, Daniels Papers. Josiah Bailey's speech on the Senate floor can be found in the *Congressional Record,* 78th

Cong., 1st sess., p. 10346. For accounts of southern discontent with the national Democratic party during this period, see Bartley, *Rise of Massive Resistance,* 31; Garson, *Democratic Party and Sectionalism,* 39–45, 94–130; and Allen Drury, *A Senate Journal, 1943–1945* (N.Y.: McGraw-Hill, 1963), 17–19, 138–41, 208. On Hill's reelection, see Hamilton, *Lister Hill,* 118–19.

24. Graves, *The Fighting South,* 242. Garson, *Democratic Party and Sectionalism,* 41–42, 129, 217–19. Harvard Sitkoff, "Harry Truman and the Election of 1948: The Coming of Age of Civil Rights in American Politics," *Journal of Southern History* 37 (Nov. 1971): 613–15. On Truman's brief honeymoon, see Drury, *Senate Journal,* 422. Only nine southern Democrats in the House voted to sustain Truman's veto of the Taft-Hartley Act, with only three southern senators (Hill, Pepper, and Sparkman) supporting the veto in the upper house.
25. Harry S Truman, *Memoirs,* Vol. II, *Years of Trial and Hope* (Garden City: Doubleday, 1956), 184, 222. On the Clifford memorandum, consult Sitkoff, "Harry Truman and the Election of 1948," 597; Hamby, *Beyond the New Deal,* 209–11; and Garson, *Democratic Party and Sectionalism,* 230–31. The Dixiecrat revolt is summarized in Key, *Southern Politics,* 317–44; and Garson, *Democratic Party and Sectionalism,* 232–313. Truman later portrayed his behavior in the 1948 election not as shrewd adoption of the strategy proposed by Clifford, but as devotion to principle. "I did not discount the handicap which the loss of a 'Solid South' presented as far as my chances of winning the election were concerned," Truman wrote in his memoirs. "I knew that it might mean the difference between victory and defeat in November. I knew, too, that if I deserted the civil-liberties plank of the Democratic Party platform I could heal the breach, but I have never traded principles for votes, and I did not intend to start the practice in 1948 regardless of how it might affect the election." See Truman, *Memoirs,* Vol. II, pp. 184.
26. Key, *Southern Politics,* 278, 317–18, 329. William C. Harvard, "From Past to Future: An Overview of Southern Politics," in Havard, ed., *The Changing Politics of the South* (Baton Rouge: Louisiana State Univ. Press, 1972), 711, 720. W. Wayne Shannon, "Revolt in Washington: The South in Congress," *ibid.,* 650–58. Hamby, *Beyond the New Deal,* 293. Bartley, *Rise of Massive Resistance,* 33, 44–50, 60–61, 152–53. See also David M. Potter, *The South and the Concurrent Majority* (Baton Rouge: Louisiana State Univ. Press, 1972).

    The later career of James F. Byrnes typified this political migration. Having healed his breach with Roosevelt to serve the country during World War II, Byrnes rose to the position of secretary of state under Truman, before breaking with the administration over what he perceived to be a dangerous concentration of power in the federal government. Byrnes returned to his native South Carolina, won the governor's chair and became the Southeastern leader of Democrats for Eisenhower in 1952.
27. Jack Bass and Walter DeVries, *The Transformation of Southern Politics* (N.Y.: Basic Books, 1976), 25–26. O. Douglas Weeks, "Texas: Land of Conservative Expansiveness," in Harvard, ed., *Changing Politics of the South,* 214.
28. Key, *Southern Politics,* 114. *Southern Patriot* 1 (Dec. 1943): 1. Daniel A. Powell, "PAC to COPE: Thirty-two Years of Southern Labor in Politics," in Gary M. Fink and Merl E. Reed, eds., *Essays in Southern Labor Hisotry* (Westport, Conn. Greenwood, 1977), 254. Bartley, *Thurmond to Wallace,* 23. Bartley breaks down the percentage of conservative voters in Georgia, by region, and by level of urbanization, for three key ballots in 1952, 1954, 1960. Hamilton, *Lister Hill,*

122. Hill drew support from organized labor, but also from defense contractors, such as the Reynolds Metals Corporation.

29. Jonathan Daniels, letter to Virginius Dabney, 26 April 1943, Jonathan W. Daniels Papers. Key, *Southern Politics,* 28–33.
30. Rupert Vance, "The Urban Breakthrough in the South," in Reed and Singal, *Regionalism and the South,* 177.
31. Edward F. Haas, *DeLesseps S. Morrison and the Image of Reform* (Baton Rouge: Louisiana State Univ. Press, 1974), 7–9, 26–34, 41, 66. Key, *Southern Politics,* 201–4. Richard E. Yates, "Arkansas: Independent and Unpredictable," in Havard, ed., *Changing Politics of the South,* 255–57. Dobb, *Selling of the South,* 155–56. Dobb, *Industrialization and Southern Society,* 102–3. [Handley?], letter to Daniels, 25 Jan. 1946.
32. Hamilton, *Lister Hill,* 141. Carl Grafton and Anne Permaloff, *Big Mules and Branchheads* (Athens: Univ. of Georgia Press, 1985), 14–17, 56–194.
33. *Congressional Record,* 87th Cong., 1st sess, Vol. 95, Part 1, pp. 569–72. Doris Kearns, *Lyndon Johnson and the American Dream* (N.Y.: Harper and Row, 1976), 103–7.
34. *Ibid.*
35. Kearns, *Lyndon Johnson and the American Dream,* 103–7, 148–49.
36. Mark Ethridge, quoted in Kesselman, *Social Politics of FEPC,* 169. Daniels, quoted in Garson, *Democratic Party and Politics of Sectionalism,* 67. John T. Kneebone, *Southern Liberal Journalists and the Issue of Race, 1920–1944* (Chapel Hill: Univ. of N.C. Press, 1985), 148–49, 196. Sosna, *In Search of the Silent South, passim.* For examples, see George F. Milton, letter to Beatrice Rothschild, 3 May 1944, Milton Papers; Mark Ethridge, letter to Jonathan Daniels, 30 Nov. 1943, with Daniels's reply of 3 Dec. 1943, Daniels Papers; and Ethridge, letter to Daniels, 11 March 1946, Daniels Papers. Consult also Graves, *The Fighting South,* 58, 125.

    Some southern New Dealers eventually came out against racial segregation. Most noteworthy among them were Aubrey Williams and Clark Foreman. Still, well into the 1940s they continued to view racial injustice as a fundamentally economic problem.
37. *Ibid.* For an example of the transformation of the southern liberal emphasis on economic exploitation into a conservative argument, see "Civil Rights in an Old Hat" (Editorial), Greensboro *Daily News,* 21 June 1963, reprinted in Committee on the Judiciary, U.S. Senate, *Civil Rights—the President's Program,* 1963, 88th Cong., 1st sess., (Washington, 1964), 257. The editorial decries the proposed civil-rights act, but asks for action "on the economic front, which is the real key to orderly racial advance." On southern liberalism in the 1940s, see Sosna, *In Search of the Silent South;* and Kreuger, *And Promises To Keep.*
38. Brooks Hays, *A Southern Moderate Speaks* (Chapel Hill: Univ. of N.C. Press, 1985), 25. Hamilton, *Lister Hill,* 162–63. Pepper, *Pepper,* 111. I am indebted to Morton Sosna for this insight. Sosna, letter to the author, 2 April 1987.
39. Alan Matusow, *The Unraveling of America* (N.Y.: Harper and Row, 1984), 6–11. Godfrey Hodgson, *America in Our Time* (N.Y.: Vintage Books, 1976), 77–89.
40. Numan V. Bartley, "The Southern Conference and the Shaping of Post-World War II Southern Politics," in Moore et al., eds., *Developing Dixie,* 179–98. Bartley concludes, persuasively, that "the fate of southern liberals was particularly crucial to the shaping of post-World War II southern politics" (p. 180).

41. On the fiscal proclivities of southern state governments, see Huber, "Fiscal History of the South," and Chapter 7 below.
42. The efforts of governors to recruit new industry are described in Cobb, *Selling of the South,* 74–75 ff. For similar interpretations of the relationship between business and southern politics, see Nicholls, *Southern Tradition and Regional Progress,* 114–23; Cobb, *Selling of the South,* 122–42; and Earl Black, *Southern Governors and Civil Rights* (Cambridge: Harvard Univ. Press, 1976), 142, 322–26.
43. Arnall, *Shore Dimly Seen,* 12–13.
44. Arnall, Foreword to Vance, *Wanted,* 4. Arnall, "Revolution Down South," *Collier's* (28 July 1945). Arnall, *Shore Dimly Seen,* 45–56. Anderson, *Wild Man from Sugar Creek,* 226–33. Jonathan Daniels, memorandum to Franklin D. Roosevelt, 18 May 1944, Jonathan W. Daniels papers. Howard Odum early identified Arnall and North Carolina governor J. Melville Broughton as indicative of "the region's new potential leadership." See Odum, letter to Jonathan Daniels, 1946, Jonathan W. Daniels Papers.
45. *Ibid.*
46. *Ibid.* See also Anderson, *Wild Man from Sugar Creek,* 75.
47. *Ibid.* He also led the popular southern crusade against "discriminatory freight rates." This movement enlisted the support of the Conference on Economic Conditions of the South, the Tennessee Valley Authority, and President Roosevelt and won some concessions from the Interstate Commerce Commission. Economic analyses of the issue, however, indicated that the rates had little effect on southern rail service, transportation in general, or economic activity overall. Nonetheless, the issue allowed politicians like Arnall to portray federal intervention as redressing old grievances. For assessments of the impact of ICC rates, see Joint Committee on the Economic Report, *Impact of Federal Policies on the South,* 31; Hoover and Ratchford, *Economic resources,* 80–81; and McLaughlin and Robock, *Why Industry Moves South,* 93. Among the many arguments against regionalized freight rates, consult Arnall, *Shore Dimly Seen,* 165–85; and three reports by the Transportation and Industrial Economics Division of the TVA Commerce Department: "Economic Consequences of Regionalized Freight Rates," July 1942; "Supplemental Phases of the Interterritorial Freight Rate Problem of the United States," 10 Jan. 1939; "Regionalized Freight Rates," March 1943. All in Records of the TVA, RG 142, TVA Board File Curtis-Morgan-Morgan 1933–57, FARC.
48. *Ibid.*
49. Ralph Eisenberg, "Virginia: Emergence of Two-Party Politics," in Havard, *Changing Politics of the South,* 46–51. Key, *Southern Politics,* 58–59, 254–55. Donald S. Strong, "Alabama: Transition and Alienation," in Havard, *Changing Politics of the South,* 446–50. Preston Edsall and J. Oliver Williams, "North Carolina: Bipartisan Paradox," *ibid.,* 370–75. Bass and DeVries, *Transformation of Southern Politics,* 61.
50. Black, *Southern Governors and Civil Rights,* 15, 31. Manning J. Dauer, "Florida: The Different State," in Havard, *Changing Politics of the South,* 157. Bass and DeVries, *Transformation of Southern Politics,* 61. Cobb, *Selling of the South,* 140–49.
51. Hodges, *Businessman in the Statehouse* (Chapel Hill: Univ. of N.C. Press, 1962), *passim.* William H. Chafe, *Civilities and Civil Rights* (N.Y.: Oxford Univ. Press, 1980), 42–70.

52. Frank E. Smith, remarks at Round-Table Discussion, Institute on Social Change, Norman, Okla., 3 May 1965, Records of the TVA, RG 142, Records of the TVA Board, Smith-Wagner-Hays-Welch, 1957–72, FARC. Key, *Southern Politics,* 675. Powell, "From PAC to COPE," 244. The liberal northeastern organization Americans for Democratic Action (ADA) misread the modest success of southern reformers in 1950. At that time, ADA planned a recruitment drive in the South, addressing economic problems. Hamby, *Beyond the New Deal,* 291. Bartley, *Rise of Massive Resistance,* 39–40.
53. Key, *Southern Politics,* Foreman and Dombrowski, quoted in Bartley, "The Southern Conference," 180.
54. Nixon, "South After the War," 325. Key, *Southern Politics,* 340. Allowing localities to commit their resources to recruitment of industry (and to the industrial subsidies and tax abatements that went with such recruitment) most appealed to footloose, highly competitive industries like textiles. Recruitment efforts strengthened rather than weakened the hold of low-paying, labor-intensive industries on the southern economy. See Cobb, *Industrialization and Southern Society,* 42–43, and Conclusion below.
55. In 1957, at the height of white resistance to the *Brown* decision, Mississippi's State Sovereignty Commission, the state's Maginot Line against the forces of integration, decided to cooperate in the construction of an integrated Veteran's Administration hospital. See Dykeman and Stokely, *Neither Black Nor White,* 311.
56. *Business Week,* 12 Feb. 1949, p. 20.
57. Alabama Business Research Council, *Transition in Alabama* (University: Univ. of Alabama Press, 1962), 1–6. Strong, "Alabama," 427.

Chapter 6

1. William Faulkner, "On Fear," in James Meriwether, ed., *Essays, Speeches and Public Letters by William Faulkner* (N.Y.: Random House, 1965), 98.
2. The path of federal activity in race relations would not have been so familiar. See Chapter 8 below.
3. Nicholls, *Southern Tradition and Regional Progress,* 98. Southern Growth Policies Board, *Halfway Home and a Long Way to Go: The Report of the 1986 Commission on the Future of the South* (Research Triangle Park: SGPB, 1986). The waning of the welfare state in the South is the subject of Chapter 7. See Chapter 7 also for a detailed discussion of minimum-wage policy.

   The assertion that the minimum wage had already achieved its effects by 1960 does not imply either that subsequent revisions of the Fair Labor Standards Act had no impact on the South or that regional wage differentials had been elminated. Rather, it refers to the fact that earlier legislation had slowed the growth of low wage industries in the South and had adjusted capital-labor (K/L) ratios so that southern industry was already more capital-intensive by 1960. Robert J. Newman, *Growth in the American South* (N.Y.: New York Univ. Press, 1984), 112–13. J. R. Moroney and J. M. Walker, "A Regional Test of the Heckscher-Ohlin Hypothesis," *Journal of Political Economy* 74 (Dec. 1966), 581–84. For a dissenting opinion on the convergence of North-South capital-labor ratios, see H. M. Douty, "Wage Differentials: Forces and Counterforces," *Monthly Labor Review* 91 (March 1968): 78, 78n.
4. Hyman Rickover, quoted in Walter McDougall, . . . *the Heavens and the Earth*

(N.Y.: Basic Books, 1985), 301. Harold Vatter, *The United States Economy in the 1950s* (N.Y.: W. W. Norton, 1963), 5–17, 72–74. On the bifurcation of the American state in the postwar era, I am indebted to Peter B. Evans et al., eds., *Bringing the State Back In* (N.Y.: Cambridge Univ. Press, 1985); William E. Leuchtenburg, "The Pertinence of Political History: Reflections on the Significance of the State in America," *Journal of American History* 73 (Dec. 1986): 585–600; and Leonard Krieger, "The Idea of the Welfare State in Europe and the United States," *Journal of the History of Ideas* 24 (Oct.–Dec. 1963): 553–68.

5. Eisenhower's views can best be found in an address delivered before the American Society of Newspaper Editors, "The Chance for Peace," 16 April 1953, quoted in Mc Dougall, . . . *the Heavens and the Earth,* 114. McDougall, . . . *the Heavens and the Earth,* 137–39. Vatter, *Economy in the 1950s,* 72–74, 98–102, 203–4. Eisenhower, despite his avowed antipathy to deficit spending, envisioned the Interstate Highway Program as a way to promote economic growth. And his economic advisers, principally those serving on the Council of Economic Advisors, hoped to use highway program disbursements as a Keynesian economic management tool. See Rose, *Interstate,* 69–83.
6. For George Humphrey's remarks, see Remarks by Secretary Humphrey, interviewed by a panel at the New York *Herald Tribune* Annual Forum, New York, 21 Oct. 1954, in *The Basic Papers of George M. Humphrey 1953–1957* (Cleveland: Western Reserve Historical Society, 1965), 423; Humphrey, speech to the Detroit Economic Club, Detroit, 9 Nov. 1953, p. 494; and speech before the Economic Club, Detroit, 8 Oct. 1956, pp. 234–35.

   Ike's liberal critics, men who would soon staff the Kennedy administration, pushed this reasoning even further. They believed wholeheartedly that defense spending was as good as nondefense demand. Economist James Tobin, for example, later a member of JFK's CEA, criticized Ike's military budget cuts for failing to embrace this Keynesian principle. See James Tobin, "Defense, Dollars, and Doctrines," in B. H. Wilkins and C. B. Friday, eds., *The Economists of the New Frontier* (N.Y.: Random House, 1963), 42–57. Consult also John K. Galbraith, "The Illusion of National Security," *ibid.*, 31–40; and Matusow, *Unraveling of America,* 10, 47.
7. Hearing Before the Select Committee on Small Business, U.S. Senate, *Impact of Defense Spending on Labor Surplus Areas-1962,* 29 Aug. 1962, 87th Cong., 2nd sess. (Washington, 1962), 7–9, 12–13, 53; and *Impact of Defense Spending on Labor-Surplus Areas-1963,* 19 Aug. 1963, 88 Cong., 1st sess. (Washington, 1963), III, 3.
8. *Ibid.* See also Humphrey's behavior in Hearing Before a Subcommittee of the Select Committee on Small Business, U.S. Senate, *Impact of Defense Spending on Labor Surplus Areas-1964,* 13 Aug. 1964, 88th Cong., 2nd sess. (Washington, 1964); and Hearings Before a Subcommittee of the Select Committee on Small Business, U.S. Senate, *The Role and Effect of Technology in the Nation's Economy,* Parts 1–4, 20 May–20 June 1963, 88th Cong., 1st sess. (Washington, 1963).
9. "New Focus on 'Control' of Military" (Editorial), New Orleans *Times-Picayune,* 4 Aug. 1963, sect. 2, p. 4. Edgar Poe, "Sen. Stennis Raps Gesell Proposal," *ibid.*, 1 Aug. 1963, p. 1. Poe, "Blasts Are Fired at Defense Dept.," *ibid.*, 8 Aug. 1963, p. 1. Ronald M. Dalfiume, *Desegregation of the United States Armed Forces* (Columbia: Univ. of Missouri Press, 1969), 210–12, 220–22. Dykeman and Stokeley, *Neither White nor Black,* 164.

Another example of the social reform activities of the military was Secretary of Defense Robert McNamara's proposal to accept for military service young men usually considered educationally unfit. Such a plan, McNamara maintained, would allow these usually impoverished men to gain experience and training that would fit them for future civilian occupations. See "Bold New Attack on Poverty" (Editorial), Atlanta *Journal-Constitution,* 24 Aug. 1966, p. 4.

10. Marvin Hurley, *Decisive Years for Houston* (Houston: Houston Chamber of Commerce, 1966), 333–34. For examples of southern opinion, see the editorial in the New Orleans *Times-Picayune* on President Kennedy's State of the Union address, 13 Jan. 1962, p. 10; see also "New Focus on 'Control' of Military," sect. 2, p. 4.
11. William Proxmire, *Report from Wasteland* (N.Y.: Praeger, 1970), 77–78, 111; Russell is quoted on p. 111. The bipartite model of state development is drawn from Krieger, "Idea of the Welfare State in Europe and the United States," 560–63.
12. On the coincidence of national trends and southern economic benefits, see David C. Perry and Alfred J. Watkins, "Regional Change and the Impact of Uneven Urban Development," in Perry and Watkins, eds., *The Rise of the Sunbelt Cities* (Beverly Hills: Sage Publications, 1977), 47–48; and, Kirkpatrick Sale, *Power Shift* (N.Y.: Vintage, 1975), 17–22. For an example of southern liberal discontent with the efforts of the defense establishment in the South, particularly federally supported research and development, see Jonathan Daniels, "An Autobiographical Adventure," in Southern Regional Education Board, *SREB: The Second 20 Years* (Atlanta: SREB, 1968). The two-stage model of postwar southern econmic development is adapted from David Goldfield, "Economic Development in the South," Appendix A to Southern Growth Policies Board, *Education, Environment and Culture* (Research Triangle Park: SGPB, 1986), 14–15.
13. The slogan which is C. Vann Woodward's, is quoted in James C. Cobb, "Y'All Come Down!," *Southern Exposure* 14 (Sept./Oct. and Nov./Dec. 1986): 23. Cobb is the leading proponent of the view that southern development efforts, even in the postwar years, remained consonant with long-standing southern traditions and patterns of growth. Consult his *Selling of the South and Industrialization and Southern Society.* See also Carl Abbott, "The End of the Southern City," in Jack Bass and Thomas E. Terill, eds., *The American South Comes of Age* (N.Y.: Knopf, 1986), 372–73.
14. Select Committee on Small Business, *Impact of Defense Spending . . . 1964.* Dept. of Defense, OASD (Comptroller), Directorate for Information Operations and Control, *Prime Contract Awards, by State, Fiscal Years 1951–1977* (Washington, n.d.); and *Prime Contract Awards, by Region and State, Fiscal Years 1978, 1979, 1980* (Washington, n.d.). Bernard Udis and Murray L. Weidenbaum, "The Many Dimensions of the Military Effort," in Udis, ed., *The Economic Consequences of Reduced Military Spending* (Lexington, Mass.: D.C. Heath, 1973), 21. NASA's Mississippi Test Facility opened in October 1961 as part of the Apollo Lunar Landing Project. The unemployment rate in Hancock County, an area of chronic unemployment and underemployment dropped from 15 percent in 1960 to 2 percent in 1966. In that same period, employment in Hancock County shifted from predominantly agricultural

employment to predominantly service and manufacturing employment. Consult Mary A. Holman, *The Political Economy of the Space Program* (Palo Alto: Pacific Books, 1974), 217–20.

15. Bolton, *Defense Purchases and Regional Growth,* 99–100. James L. Clayton, ed., *The Economic Impact of the Cold War* (N.Y.: Harcourt, Brace & World, 1970), Table 7. Udis and Weidenbaum, "Many Dimensions of the Military Effort," 28–30. Roger Riefler and Paul Downing, "Regional Effect of Defense Effort on Employment," *Monthly Labor Review* 91 (July 1968): 2–6.
16. "Southern Militarism," 60. *Our Own Worst Enemy,* 111.
17. "Southern Militarism," 73. George W. Hopkins, "From Naval Pauper to Naval Power," in Roger W. Lotchin, ed., *The Martial Metropolis* (N.Y.: Praeger, 1984), 18–19.
18. Clayton, *Economic Impact,* 81, 248–49. Roger E. Bolton, "Defense Spending: Burden or Prop," in Bolton, ed., *Defense and Disarmament: The Economics of Transition* (Englewood Cliffs, N.J.: Prentice-Hall, 1966), 8. Charles M. Tiebout, "The Regional Impact of Defense Expenditures," *ibid.,* 127. John H. Cumberland, "Dimensions of the Impact of Reduced Military Expenditures on Industries, Regions and Communities, in Udis, ed., *Economic Consequences,* 79–83. Holman, *Political Economy of Space Program,* 51–53.
19. Murray L. Weidenbaum, "The Transferability of Defense Industry Resources to Civilian Use," in Bolton, ed., *Defense and Disarmament,* 106–8. William L. Baldwin, *The Structure of the Defense Market 1955–1964* (Durham: Duke Univ. Press, 1967), 3, 179. *Our Own Worst Enemy,* 157–66. In the 1980s, a few southern textile manufacturers, many of them under government contract, adapted similar automated processes to their mills. See "Burlington Industries: Plant Modernization and Production Innovation," *Appalachia* 17 (May–Aug. 1984): 9–10. On automation in the aircraft industry, see David C. Mowery and Nathan Rosenberg, "Technical Change in the Commercial Aircraft Industry, 1925–1975," in Nathan Rosenberg, *Inside the Black Box* (N.Y.: Cambridge Univ. Press, 1982), 163, 171–74; and, David F. Noble, *Forces of Production* (N.Y.: Oxford Univ. Press, 1986).
20. Proxmire, *Report from Wasteland,* xi–xii, 109. Derek Shearer, "Converting the War Machine," *Southern Exposure* 1 (Spring 1973): 40–42. On the effects of the McNamara base closings, see John E. Lynch, *Local Economic Development After Military Base Closures* (N.Y.: Praeger, 1970).
21. *Our Own Worst Enemy,* 10–11, 51. For an analysis of the South's role as the nation's commissary, see below.
22. *Ibid.* See also McDougall, . . . *the Heavens and the Earth;* two works by Seymour Melman, *Our Depleted Society* (N.Y.: Holt, Rinehart and Winston, 1965); and *The Permanent War Econmy* (N.Y.: Simon and Schuster, 1974).
23. John Crowe Ransom, "Reconstructed But Unregenerate," in Twelve Southerners, *I'll Take My Stand* (Baton Rouge: Louisiana State Univ. Press, 1983), 1–27. Barry Bluestone and Bennett Harrison, *The Deindustrialization of America* (N.Y.: Basic Books, 1982), 86–88.
24. See Table 5-1. *New York Times,* 8–13 Feb. 1976. Sale, *Power Shift,* 28–29.
25. Select Committee on Small Business, *Impact of Defense Spending . . . 1962,* 169; *Impact of Defense Spending* [1963], III. Clayton, *Economic Impact,* 51–52. Proxmire, *Report from Wasteland,* 13–14.
26. Charles D. Liner, "The Sunbelt Phenomenon—A Second War Between the States?," *Popular Government* 43 (Summer 1977): 16–24. "Southern Militarism," 70. I am indebted to an anonymous reviewer for this insight. In addi-

tion to obsolete plant, Frostbelt industries suffered from growing diseconomies of scale in the 1950s and 1960s.

27. Luther H. Hodges, remarks prepared for delivery by Secretary of Commerce Luther H. Hodges, at Jet Age Cavalcade, Galveston, 26 Aug. 1962, Luther H. Hodges Papers. Clayton, *Economic Impact,* 51–52. Bolton, *Defense Purchases and Regional Growth,* 2. The reference to the "new military aircraft industry" refers to the fact that while one-third of American aircraft went to the military in 1939 and one-half in 1946, more than 90 percent of U.S. aircraft production was purchased by the military in 1953. Noble, *Forces of Production,* 6.
28. Select Committee on Small Business, *Impact of Defense Spending . . . 1962,* 157–60; and . . . *1964,* 24; *Impact of Defense Spending* [1963], 12–13. Dept. of Defense, *Armed Services Procurement Regulation* (Washington, 1976), 1-801.1, 1–804. For a list of the ten Defense Dept. directives pursuant to these regulations, and reproductions of those directives, consult Armed Services Committee, U.S. Senate, *Hearings Before a Subcommittee of the Armed Services Committee, U.S. Senate on S. 500, S. 1383, and S. 1875,* 86th Cong., 1st sess., 13–31 July 1959 (Washington, 1959), 216–70. Consult also Bolton, *Defense Purchases and Regional Growth,* 141–46.
29. *Ibid.*
30. Armed Services Committee, *Hearings Before a Subcommittee of the Armed Services Committee, U.S. Senate on S. 500, S. 1383, and S. 1875,* pp. 25–26, 72–75, 81, 103. The 17-member Armed Services Committee had eleven Democrats; six, including Chairman Richard Russell of Georgia and the five highest ranking Democrats, hailed from the South. The six-person subcommittee investigating the Javits-Keating-Case bill had three southerners, including the chairman, on its rolls. During the hearings, Keating accepted a temporary ban on the payment of price differentials. Subcommittee chairman Strom Thurmond (D-SC) forced the same concession from Senator Jacob Javits (R-NY). See also Bolton, *Defense Purchases and Regional Growth,* 141–42.
31. U.S. Congress, Joint Committee on Defense Production, *Third Annual Report,* 83rd Cong., 1st sess. (Washington, 1953), 19–20. Select Committee on Small Business, *Impact of Defense Spending . . . 1962,* 50–51; . . . *1964,* 16; *Impact of Defense Spending* [1963], 3–8. U.S. Dept. of Commerce, Economic Development Administration, *Report of the Independent Study Board on the Regional Effects of Government Procurement and Related Policies* (Washington, 1967), ix–xi, 9. In that 1967 report, the Commerce Dept. recommended against reallocation of procurement and science expenditures to aid lagging regional economies.
32. *Ibid.* For the commentary on midwestern technical resources, consult Select Committee on Small Business, *Impact of Defense Spending* [1963], pp. 3–8. In 1965, Illinois governor Otto Kerner complained that a lack of government R&D contracts drained Illinois of much of the technical talent it trained. See hearings before the Subcommittee on Employment and Manpower of the Committee on Labor and Public Welfare, U.S. Senate, *Impact of Federal Research and Development Policies on Scientific and Technical Manpower,* 89th Cong., 1st sess., 2 June–22 July 1965 (Washington, 1965), 7–11. In 1964 each state's share of Defense Dept. research contracts reflected its share of the nation's distinguished scientists. Only the South was an exception to that rule. Texas, North Carolina, Florida, and Georgia ranked higher in contracts

than in National Academy of Sciences members. *Impact of Federal Research and Development Policies on Scientific and Technical Manpower,* 484.

33. The full committee contained three southerners—George Smathers (D-Fla.), Russell Long (D-La.), and the chairman, John Sparkman (D-Ala.). Humphrey's subcommittee had no southern members. Select Committee on Small Business, *Impact of Defense Spending . . . 1964.* Select Committee on Small Business, *Hearing Before a Subcommittee of the Select Committee on Small Business on the Implementation of Defense Manpower Policy No. 4,* 82nd Cong., 2nd sess., 20 March 1952 (Washington, 1952). For Senator Hart's resolution, consult Armed Services Committee, *Hearings Before a Subcommittee of the Armed Services Committee, U.S. Senate on S. 500, S. 1383, and S. 1875,* p. 312.
34. Bass and DeVries, *Transformation of Southern Politics,* 138–39, 369–91. "Federal Spending: The North's Loss Is the Sunbelt's Gain," *National Journal* 8 (26 June 1976): 878. W. Wayne Shannon, "Revolt in Washington, the South in Congress," in Havard, ed., *Changing Politics,* 637–687. "Southern Militarism," 60–62.
35. *Ibid.* Marc Miller, "The Low Down on High Tech," *Southern Exposure* 14 (Sept./Oct. and Nov./Dec. 1986): 39. Carl N. Degler, "Thesis, Antithesis, Synthesis: The South, the North, and the Nation," *Journal of Southern History* 53 (Feb. 1987): 17. Hopkins, "From Naval Pauper to Naval Power," 13–14.
    Joseph Petit, Dean of Engineering at Stanford Univ. left California for Georgia Tech for much the same reason. See Robert C. McMath, Jr., et al., *Engineering the New South: Georgia Tech, 1885–1985* (Athens, Univ. of Georgia Press, 1985), 409–12
36. Bass and DeVries, *Transformation of Southern Politics,* 138–39, 369–91. "Federal Spending: The North's Loss Is the Sunbelt's Gain," 878. W. Wayne Shannon, "Revolt in Washington, the South in Congress," 637–87. Rivers's career is analyzed in Hopkins, "From Naval Pauper to Naval Power," 18–23; and Marshall Frady, "The Sweetest Finger This Side of Midas," *Life,* 27 Feb. 1970, pp. 50–57. On the South's control of Congressional seniority, see Potter, *The South and the Concurrent Majority.*
37. *Ibid.*
38. *Ibid.* See also Proxmire, *Report from Wasteland,* 100–03; David R. Goldfield, *Promised Land: The South Since 1945* (Arlington Heights, Ill.: Harlan Davidson, 1987), 32; and "Southern Militarism," 72.
39. First and foremost among southern politicians who recognized the regional development possibilities in the military budget was Lyndon Johnson. See Kearns, *Johnson and the American Dream, passim.* I am also indebted to the forthcoming biography of Johnson by Robert Dallek. Bass and DeVries, *Transformation of Southern Politics,* 138–39, 369–91. "Federal Spending: The North's Loss Is the Sunbelt's Gain," 878. W. Wayne Shannon, "Revolt in Washington, the South in Congress," 637–87.
40. Alex Roland, *Model Research: The National Advisory Committee for Aeronauntics 1915–1958,* 2 vols. (Washington: NASA, 1985), 199, 221, 300. McDougall, *. . . the Heavens and the Earth,* 148–56, 162, 172–74. The Defense Dept. official, Assistant to the Secretary Oliver Gale, is quoted on p. 148. Holman, *Political Economy of Space Program,* 48. The frenzy over *Sputnik* also led to the enactment of the National Defense Education Act. Piloted by Senators J. William Fulbright (D-Ark.) and Lister Hill (D-Ala.), the long-delayed aid-to-education bill was passed on southern terms, without attention to integration.

41. McDougall, . . . *the Heavens and the Earth,* 376 *passim.* Hearings before the Committee on Aeronautical and Space Sciences, U.S. Senate, *NASA Authorization for Fiscal Year 1964,* 88th Cong., 1st sess., 24–30 April 1963 (Washington, 1963), 903. Subcommittee on NASA Oversight of the Committee on Science and Aeronautics, U.S. House of Representatives, *The National Space Program—Its Values and Benefits,* 90th Cong., 1st sess. (Washington, 1967), 9.
42. Select Committee on Small Business, *Impact of Defense Spending . . . 1964,* 29–31. Hearings before the Committee on Aeronautical and Space Sciences, U.S. Senate, *NASA Authorization for Fiscal Year 1965,* 88th Cong., 2nd sess., 4–18 March 1964 (Washington, 1964), 323. During the 1965–66 academic year, 25 percent of NASA's pre-doctoral scholarship students in training studied in the South. See House Subcommittee on NASA Oversight, *National Space Program,* 60. The research contracts are detailed in the quarterly NASA, *NASA's University Program.* On improvements in primary and secondary education in Huntsville and Brevard County, Fla., see Mary A. Holman and Ronald Konkel, "Manned Space Flight and Employment," *Monthly Labor Review* 91 (March 1968): 34.
43. NASA, *Marshall Space Flight Center 1960–1985* (Washington, 1985), chronology. UAH received several hundred thousand dollars in NASA contracts each year during the 1970s and early 1980s. Holman, *Political Economy of the Space Program,* 204–10. Steven Beschloss, "Prosperity's Broken Promise," *Southern Exposure* 15 (Fall and Winter 1987): 58.
44. Hurley, *Decisive Years for Houston,* 207, 229–33. McDougall, . . . *the Heavens and the Earth,* 374. Perry and Watkins, "Introduction," in Perry and Watkins, eds., *Rise of the Sunbelt Cities,* 105. On site selection for Manned Space Center, see Holman, *Political Economy of the Space Program,* 48–50. The oil boom of the 1970s reversed the trend toward diversification and independence from oil. And even though the boom accelerated the rapid growth of Houston in the 1970s, it left the city vulnerable to the oil shock of the early 1980s. Eventually, the city recovered and the local economy continued to diversify. See Roberto Suro, "How Houston Shaped Up with a Rigid No Oil Diet," *New York Times,* 30 April 1989, p. E5. For the anecdote about the first words from the moon, see Joel Garreau, *The Nine Nations of North America* (Boston, 1981).
45. NASA, *Marshall Space Center,* 15. McDougall, . . . *the Heavens and the Earth,* 373–74. Webb is quoted in House Subcommittee on NASA oversight, *National Space Program,* 12. The quotation is a reference to Russia, but the report applies that analysis to the United States as well; see pp. 8–12. On Webb's views in general, see McDougall.
46. The heaviest populations of active army personnel served in the Third and Fourth armies, housed in the Southeast and Southwest, respectively. See, for example, Military Construction Subcommittee of the Committee on Armed Services, U.S. Senate, *Military Construction Authorization, Fiscal Year 1963,* 87th Cong., 2nd sess., 8 March–2 April 1962 (Washington, 1962), 48. John E. Lynch, *Local Economic Development After Military Base Closures,* 261, 291–93. Advisory Commission on Intergovernmental Relations, *The Federal Influence on State and Local Roles in the Federal System* (Washington, 1981), 26. Holman and Konkel, "Manned Space Flight and Employment," 34. C. L. Jusenius and L. C. Ledebur, "A Myth in the Making," in E. Blaine Liner and

Lawrence K. Lynch, eds., *The Economics of Southern Growth* (Durham: SGPB, 1977), 170.

47. John E. Lynch, *Local Economic Development After Military Base Closures,* 81, 145–46, 231–32; Greenville (S.C.) *News* quoted on p. 81. A major employer is one employing more than 100 persons. Edward F. Haas, "The Southern Metropolis, 1940–1976," in Blaine Brownell and David R. Goldfield, eds., *The City in Southern History* (Port Washington, N.Y.: Kennikat Press, 1977), 174. Edward H. Kolchum, "Georgia Tech Facility Is Key to Growth," *Aviation Week and Space Technology,* 28 Feb. 1983, p. 72. Another example is Tampa, Fla., a city one urban historian called a "metropolitan-military complex." See Gary R. Mormino, "Tampa: From Hell Hole to the Good Life," in Richard M. Bernard and Bradley R. Rice, eds., *Sunbelt Cities: Politics and Growth Since World War II* (Austin: Univ. of Texas Press, 1983), 147.
48. Dept. of Defense, *Prime Contracts, By State. Our Own Worst Enemy,* 10–11, 51. Walt Whitman Rostow, "Regional Change in the Fifth Kondratieff Upswing," in Perry and Watkins, eds., *Rise of the Sunbelt Cities,* 95–96. Bolton, *Defense Purchases and Regional Growth.* Among the journalistic accounts that emphasize miltary spending is Sale, *Power Shift.* For examples of economists who downplay the importance of defense disbursements, see Douglas E. Booth, "Long Waves and Uneven Regional Growth," *Southern Economic Journal* 53 (Oct. 1986): 458-59; and, Bernard Weinstein and Robert E. Firestine, *Regional Growth and Decline in the United States* (N.Y.: Praeger, 1978), 43. On the influence of the space program in Los Angeles, see Holman, *Polital Economy of Space Program,* 229–37. On New England, consult Bennett Harrison, "Regional Restructuring and 'Good Business Climates': The Economic Transformation of New England Since World War II," in Larry Sawers and William K. Tabb, eds., *Sunbelt/Snowbelt* (N.Y.: Oxford Univ. Press, 1984), 48–98.
49. Reubin Askew, "Introduction: The South and the Nation," in Liner and Lynch, eds., *The Economics of Southern Growth,* 2–3. Perry and Watkins, "Regional Change," 50. "Federal Spending: North's Loss Is Sunbelt's Gain," 880–84.
50. Henry B. Kline, "Economic Motives for the Diffusion of National Defense Production in Depressed Agricultural Areas," Dec. 1940, Records of the TVA, RG 142, Records of the TVA Commerce Dept., FARC, pp. 1–2.
51. Dewey W. Grantham, Jr., "The New Southern Region: An Essay in Contemporary History," in Southern Regional Education Board, *The Future South and Higher Education* (Atlanta: SREB, 1968), 1. Daniels, "An Autobiographical Adventure," 25. Coclanis and Ford, "The South Carolina Economy Reconstructed and Reconsidered." For data on convergence by the mid-1950s, consult Tables 3-1, 4-4, and 4-5 above.
52. Weinstein and Firestine, *Regional Growth and Decline,* 11–19. Select Committee on Small Business, *Impact of Defense Spending . . . 1962,* 22–23. Newman, *Growth in the American South,* 5,12. Lawrence Lynch and E. Evan Brunson, "Comparative Growth and Structure," in Liner and Lynch, eds., *The Economics of Southern Growth,* 20–21. Douty, "Wage Differentials: Forces and Counterforces," 80. John McKinney and Linda Brookover Borque, "The Changing South: National Incorporation of a Region," *American Sociological Review* 36 (June 1971): 399–412. Philip Rones dissents from the prevailing wisdom that manufacturing energized Sunbelt growth. See Rones, "Moving to the

Sun: Regional Job Growth, 1968 to 1978," *Monthly Labor Review* 103 (March 1980): 15.

53. Lynch and Brunson, "Comparative Growth and Structure," 20–24. William H. Miernyk, "The Changing Structure of the Southern Economy," in Liner and Lynch, eds., *Economics of Southern Growth,* 41–45. Both of those authors use a fifteen-state South which includes West Virginia and Maryland. Miernyk, "Changing Structure," provides Employment Location Quotients, for 1940 and 1975, for the South (plus W.Va and Md.) on p. 45 (1.0 indicates percentage of employment in that sector equal to national average). Agriculture declined from 1.73 to 1.00; manufacturing rose from 0.63 to 0.90; transportation, 0.84 to 1.02; wholesale and retail trade from 0.81 to 0.98 and 0.99, respectively; government rose from 0.90 to 1.05; and services from 0.91 to 0.99.
54. Faulkner, "On Fear," 98. Vatter, *United States Economy in the 1950s,* 250–53. Daniel, "Transformation of the Rural South," 243–44. Price supports and acreage reduction also encouraged the trend toward larger farm units.
55. Daniel, *Breaking the Land,* 248–66. Daniel, "Transformation of the Rural South," 232–48. See also Jack Temple Kirby, *Rural Worlds Lost: The American South, 1920–1960* (Baton Rouge: Louisiana State Univ. Press, 1986). On the conversion from cotton to food production, consult Fornari, "The Big Change: Cotton to Soybeans." In combination with the changing tastes of American consumers, those changes in crop choice reflected the influence of industrialization in general and defense spending in particular. In 1940 the TVA predicted that a shift from staple to food production would result from defense production in the South. Industrialization, the TVA argued, would stem out-migration and create food markets. On the relationship between industrialization and conversion from cotton to fresh food production, see Kline, "Economic Motives for the Diffusion of National Defense Production," 21, and Sale, *Power shift,* 22.
56. Webb, quoted in Grantham, "The New Southern Region," 17. Luther H. Hodges, preliminary text of an address in connection with the North Carolina Trade and Industry Mission, to be delivered at six European cities, 1–15 Nov. 1959, Records of Gov. Luther H. Hodges, Governors' Papers, N.C. Dept. of Archives and History, Raleigh, Box 465. Newman, *Growth in American South,* 14. On the South's performance during the recessions and expansions of 1945–61, see L. Randolph McGee, *Income and Employment in the Southeast: A Study in Cyclical Behavior* (Lexington: Univ. Press of Kentucky, 1967), 2, 27–34. For the South's performance in the divestment period of the 1970s, consult, Bluestone and Harrison, *Deindustrialization,* Table A-4.
57. Vance, "Education and the Southern Potential," 278. U.S. Advisory Commission on Intergovernmental Relations, *State-Local Taxation and Industrial Location* (Washington, 1967) 11–12. Newman, *Growth in the American South,* 112–13. Pratt, *Growth of a Refining Region,* 179–81. Moroney and Walker, "Regional Test of the Heckscher-Ohlin Hypothesis," 581–84. U.S. Dept. of Commerce, Office of Area Development, "Mechanization in U.S. Manufacturing: 1939–1954," *Area Development Bulletin* 3 (Oct.–Nov. 1957): 1,6. For a description of the competing models of regional development, see Newman, *Growth in the American South,* 95–99; and Weinstein and Firestine, *Regional Growth and Decline,* 51–65.
58. Walt Whitman Rostow, *The Process of Economic Growth* (N.Y.: W. W. Norton,

1962), 261–65. See also Rostow's classic, *The Stages of Economic Growth,* 2nd ed. (Cambridge: Cambridge Univ. Press, 1971). On the South, see Advisory Commission on Intergovernmental Relations, *State-Local Taxation and Industrial Location,* 11–12; Cobb, *Industrialization and Southern Society,* 50; and Bartley, "Another New South," 131–32.

59. The largest absolute employment gains, however, came in the trades, government, services, and construction—the United States leaders. Ray Marshall, "The Old South and the New," in Ray Marshall and Virgil L. Christian, eds., *Employment of Blacks in the South: A Perspective on the 1960's* (Austin: Univ. of Texas Press, 1978), 7–8. Select Committee on Small Business, *Impact of Defense Spending . . . 1962,* 7. Newman, *Growth in the American South,* 12, 20–21. Vatter, *United States Economy in the 1950s,* 167. Despite the rapid rise of the southern machinery industry, it never became a leader in the adoption of automated production technology. Consult, J. Rees, et al., "New Technology in the United States Machinery Industry," in A. T. Thwaites and R. P. Oakey, eds., *The Regional Impact of Technological Change* (London: Francis Pinter, 1985), 192–93.

60. Bluestone and Harrison, *Deindustrialization,* 39–40, 6–9.

61. Douglass C. North, "A Reply," *Journal of Political Economy* 64 (April 1956): 166. Noble, *Forces of Production,* 229, 339. Ron Linton, director of the Office of Economic Utilization, Dept. of Defense, testified in 1962 that each defense production job created three service-sector jobs. Select Committee on Small Business, *Impact of Defense Spending . . . 1962,* 55.

62. Timothy O'Rourke, "The Demographic and Economic Setting of Southern Politics," in James F. Lea, ed., *Comtemporary Southern Politics* (Baton Rouge: Louisiana State Univ. Press, 1988), 18–19. Peter A. Morrison with Judith P. Wheeler, "Rural Renaissance in America? The Revival of Population Growth in Remote Areas," *Population Bulletin* 31 (1976): 3. Jeanne C. Biggar, "The Sunning of America: Migration to the Sunbelt," *Population Bulletin* 34 (March 1979): 26. Advisory Commission on Intergovernmental Relations, *State-Local Taxation and Industry Location,* 17–18, 24–26. The comparable figure for Pennsylvania was 25 percent; for Illinois, 27 percent.

63. Summary of the Southern Governors Conference, Nashville Meeting, 24–27 Sept. 1961, Records of the Governor's Office, Records of Gov. Burford Ellington (1959–63), Tennessee State Library and Archives, Nashville, Box 81, p. 42. Stefan H. Robock, "Report on Defense Expansion in the Valley Region," 2 July 1952, Records of the TVA, RG 142, TVA Board File Curtis-Morgan-Morgan 1933–57, FARC. Beschloss, "Prosperity's Broken Promise," 57. This "rural Renaissance," however, would prove short-lived; by the 1980s, urban areas accounted for most job growth in the South.

64. *Ibid.* Lilah Watson, quoted in Ginger Varney, "The Super Pig-Out," *LA Weekly,* 4–10 March 1988, p. 10. "One of the major reasons for the shift in net immigration in the South and net out-migration in the Northeast," the director of the Census Bureau concluded in 1976, "has been the nation's reliance on truck transportation." Vincent Barabba, quoted in "The Second War Between the States," *Business Week,* 17 May 1976, p. 97. Thomas D. Clark, quoted in Grantham, "The New Southern Region," 7n. In 1956, Secretary of the Treasury George Humphrey stressed the role of the Defense Highway System in economic growth. See Humphrey, statement before the Senate Finance Committee, 17 May 1956, in Humphrey, *Basic Papers,* 526.

On the contours of the federal highway program in the 1960s and 1970s, see Federal Highway Administration, *America's Highways,* 261–62, *passim.* On the development role of interstate highways in San Antonio, Texas, consult David R. Johnson, "San Antonio: The Vicissitudes of Boosterism," in Bernard and Rice, eds., *Sunbelt Cities,* 241.

65. Stuart Rosenfeld, "A Divided South," *Southern Exposure* 14 (Sept./Oct. and Nov./Dec. 1986): 11. Beschloss, "Prosperity's Broken Promise," 57. Jane Jacobs argues that the dispersal of federal investment through the nonmetropolitan South accounts for the failure of regional economic development efforts. See Jacobs, "Why TVA Failed" and *Cities and the Wealth of Nations.*
66. Holman, *Political Economy of the Space Program,* 205. Holman and Konkel, "Manned Space Flight and Employment," 31. Daniel Charles, "Star Wars Fell on Alabama," *The Nation,* 19 Dec. 1987, pp. 748–49. Rones, "Moving to the Sun," 16.
67. John Herbers, "A Different Dixie," *New York Times,* 6 March 1988, sect. 4, p. 1. Miller, "Low Down on High Tech," 36. Thomas D. Clark, *The Emerging South,* 2nd ed. (N.Y.: Oxford Univ. Press, 1968), 272. Don Bellante, "The North-South Differential and the Migration of Heterogenous Labor," *American Economic Review* 69 (March 1979): 168. Wright, *Old South, New South,* 255–56. The author thanks Siecor Corporation for providing personnel information.
68. Between 1970 and 1976 the South experienced a net in-migration of 2.6 million people, the West 1.4 million. U.S. Dept. of Commerce, Bureau of the Census, *Current Population Reports,* Series P-25, no. 640 (Nov. 1976). James G. Maddox, "The Growing Southern Economy," in SREB, *The Future South and Higher Education,* 60. Weinstein and Firestine, *Regional Growth and Decline,* 5. Wright, *Old South, New South,* 255–56.
69. Census Bureau, *Current Population Reports,* Series P-25, no. 640. Rones, "Moving to the Sun," 16. Richard Raymond, "Determinants of Non-White Migration During the 1950s: Their Regional Significance and Long-Run Implications," *American Journal of Economics and Sociology* 31 (Jan. 1972): 15–16. Tommy Rogers, "Migration Attractiveness of Southern Metropolitan Areas," *Social Science Quarterly* 50 (Sept. 1969): 335.
70. Select Committee on Small Business, *The Role and Effect of Technology in the Nation's Economy,* 33–34. Charles C. Killingsworth, "The Continuing Labor Market Twist," *Monthly Labor Review* 91 (Sept. 1968): 12–17. Dennis F. Johnston, "Education and the Labor Force," *ibid.,* 2–4. Dennis F. Johnston, "The Labor Market Twist 1964–1969," *ibid.,* 94 (July 1971): 35.
71. These issues form the subject of Chapter 7 below. See also John Cogan, "The Decline in Black Teenage Employment, 1950–1970," *American Economic Review* 72 (Sept. 1982): 621–38.
72. Weinstein and Firestine, *Regional Growth and Decline,* 23–24. Clayton, *Economic Impact,* 63–64, 73. Teibout, "Regional Impact of Defense Expenditures," 128.
73. Select Committee on Small Business, *Impact of Defense Spending . . . 1962,* 30. Bluestone and Harrison, *Deindustrialization,* 117. Noble, *Forces of Production,* 229, 339. Wage patterns are discussed in Chapter 7 below. See also Wright, *Old South, New South,* 250.
74. Hodges, *Businessman in the Statehouse,* 61–62. Contradictory assessments of the impact of unions on regional development include: Booth, "Long Waves

and Uneven Regional Growth," who asserts that the rate of unionization has no effect on new incorporation rates; Cobb, *Selling of the South,* who views anti-unionism as essential to the maintenance of the traditional low wage industry of the region; and Newman, *Growth in the American South,* who attributes the greatest part of regional industrial growth to anti-union legislation.

75. Luther H. Hodges, "A Letter to 'Family and Friends,' Giving Highlights of Trip," 13–30 July 1957, Luther H. Hodges Papers. Samuel J. Ervin, remarks prepared for delivery before the Virginia State Bar Association, 28 Jan. 1966, Samuel James Ervin, Jr., Papers, Southern Historical Collection, Univ. of N.C., Chapel Hill, Series II-Subject Files. Consult also the high volume of consituent mail regarding the Kennedy-Ervin labor bill in 1959, Samuel J. Ervin Papers, Series I-Correspondence Files. See especially the exchange J. E. Dowd, letter to Sam Ervin, 23 Jan. 1959, and Ervin, letter to Dowd, 27 Jan. 1959. On the J. P. Stevens affair, see Bruce Raynor, "Unionism in the Southern Textile Industry: An Overview," in Gary M. Fink and Merle E. Reed, *Essays in Southern Labor History* (Westport, Conn.: Greenwood, 1977), 81.
76. On unionization rates, see Newman, *Growth in the American South,* 59–62. Consult also Marshall, *Labor in the South,* 267–68, 303, 317. On the 1970s, Cobb, *Selling of the South,* 256, 259; and Bluestone and Harrison, *Deindustrialization,* 165–67. The regional migration of industry also widened the union/non-union wage gap over time. Consult H. Gregg Lewis, *Union Relative Wage Effects: A Survey* (Chicago: Univ. of Chicago Press, 1986).
77. *Our Own Worst Enemy,* 29–30, 37–41. Noble, *Forces of Production,* 26–29, *passim.* Rosenfeld, "Divided South," 13. Bluestone and Harrison, *Deindustrialization,* 117. "Burlington Industries: Plant Modernization and Production Innovation," 9–10. Democratic National Committee, press release: Summary of Remarks by Secretary of Commerce Luther H. Hodges to Annual Meeting, Democratic Women's Clubs of Kentucky, 23 Sept. 1961, Luther H. Hodges Papers.
78. Mancur Olson, "The Causes and Quality of Southern Growth," in Liner and Lynch, eds., *Economics of Southern Growth,* 121. Wright, *Old South, New South,* 263–64. Thomas Ferguson and Joel Rogers, *Right Turn* (N.Y.: Hill and Wang, 1986), 51–61. David M. Gordon, et al., *Segmented Work, Divided Workers* (N.Y.: Cambridge Univ. Press, 1982), 187–88. Noble, *Forces of Production,* 28–29. Lichtenstein, *Labor's War at Home,* 238–41.
79. Southern Governors Conference, "The South's Competitive Postion: A Report to the Southern Governors Conference Presented by the Committee on Industrial Development," Asheville, N.C., 13 Oct. 1959, in Records of the Governor's Office, Records of Gov. Burford Ellington (1959–63), Tennessee State Library and Archives, Nashville, Box 81. Cobb, *Selling of the South,* 225. Kenneth Johnson and Marilyn Scurlock, "The Climate for Workers," *Southern Exposure* 14 (Sept./Oct. and Nov./Dec. 1986): 29.
80. On right-to-work, consult Southern Governors Conference, "The South's Competitive Postion." Newman, *Growth in the American South,* 52–53, 94, 100. On taxes, see Bluestone and Harrison, *Deindustrialization,* 186, and the tables appended to Huber, "Fiscal History of the South." On industrial promotion, see Cobb, "Y'All Come Down," 18–23; Cobb, *Industrialization and Southern Society,* 31, 58; and Cobb, *Selling of the South.* On Hodges's fishing

expeditions, see his *Businessman in the Statehouse,* 57–58, and his "Report from Europe," 8 Nov. 1959, Luther H. Hodges Papers, Series E.

81. On right-to-work, William J. Moore et al., "Do Right to Work Laws Matter? Comment," *Southern Economic Journal* 53 (Oct. 1986): 522–23. See also Newman, *Growth in the American South,* 174–75; and Marshall, *Labor in the South,* 321, 329. On the effects of development programs, see Cobb, *Sellingof the South,* 36–39, 222, *passim;* and Cobb, *Industrialization and Southern Society,* 43. On taxation and fiscal policy, consult Advisory Commission on Intergovernmental Relations, *State-Local Taxation and Industrial Location,* 70, 38; and Huber, "Fiscal History of the South," 2–13, 31.
82. Richard J. Cebula, "Local Government Policies and Migration," *Public Choice* 19 (Fall 1974): 85–93. Officials in Texas, Arkansas, Florida, and Mississippi all cited the desire to recruit industry as the motivation for right-to-work legislation. See Marshall, *Labor in the South,* 319–20.
83. Olson, "Causes and Quality of Southern Growth," 107–19. Olson, "The South Will Fall Again: The South as Leader and Laggard in Economic Growth," *Southern Economic Journal* 50 (April 1983): 917–32. Rostow, "Regional Change and the Fifth Kondratieff Upswing," 88–91. Booth, "Long Waves and Uneven Regional Growth," 448–51. Stanley Engerman, "Regional Aspects of Stabilization Policy," in Richard Musgrave, ed., *Essays in Fiscal Federalism* (Washington: Brookings Institution, 1965), 30–31.
84. U.S. Advisory Commission on Intergovernmental Relations, *Trends in Fiscal Federalism 1954–1974* (Washington, 1975), 27. Grantham, "The New Southern Region," 19. Cobb, *Selling of the South,* 37–44. Michael I. Luger, "Federal Tax Incentives as Industrial and Urban Policy," in Sawers and Tabb, eds., *Sunbelt/Snowbelt,* 201–34. This analysis has neglected the argument that air-conditioning and climate-control technology was central to the rise of the Sunbelt South. That argument is advanced in Raymond Arsenault, "The End of the Long Hot Summer: The Air Conditioner and Southern Culture," *Journal of Southern History* 50 (Nov. 1984): 597–628. It should be noted, however, that the federal government introduced air conditioning to the public buildings of the South in the 1940s.
85. Hurley, *Decisive Years for Houston,* 33–34.
86. David E. Lilienthal, "The Restoration of Economic Equality Among the Regions of the United States," address before the Southern Newspaper Publisher's Association, Mineral Wells, Texas, 21 May 1940. Committee of Southern Regional Studies and Education of the American Council on Education, "Preparation, Distribution and Use of Instruction Materials Relating to the Use of the Resources of the Southern Region: A Report of the Work-Conference on Southern Regional Studies and Education," Gatlinburg, Tenn., 2–13 Aug. John Ferris, "Industrial Development in the Southeast," address before the Spring Meeting of the American Society of Mechanical Engineers, Chattanooga, 3 April 1946, p. 14. all in Records of the TVA, RG 142, TVA Board File Curtis-Morgan-Morgan 1933–57, FARC. See also TVA Commerce Dept., "Proposed Regional Industrial Research Program of the TVA and Other Associated Agencies in the Tennessee Valley Region," July 1941, Records of the TVA, RG 142, Records of the TVA Commerce Dept.; John P. Ferris, introductory talk at the Meeting of Directors of the Valley States Planning and Development Agencies and Other Organizations Interested in Industrial-Business Development, Fontana, N.C., 9 July 1947, Rec-

ords of the TVA, RG 142, TVA Board File Curtis-Morgan-Morgan 1933–57, p. 5. Both in FARC. World War II did not alter the South's share of research facilities. It stood at 6.5 percent of the national total in 1946. Hoover and Ratchford, *Economic Resources,* 77.

87. Census Bureau, *Statistical Abstract,* 1986 (Washington, 1986), 577. Select Committee on Small Business, *Role and Effect of Technology on the Nation's Economy,* 32–33. Luther H. Hodges, summary of remarks by Secretary of Commerce Luther H. Hodges, Conference of the American Industrial Development Council, 2 April 1962, Hodges Papers, p. 10. Clayton, *Economic Impact,* 129. Vatter, *United States Economy in the 1950s,* 165.
88. Census Bureau, *Statistical Abstract, 1986,* 577. While the Defense Dept.'s share of federally sponsored and total research and development expenditures fell off after the late 1960s, it resurged during the 1980s. It is also important to note that much privately funded research is oriented toward winning defense contracts. One study estimated that 30 percent of such private research was defense-oriented. Daniel Kevles, "The Remilitarization of American Science," lecture delivered at Stanford University, 29 April 1987. On the attitudes of American scientists, see McDougall, . . . *the Heavens and the Earth,* 160.
89. Select Committee on Small Business, *Role and Effect of Technology on the Nation's Economy,* 19, 28–33. General Accounting Office, *The Federal Role in Fostering University-Industry Cooperation* (Washington, 1983).
90. Select Committee on Small Business, *Impact of Defense Spending . . . 1962,* 28. Hearings Before the Subcommittee on Defense Procurement of the Joint Economic Committee, U.S. Congress, *Impact of Military Supply and Service Activities ion the Economy,* 88th Cong., 1st sess. (Washington, 1963), 147–48, 163. Seymour Melman, *Our Depleted Society* (N.Y.: Holt, Rinehart and Winston, 1965), 4, 72–73. Weidenbaum. "Transferability of Defense Industry Resources."
91. Summary of the Southern Governors Conference, Nashville Meeting, 24–27 Sept. 1961, Records of Gov. Burford Ellington, Tennessee State Library and Archives, Box 81, pp. 34–37. Luther H. Hodges, "Educations's Race with Catastrophe," address by Secretary of Commerce Luther H. Hodges at Wake Forest College Commencement, Winston-Salem, N.C., 3 June 1963, Hodges Papers, pp. 4–5. Luther H. Hodges, summary of remarks by Secretary of Commerce Luther H. Hodges, Conference of American Industrial Development Council, 2 April 1962, Hodges Papers.
92. SREB, "Within Our Reach: Report of the Commission on Goals for Higher Education in the South," Nov. 1961, in Records of Gov. Ellington, Tennessee State Library and Archives, Box 76, p. 6. Southern Regional Education Board, "The Southern Regional Education Program" (pamphlet), Jan. 1955, in Records of the Governor's Office, Records of Gov. Frank Goad Clement, Tennessee State Library and Archives, Nashville, Box 39. Board of Control for Southern Regional Education, minutes of meeting, 21 Nov. 1949, Biloxi, in Records of the Governor's Office, Records of Gov. Gordon Browning, Tennessee State Library and Archives, Nashville, Box 111. SREB, *The Academic Common Market* (Atlanta: SREB, 1976), 6. Regional Council for Education, minutes of 11 Oct. 1948 meeting at Atlanta, in Records of Gov. Browning, Tennessee State Library and Archives, Nashville, Box 111. Although the South pioneered this system, the Western and the New England

Regional Education Boards followed within five years. See Askew, "Introduction," 4.

93. SREB, *Statistics for the Sixties: Higher Education in the South* (Atlanta: SREB, 1963), 5, 63, 79. A 1951 survey revealed that almost half of the business leaders born in the South had left the region. By the time of a follow-up survey in 1977, only one-fifth of the nation's southern-born business leaders pursued their careers outside the region. Cobb, *Industrialization and Southern Society,* 56.
94. SREB, *Statistics for the Sixties,* 53. Summary of the Southern Governors Conference, Nashville Meetine, 24–27 Sept. 1961, p. 35. In 1986, the Southern Growth Policies Board declared: "The South still spends half the national average on research and development, *mostly because it has not yet successfully attracted federal funds.* Southerners are justifiably proud of high technology centers such as the Research Triangle Park, Cape Canaveral, Huntsville, or the technology corridor in Tennessee. Yet these are only models for the technological revolution which the South needs to spread more generally across the region [emphasis added]." SGPB, *Halfway Home and a Long Way to Go,* 22. On the region's deficiency in scientists, see Subcommittee on Employment and Manpower, *Impact of Federal Research and Development Policies on Scientific and Technical Manpower,* 32–33.
95. SGPB, *1986 Commission on the Future of the South: The Report of the Committee on Government Structure and Fiscal Capacity* (Research Triangle Park: SGPB, 1986), 14. SGPB, "Within Our Reach," Nov. 1961, pp. 1–2. Other regions linked R&D with economic development. The U.S. Dept. of Commerce asserted in 1967 that "There is little doubt in the minds of most development officials that the level of scientific and technical activity in any region, particularly in a lagging region, must be upgraded if the region is to improve its economic status." U.S. Dept. of Commerce, EDA, *Impact of Science and Technology on Regional Development* (Washington, 1967), 70. Hodges's files contained promotional literature from Princeton Research Park that served as a model for his own efforts. See, for example, "Princeton Research Park: The Ideal Location for Your Advanced Research Laboratory," in Records of Gov. Luther H. Hodges, Governors' Papers, North Carolina Dept. of Archives and History, Raleigh, Box 224.
96. Hodges, *Businessman in Statehouse,* 203–16. Luther H. Hodges, address by Secretary of Commerce Luther H. Hodges, 6th International Congress on Glass, Washington, D.C., 9 July 1962, Hodges Papers. George Esser, memorandum to George Simpson, 7 May 1957, Records of Gov. Luther H. Hodges, Governors' Papers, North Carolina Dept. of Archives and History, Box 224.
97. David Sylvester, "Cultivating High Tech," San Jose *Mercury News,* 14 May 1984, pp. 1D, 4D. W. John Moore, "High-Tech Hopes," *National Journal,* 15 Nov. 1986, p. 2771. Cobb, *Selling of the South,* 174. General Accounting Office, *Federal Role in University-Industry Cooperation,* 20.
98. Summary of remarks of Secretary of Commerce Luther H. Hodges to Huntsville Industrial Expansion Committee, Huntsville, Ala., 5 May 1962, Hodges Papers. The list of facilities can be found in the Records of Gov. G. Browning, Tennessee State Library and Archives, Nashville. NASA, *NASA's University Program.* See also Miller, "Low Down on High Tech," 39. Jim Cotham, Direc-

tor of Tennessee's Economic and Community Development Dept. in 1981 and 1982, explained that defense was "a very large part of the high tech recruiting campaign" that Gov. Lamar Alexander pursued in Tennessee. Cotham, quoted in *Our Own Worst Enemy,* 12.

99. Robert C. McMath, Jr., et al., *Engineering the New South: Georgia Tech, 1885–1985,* 212–14, 237, 360, 398, 409–12, 441–47.

100. *Ibid.* Kolchum, "Georgia Tech Facility," 73. State of Georgia, Office of Gov. Joe Frank Harris, press release, "Remarks of Governor Joe Frank Harris at the Dedication of the Advanced Technology Development Center," Atlanta, 14 May 1984.

101. Moore, "High-Tech Hopes," 2769–72. Kolchum, "Georgia Tech Facility," 74. "Microelectronics Center of North Carolina," *Appalachia* 17 (May–Aug. 1984): 7.

102. Miller, "Lowdown on High Tech," 36. Moore, "High-Tech Hopes," 2772. Sylvester, "Cultivating High Tech," 4D. Rosenfeld, "Divided South," 12–13. Cobb, "Y'All Come Down," 21. Southern leaders merely followed the prevailing wisdom that technological leadership had always fueled American growth. That assumption reflected the conclusions of several economic historians of 20th-century American growth, all of whom linked productivity growth to advances in education and technology. Consult Robert Solow, "Technical Change and the Aggregate Production Function," in Nathan Rosenberg, ed., *The Economics of Technological Change* (Middlesex, Eng.: Penguin Books, 1971), 344–62; Edward F. Denison, "United States Economic Growth," *ibid.,* 363–81; and Robert Solo, "The Capacity to Assimilate an Advanced Technology," *ibid.,* 480–88. In 1966 the EDA went so far as to publish a guide for research-hungry cities. U.S. Dept. of Commerce, Economic Development Administration, *A Research Park . . . Is It for Your Community?* (Washington, 1966).

103. SGPB, *Halfway Home and a Long Way to Go,* 20–22. Billings, *Planters and the Making of a "New South,"* 208. The analysis of federal aid to education is based on my currently unpublished manuscript, "State and Society in Postwar America: Federal Aid to Education as a Case Study, 1937–1970." Sources include the Frank P. Graham Papers (especially the records of Graham's service on the President's Advisory Commission on Education), the *Congressional Record,* transcripts of Congressional hearings, and the Governor's Office Records in the Tennessee State Archives. See also Chapter 7 below. For a summary of the political battles over federal aid to education, consult Diane Ravitch, *The Troubled Crusade* (N.Y.: Basic Books, 1983), 28–42.

104. Daniels, "An Autobiographical Adventure," 28. Monroe Kimbrel, "Non-Economic Dividends from Education," in *SREB: The Next 20 Years,* 18. Rupert Vance reiterated the aimlessness of the onetime southern New Dealers in 1960: "As the affluent society crosses the Mason-Dixon line, the regionalist of the 1930s turns up as just another 'liberal without a cause.'" See Vance, "The Sociological Implications of Southern Regionalism" in Reed and Sigal, *Regionalism and the South,* 211.

105. SGPB, *Halfway Home and a Long Way to Go,* 7, 29.

Chapter 7

1. Thomas D. Clark, *The Emerging South,* 2nd ed. (N.Y.: Oxford Univ. Press,

1968), 124–37. See also Pete Daniel, *Standing at the Crossroads* (N.Y.: Hill and Wang, 1986).

2. Committee on Education and Labor, U.S. House of Representatives, *Malnutrition and Federal Food Service Programs,* Hearings before the Committee on Education and Labor, 90th Cong., 2nd sess., Part 2, 4–10 June 1968 (Washington, 1968), 968, 980–81. Jack Bass, "Hunger? Let Them Eat Magnolias," in H. Brandt Ayers and Thomas H. Naylor, eds., *You Can't Eat Magnolias* (N.Y.: McGraw-Hill, 1972), 273.
3. SGPB, *Halfway Home and a Long Way to Go,* 15. On the tendency of rapid regional growth to leave those at the bottom of the economic ladder in distress, see Olson, "Causes and Quality of Southern Growth."
4. James L. Walker, *Economic Development and Black Employment in the Nonmetropolitan South* (Austin: Board of Regents of the Univ. of Texas, 1977), 41. For the debate over the applicability of the term "Sunbelt" to the South, see the following books: Sale, *Power Shift;* Kevin Phillips, *The Emerging Republican Majority* (Garden City: Anchor Books, 1970); Phillips, *Post-Conservative America* (N.Y.: Vintage, 1982); Tindall, *The Ethnic Southerners,* xi; Cobb, *Selling of the South,* 187; Woodward, *Thinking Back,* 140; SGPB, *1986 Commission on the Future of the South: Report of the Committee on Government Structure and Fiscal Capacity* (Research Triangle Park: SGPB), 9; David R. Goldfield, *Cotton Fields and Skyscrapers* (Baton Rouge: Louisiana State Univ. Press, 1982), 192. In the popular press, the issue was joined by Nicholas Leeman, "Searching for the Sunbelt," *Harper's,* Feb. 1982, pp. 14ff; "The Dark Side of the Sunbelt," *Newsweek,* 19 July 1982; and "The Second War Between the States," *Business Week,* 17 May 1976, pp. 92–114.
5. David C. Perry and Alfred J. Watkins, "People, Profit, and the Rise of the Sunbelt Cities," in Perry and Watkins, eds., *Rise of the Sunbelt Cities,* 295–96. The subemployment rate included the unemployed, discouraged workers (the unemployed no longer actively seeking work), involuntary part-time workers, and full-time workers earning hourly wages of less than $3.50. Cobb, *Industrialization and Southern Society,* 137. Bass and DeVries, *Transformation of Southern Politics, 218–19.*
6. Paul Recer, "The Texas City That's Bursting Out All Over," *U.S. News and World Report,* 27 Nov. 1978, p. 47. Bluestone and Harrison, *Deindustrialization,* 83–92. Barry J. Kaplan, "Houston: The Golden Buckle of the Sunbelt," in Richard M. Bernard and Bradley R. Rice, eds., *Sunbelt Cities: Politcs and Growth Since World War II* (Austin: Univ. of Texas Press, 1983), 196–212. Carl Abbott, "The American Sunbelt: Idea and Reality," *Journal of the West* 18 (July 1979): 5–18.
7. George Stamas, "The Puzzling Lag in Southern Earnings," *Monthly Labor Review* 104 (June 1981): 27–28. H. M. Douty, "Wage Differentials: Forces and Counterforces," *Monthly Labor Review* 91 (March 1968): 74–77. U.S. Senate, Subcommittee on Labor of the Committee on Labor and Public Welfare, *Amending the Fair Labor Standards Act of 1938,* 84th Cong., 1st sess. (Washington, 1955), 406–7. Robert S. Goldfarb and Anthony M. J. Yezer, "Evaluating Alternative Theories of Intensity and Regional Wage Differentials," *Journal of Regional Science* 16 (Autumn 1976): 353–61. Naylor and Clotfelter, *Strategies for Change in the South,* 31.
8. Goldfarb and Yezer, "Evaluating Alternative Theories, 353–61. Philip R. P. Coelho and Moheb A. Ghali, "The End of the North-South Wage Differen-

tial," *American Economic Review* 61 (Dec. 1971): 932–37. Bellante, "The North-South Differential and the Migration of Heterogenous Labor," 166–74.

The aggregate regional wage differential also declined between 1955 and 1980. Newman claims that wage effects—actual changes in southern wages relative to the rest of the nation—rather than composition effects—changes in the composition of the regional and national labor forces—account for the convergence and for the remaining differentials in 1980. Newman, *Growth in the American South,* 172. Michael Bradfield argues that the Capital/Labor ratio and the occupation mix reflect the same economic factors as the wage rate. In his mind, it is tautological to use the industry mix to explain the wage rate. See Bradfield, "Necessary and Sufficient Conditions To Explain Equilibrium Regional Wage Differentials," *Journal of Regional Science* 16 (1976): 247–55.

9. Philip L. Rones, "Moving to the Sun: Regional Job Growth, 1968 to 1978," *Monthly Labor Review* 103 (March 1980): 14. On the relative weights of the wage and composition effects on regional differentials, see Newman, *Growth in the American South,* 172. On the appeal of industrial promotion to low wage rather than high wage industries, see Cobb, *Selling of the South,* 219, 227.
10. Stamas, "Puzzling Lag," 28–30. Office of Economic Opportunity, Texas Dept. of Community Affairs, *Poverty in Texas 1973* (Austin: State of Texas, 1974), 93, 168. E. Walton Jones, "Farm Labor and Farm Policy," *Monthly Labor Review* 91 (March 1968): 12. Jones notes that returns to farm labor as compared with industry remained lowest in the South in 1968. William J. Stober, "Employment and Economic Growth: Southest," *Monthly Labor Review* 91 (March 1968): 16. SGPB, *Halfway Home and a Long Way to Go,* 7.
11. *Ibid.,* 7, 11. Krumm, *The Impact of the Minimum Wage on Regional Labor Markets,* 33, 58. The minimum wage augmented the South's demand for skilled labor, lessening the demand for unskilled labor. It also enhanced economic returns for skill.
12. James P. Mitchell, *The Skilled Work Force of the United States* (Washington, 1955), in Records of the Dept. of Labor, RG 174, Records of Secretary James P. Mitchell, NA. Melman, *Our Depleted Society,* 115. Huber, "Fiscal History of the South," Table 30. Krumm, *Impact of Minimum Wage on Regional Labor Markets,* 33, 58.
13. One study of black returnees to Birmingham listed social factors—older people on fixed incomes, people returning home, preference for southern style of life—as the dominant reasons for return. Economic motivations were cited much less often. Rex R. Campbell et al., "Return Migration of Black People to the South," *Rural Sociology* 39 (Winter 1974): 515–26. Marcus E. Jones, *Black Migration in the United States* (Saratoga, Calif.: Century Twenty-One Publishing, 1980), 8–98. Cobb, *Industrialization and Southern Society,* 55. David E. Kaun, "Negro Migration and Unemployment," *Journal of Human Resources* 5 (Spring 1970): 191–93. Newman, *Growth in the American South,* 145–50. Emory F. Via, "Discrimination, Integration, and Job Equality," *Monthly Labor Review* 91 (March 1968): 83–84. Cobb, *Selling of the South,* 118. Wage conditions slowly improved in the 1970s. See "Black-White Pay Gap Narrows as Skill Levels Converge," *Monthly Labor Review* 101 (Aug. 1978): 45. According to a recent *New York Times* report, reverse migration has accelerated during the late 1980s. See Kenneth R. Weiss, "80-Year Tide of Migra-

tion by Blacks Out of the South Has Turned Around," *New York Times,* 11 June 1989, p. 29.

14. Victor Perlo, *Economics of Racism U.S.A.* (N.Y.: International Publishers, 1975), 2–3, 33–36. Arnold Strasser, "Differentials and Overlaps in Annual Earnings of Blacks and Whites," *Monthly Labor Review* 94 (Dec. 1971): 21.
15. Frank E. Smith, address before the Huntsville Area Companies Community Resources Development Meeting, Huntsville, 16 Nov. 1967, in Records of the TVA, RG 142, Records of the TVA Board of Directors, Smith-Wagner-Hays-Welch 1957–72, FARC, Box 9. Luther H. Hodges, address by Secretary of Commerce Luther H. Hodges at Florida State University, Tallahassee, 20 April 1963, Hodges Papers.
16. Persky and Kain, "Migration, Employment, and Race in the Deep South." Walker, *Economic Development and Black Employment,* 4, 41–48.
17. *Ibid.,* 41–48.
18. David F. Ross, "State and Local Governments," in Marshall and Christian, eds., *Employment of Blacks in the South,* 79. Cobb, *Selling of the South,* 118. Rosenfeld, "Divided South," 15.
19. Ray Marshall and Virgil L. Christian, Jr., "Some South and Non-South Comparisons," in Marshall and Christian, eds., *Employment of Blacks in the South,* 181–82, 192–93, 200–201. Walker, *Economic Development and Black Employment,* 15.
20. Maynard Jackson, "Glory Hallelujah, While They're Trying to Sock It to You," in Ayers and Naylor, eds., *You Can't Eat Magnolias,* 127.
21. Ernest Hollings, quoted in Bass, "Hunger? Let Them Eat Magnolias," 276–77.
22. For committee memberships, consult the title pages of Joint Hearings Before the Committee on Education and Labor, U.S. Senate and the Committee on Labor, House of Representatives, *Fair Labor Standards Act of 1937,* 75th Cong., 1st sess. (Washington, 1937); Hearings Before the Subcommittee on Labor of the Committee on Labor and Public Welfare, U.S. Senate, *Amending the Fair Labor Standards Act of 1938,* 84th Cong., 1st sess. (Washington, 1955); Hearings Before the Subcommittee on Labor of the Committee on Labor and Public Welfare, U.S. Senate, *Amendments to the Fair Labor Standards Act,* 89th Cong., 1st sess. (Washington, 1965); and, Hearings Before the Subcommittee on Labor of the Committee on Labor and Public Welfare, U.S. Senate, *Fair Labor Standards Act Amendments of 1971,* 92nd Cong., 1st sess. (Washington, 1971). See also Robert X. Browning, *Politics and Social Welfare Policy in the United States* (Knoxville: Univ. of Tennessee Press, 1986), 32–36.
23. Jackson, "Glory Hallelujah," 129. Robert Coles, *Farewell to the South* (Boston: Little, Brown, 1972), 99–100. Frank E. Smith, *Look Away from Dixie* (Baton Rouge: Louisiana State Univ. Press, 1965) 41–42.
24. Numan V. Bartley ties this pattern to the peculiar "regional framework" of southern urbanization. See Bartley, "The Era of the New Deal As a Turning Point in Southern History," in Cobb and Namoroto, eds., *The New Deal and the South,* 145.
25. E. C. Gathings, quoted in U.S. House of Representatives, Hearings Before the Committee on Agriculture, U.S. House of Representatives, *Food Stamp Plan,* 88th Cong., 1st sess., 10–12 June 1963, Serial O (Washington, 1963), 35. "U.S. Government, Too Big?" (Editorial), New Orleans *Times-Picayune,* 11 Feb. 1962, sect. 2, p. 12. Charles Murray, *Losing Ground* (N.Y.: Basic

Books, 1984), 19. For a sampling of southern attitudes on poverty and anti-poverty programs, see the New Orleans *Times-Picayune,* 16 July 1961, 9 Feb. 1962, 4 July 1962, 24 Jan. 1964, and, the Atlanta *Journal and Constitution,* 23 and 28 July 1965, 22 Sept. 1965, and 28 Sept. 1966. The New Orleans *Times-Picayune* presented consistently strong opposition to the War on Poverty. The Atlanta *Journal and Constitution,* under the direction of Ralph McGill, the noted southern liberal, was generally supportive of such measures, although its news pages faithfully reported concerted opposition to them by ordinary Georgians and by Georgia politicians such as Eugene Talmadge and Bo Callaway.

26. Martin V. Melosi, "Dallas-Fort Worth: Marketing the Metroplex," in Bernard and Rice, eds., *Sunbelt Cities,* 162–65, 176–85.
27. On the national concern with "impacted" poverty and chronic unemployment, see U.S. Dept. of Commerce, EDA, *Regional Economic Development in the United States* (Washington, 1967), II-15, II-23, IV-20.
28. *Ibid.* Smith, *Look Away from Dixie,* 9–10. William L. Batt, Jr., Oral History Interviews, 26 July 1966, conducted by Jerry N. Hess, Harry S Truman Presidential Library, Independence, Mo., pp. 57–58. See also the related commentary of Alabama Representative Carl Elliott in 1955, in Dept. of Labor, Notes on Hearings Before the House Committee on Education and Labor on Amendments to the Fair Labor Standards Act, Records of the Dept. of Labor, RG 174, Records of Secretary of Labor James P. Mitchell, No. 14, 23 June 1955, NA.
29. David E. Lilienthal, "The Partnership of the Federal Government and Local Communities in the Tennessee Valley," address before the Decatur Chamber of Commerce, Decatur, Ala., 30 July 1942, in Records of the TVA, RG 142, TVA Board File Curtis-Morgan-Morgan 1933–57, Box 26. Frank E. Smith, letter to R. Sargent Shriver, 11 Sept. 1964, Records of the TVA, RG 142, Records of the TVA Board of Directors, Smith-Wagner-Hays-Welch 1957–72. Both in FARC. On the Kennedy program in general, see Matusow, *Unraveling of America,* 19–20, *passim.*
30. David E. Lilienthal letter to B. T. Gregory, 6 Aug. 1945, with Gregory's reply, 10 Aug. 1945, in Records of the TVA, RG 142, TVA Board File Curtis-Morgan-Morgan 1933–57, Box 72. Lawrence Durisch, memorandum to General Manger, 7 Feb. 1962, in Records of the TVA, RG 142, Records of the TVA Board of Directors, Smith-Wagner-Hays-Welch 1957–72, Box 12. TVA Government Relations and Economics Staff, "Industrial Development in the TVA Area," draft report, Feb. 1962, in Records of the TVA, RG 142, Records of the TVA Board of Directors, Paty-Pope 1939–57, Box 48. TVA Power Utilization Division, "Industrial Development in the TVA Area During 1955," Feb. 1956, in Records of the TVA, RG 142, Records of the TVA Board of Directors, Smith-Wagner-Hays-Welch 1957–72, Box 12. All in FARC.
31. Dewey W. Grantham, "TVA and the Ambiguity of American Reform," in Hargrove and Conkin, *TVA,* 318–19. See also Droze, "TVA, 1945–80: The Power Company," 66–88; Ruttan, "The TVA and Regional Development," 160; and Wheeler and McDonald, "The New Mission," 168–70.
32. John M. Peterson, memorandum to Lawrence Durisch, 15 Sept. 1955, in Records of the TVA, RG 142, Records of the TVA Board of Directors, Paty-Pope 1939–57, Box 47, FARC. Richard A. Couto, "New Seeds at the Grassroots," in Hargrove and Conkin, *TVA,* 233–37. Droze, "TVA, 1945–80: The

Power Company," 66–88. Ruttan, "The TVA and Regional Development," 160. Avery Leiserson, "Administrative Management and Political Accountability," in Hargrove and Conkin, *TVA,* 122–49. Wheeler and McDonald, "The New Mission," 168–70. Frank Goad Clement, speech at Decatur, Ala., 20 Oct. 1954, Records of Gov. Clement, Tennessee State Library and Archives, Nashville, Box 97. On the diminution of TVA's reform role, see Dewsey W. Grantham, "TVA and the Ambiguity of American Reform," in Hargrove and Conkin, *TVA,* 316–35.

33. EDA, *Regional Economic Development,* IV-45, IV-49, V-3–V-14. Advisory Commission on Intergovernmental Relations, *Multistate Regionalism,* 16–17. On the debate over how to define distressed areas, see James P. Mitchell, letter to Rep. William L. Dawson, 6 Feb. 1957, Records of the Dept. of Labor, RG 174, Records of Secretary James P. Mitchell, NA. See also Gerald Kraft et al., "On the Definition of a Depressed Area," in Kain and Meyer, eds., *Essays in Regional Economics,* 58–106; and James T. Patterson, *America's Struggle Against Poverty 1900–1980* (Cambridge: Harvard Univ. Press, 1981), 95.
34. *Ibid.* U.S. Dept. of Commerce, EDA, *Federal Activities Affecting Location of Economic Development,* Vol. 1 (Washington, 1970), I-12. EDA, *1979 Annual Report,* 51. EDA, *Regional Economic Development,* IV-42. Advisory Commission on Intergovernmental Relations, *Multistate Regionalism,* 42, 59. William H. Miernyk, "Local Labor Market Effects of New Plant Locations," in Kain and Meyer, *Essays in Regional Economics,* 168–70. The ARA, of course, was not entirely ineffective. James Nutler, Jr., Director of the Georgia Dept. of Industry and Trade, credited the ARA for having "created or salvaged" 3,625 jobs in Georgia. See "Georgian Goes To Bat for ARA," Atlanta *Journal and Constitution,* 1 May 1965, p. 5.
35. *Ibid.*
36. *Ibid.*
37. Advisory Commission on Intergovernmental Relations, *Multistate Regionalism,* 18–19. EDA, *Regional Economic Development,* V-5, V-6. Section 2 of the Appalachian Regional Development Act, reprinted in *Appalachia* 18 (March 1985): 25. David S. Walls and John Stephenson, Preface, in Walls and Stephenson, eds., *Appalachia in the Sixties* (Lexington: Univ. Press of Kentucky, 1972), xii–xiii. James C. Millstone, "East Kentucky Coal Makes Profits for Owners, Not Region," *ibid.,* 69–76. Dan Wakefield, "In Hazard," *ibid.,* 10–25. Harry Ernst and Charles H. Drake, "The Lost Appalachians," *ibid.,* 3–10. On the parallels between Kennedy's concern for Appalachia and Roosevelt's for the South, see, for example, the editorial "Appalachia Bill Is Aimed at Distress," Atlanta *Journal and Constitution,* 6 Feb. 1965, p. 4.
38. *Appalachia* 18 (March 1985): 25, 34–35. The original act authorized a budget of approximately one billion dollars over six years. Section 2 of the Appalachian Regional Development Act is reprinted on p. 25. Walls and Stephenson, eds., *Appalachia in the Sixties,* Preface, pp. xii–xiii. "Appalachian Governors Propose Finish-Up Program for ARC," *Appalachia* 15 (Nov./Dec. 1981–Jan./Feb. 1982): 1–3.
39. "Interview with Henry Krevor," *Appalachia* 13 (Jan–Feb. 1980): 2. President's Appalachian Regional Commission Staff, Summary of Recommendations of PARC Staff to the President's Appalachian Regional Commission, 30 Oct. 1963. John L. Sweeney (Executive Director of PARC), memorandum to Participants in PARC Conferences, 17–22 Nov. 1963. Both in Records of the

TVA, RG 142, Records of the TVA Board of Directors, Smith-Wagner-Hays-Welch 1957–72, FARC. Advisory Commission on Intergovernmental Relations, *Multistate Regionalism,* 25. "Sanders Wins Appalachia Job," Atlanta *Journal and Constitution,* 20 April 1965, p. 1. James S. Brown, "A Look at the 1970 Census," in Walls and Stephenson, eds., *Appalachia in the Sixties,* 143–44.

40. "Interview with Henry Krevor," 3. Appalachian Regional Commission, "Appalachia Governors Propose," 2. Advisory Commission on Intergovernmental Relations, *Multistate Regionalism,* 14, 34. EDA, *Regional Economic Development,* V-5, V-6.
41. John B. Waters, memorandum to Sam J. Ervin, 2 July 1970. Waters, letter to Ervin, 22 July 1970. Both in Samuel J. Ervin Papers, Southern Historical Collection, Univ. of N.C., Chapel Hill, Series II-Subject Files, Appalachian Regional Commission. Brown, "A Look at the 1970 Census," 141. Jerome Pickard, "A Decade of Change for Appalachia," *Appalachia* 14 (July–Aug. 1981): 1. Luther H. Hodges, address by Secretary of Commerce Luther H. Hodges at Dedication, Interstate Highway 40, Hickory, N.C., 26 May 1961, Hodges papers.
42. Advisory Commission on Intergovernmental Relations, *Multistate Regionalism,* 19, 42–43. Walls and Stephenson, eds., *Appalachia in the Sixties,* Preface, p. xiii.
43. *Ibid.* Coles, *Farewell to the South,* 95–105.
44. U.S. House, Hearings Before the Subcommittee on the War on Poverty Program of the Committee on Education and Labor, *Examination of the War on Poverty Program,* 89th Cong., 1st sess., 12–30 April 1965 (Washington, 1965), 17. Robert D. Reischauer, "Fiscal Federalism in the 1980s: Dismantling or Rationalizing the Great Society," in Marshall Kaplan and Peggy Cuciti, eds., *The Great Society and Its Legacy* (Durham: Duke Univ. Press, 1986), 182–83. Frances Fox Piven and Richard Cloward, *Regulating the Poor: The Functions of Public Welfare* (N.Y.: Vintage Books, 1971), 248–50, 270–71. Per capita federal budget outlays for national security dropped from 72 percent of the budget in 1954 to 63 percent in 1965 and also declined in absolute dollars. Per capita spending on domestic programs, on the other hand, increased absolutely and as a share of total expenditures. See U.S. House, *Examination of the War on Poverty Program,* 761.
45. U.S. Office of Economic Opportunity, *A Nation Aroused: 1st Annual Report* (Washington, 1965), 15–19. Office of Economic Opportunity, *The Quiet Revolution* (Washington, 1967), 11. Matusow, *Unraveling of America,* 247–50, *passim.*
46. Roger H. Davidson, "Poverty and the New Federalism," in Sar A. Levitan and Irving H. Siegel, eds., *Dimensions of Manpower Policy* (Baltimore: Johns Hopkins Univ. Press, 1966), 69–70. Piven and Cloward, *Regulating the Poor,* 126.
47. U.S. House, *Examination of the War on Poverty Program,* 113–14, 716. Office of Economic Opportunity. *The Quiet Revolution,* 13.
48. U.S. House, *Examination of the War on Poverty Program,* 116–24.
49. *Ibid.,* 45. Davidson, "Poverty and the New Federalism," 69–72.
50. Matusow, *Unraveling of America,* 255–56, 269–70. Patterson, *America's Struggle Against Poverty,* 147–48. "Atlanta Points the Way on Poverty War" (edi-

torial), Atlanta *Journal and Constitution,* 16 April 1965, p. 4. U.S. House, *Examination of the War on Poverty Program,* 400–414.

51. U.S. Senate, Hearings Before the Subcommittee on Intergovernmental Relations of the Committee on Government Operations, *Intergovernmental Cooperation Act of 1967 and Related Legislation,* 90th Cong., 2nd sess., 9–29 May 1968 (Washington, 1968), 321–22. U.S. House, *Examination of the War on Poverty Program,* 3, 6, 24–25.
52. U.S. House, *Malnutrition and Federal Food Service Programs,* 974–75. U.S. House, Hearings Before the Committee on Education and Labor, *Elementary and Secondary Education Amendments of 1967,* Part 2, 90th Cong., 1st sess., 9–20 March 1967 (Washington, 1967), 952.
53. U.S. Senate, Hearings Before a Subcommittee of the Committee on Agriculture and Forestry, *Food Distribution Programs,* 86th Cong., 1st sess., 4–8 June 1959 (Washington, 1959), 9–10, 33, 90. U.S. Senate, Hearings Before the Committee on Agriculture and Forestry, *Food Stamp Act of 1964,* 88th Cong., 2nd sess., 18–19 June 1964 (Washington, 1964), 69. U.S. Senate, Committee on Agriculture, Nutrition, and Forestry, *The Food Stamp Program: History, Description, Issues, and Options* (Washington, 1985), 3–18. Among the cities participating in the 1939–43 pilot program were Memphis, New Orleans, and Birmingham. See Smith, *New Deal in the Urban South,* 151.
54. *Ibid.*
55. U.S. Senate, *The Food Stamp Program,* 21–22. U.S. Senate, Hearings Before the Committee on Agriculture and Forestry, *Food Stamp Program and Commodity Distribution,* 91st Cong., 1st sess., 22–27 May 1969 (Washington, 1969), 131–32. Sar A. Levitan and Clifford M. Johnson, "Did the Great Society and Subsequent Initiatives Work," in Kaplan and Cuciti, eds., *The Great Society and Its Legacy,* 88. U.S. House, *Food Stamp Plan,* 14. U.S. Senate, *Food Stamp Act of 1964,* 11. U.S. House, *Malnutrition and Federal Food Service Programs,* 968.
56. U.S. Senate, *Food Stamp Program and Commodity Distribution,* 132, 139.
57. *Ibid.,* 199. U.S. Senate, *The Food Stamp Program,* 21–22, 266, 274–75. U.S. Senate, *Food Stamp Act of 1964,* 16–17, 32, 51. U.S. House, *Food Stamp Plan,* 14. Secretary Freeman also assured Senators Talmadge (D-Ga.), Johnston (D-S.C.), and Ellender (D-La.) that he would crack down on merchants who redeemed food stamps for alcohol and would see to it that government nutritionists considered the peculiar dietary needs of southerners.
58. U.S. House, *Malnutrition and Federal Food Service Programs,* 1053–54. "Food Stamps Prove Popular" (Editorial), Atlanta *Journal and Constitution,* 22 Oct. 1966, p. 4. Maurice MacDonald, *Food, Stamps, and Income Maintenance* (N.Y.: Academic Press, 1977), 91–96. Gene D. Sullivan, "Food Stamps: A Boost to the Southeastern Economy," *Monthly Review of the Federal Reserve Bank of Atlanta,* 58 (June 1973): 86–91. U.S. Senate, *Food Stamp Program and Commodity Distribution,* 196–97, 352.
59. U.S. Senate, *The Food Stamp Program,* 31–48. MacDonald, *Food, Stamps, and Income Maintenance,* ix, 5–10. U.S. Senate, *Food Stamp Program and Commodity Distribution,* 1–3.
60. U.S. Senate, *The Food Stamp Program,* 266, 274–75. One study of the "poverty reduction effect" of food stamps, the percentage of impoverished people who were raised out of poverty by the stamps, revealed lower than U.S. average poverty reduction effects in 9 of the 13 southern states. Kentucky, Lou-

isiana, South Carolina, and Texas were the exceptions. Of course, the depth of southern poverty and the niggardliness of southern welfare benefits were responsible for these low poverty reduction effects. See MacDonald, *Food, Stamps, and Income Maintenance,* 80–87.

61. Vance, *All These People,* 383–417. *Congressional Record,* 80th Cong., 1st sess., pp. 365–67. Harry L. Case, memorandum to George F. Gant, 7 Sept. 1949, Records of the TVA, TVA Board File Curtis-Morgan-Morgan 1933–57, FARC. The indices included median school years completed, teacher salaries, per pupil expenditures, value of facilities, and school-term lengths. President Roosevelt had personally witnessed the penury of southern public education during his years at Warm Springs. It so distressed him that he recited one discouraging anecdote in many of his official pronouncements. One day in 1924, FDR often recalled, a young man asked him to deliver the diplomas at a nearby high school commencement. "I said, 'Yes'," the president remembered, "and then asked, 'Are you the president of the graduating class?' He said, 'No, I am the principal of the school.'" The president subsequently discovered that the principal was only nineteen years old, had but one year of training at the University of Georgia, and was attending college every other year on the proceeds of his duties. Roosevelt, informal extemporaneous remarks at the dedication of a schoolhouse, Warm Spings, Ga., in Rosenman, ed., *Public Papers,* 1937 Vol., p. 136. He repeated the anecdote in his remarks before the Conference on Rural Education in 1944; see *Public Papers,* 1944 Vol., pp. 312–15.
62. *Congressional Record,* 78th Cong., 1st sess., pp. 8399, 8510. *Ibid.,* 80th Cong., 2nd sess., p. 3952. Vance, *All These People,* 380–417.
63. Paul T. David, memoranda to members of President's Advisory Commission on Education, 7 March, 29 March 1938, and 11 May 1939, Graham Papers. *Congressional Record,* 76th Cong. 1st sess., pp. 1345–46; 78th Cong., 1st sess., p. 8510; 75th Cong., 1st sess., pp. 1872–73. U.S. Senate, 76th Cong., 1st sess., Report No. 244, Part 2, Individual Views to Accompany S. 1305. Swain, *Pat Harrison,* 210–18. "Estimated Allotment of Federal Aid Funds Under the Provisions of the Harrison-Thomas-Fletcher Bill," 30 April 1938, Graham Papers. Title I of the 1938 Harrison-Fletcher bill earmarked 63.7 percent of its funds to the southern states.
64. *Congressional Record,* 78th Cong., 1st sess. pp. 8302–3; 80th Cong., 1st sess., p. 718; 80th Cong., 2nd sess., pp. 3941, 3951–52, 3958; 81st Cong., 1st sess., pp. 5686–87. Drury, *Senate Journal,* 368–69. When Senator Taft, saddled with the new responsibility of leading the majority party and harboring presidential ambitions, reversed field during the 1946 session and co-sponsored the federal aid to education measure, southern senators finally secured a roll-call vote on the proposal. The 1946 bill passed 58 to 22, with all but four southern senators in favor. The next year, with Lyndon Johnson inheriting the seat of the vehemently anti-New Deal Lee O'Daniel, the South cast only two nay votes.
65. *Congressional Record,* 75th Cong., 3rd sess., pp. 1938, 6602–3; 76th Cong., 3rd sess., pp. 4707–8; 80th Cong., 2nd sess., pp. 3941, 3951–52. David, memorandum to members, 29 March 1938, Graham Papers. U.S. Senate, 76th Cong., 1st sess., Report No. 244, Part 2, Individual Views to Accompany S. 1305. Holmes, *New Deal in Georgia,* 41–43. In the South Carolina state legislature, the House passed a resolution endorsing the Harrison federal aid

bill in 1938, while the Senate, bowing to the traditions of that body, opposed the bill as an infraction on state's rights. See James C. Derieux, letter to Frank P. Graham, 10 May 1938, with attached editorial from the Columbia (S.C.) *State,* Graham Papers. See also Dorough, *Mr. Sam,* 83.

66. *Congressional Record,* 78th Cong., 1st sess., 8306–7, 8488; 78th Cong., 1st sess., Appendix, pp. A4467, A4470. Frank P. Graham, quoted in Raleigh *News and Observer,* 7 June 1950, Frank P. Graham Papers. Lister Hill also was forced to retreat on this issue. See Hamilton, *Lister Hill,* 103. The statistical consequences of segregation almost assured the raising of the issue. In 1940, the southern schools showed the largest gap between black and white educational attainments in the nation, despite the low level of white achievement. In the academic year 1935–36, black schools posessed only one-fifth the value of property per pupil as did white schools in reporting states. In the Southeast, per pupil expenditures for blacks accounted for but 15 percent of the national average and only a quarter of the paltry sum distributed to southeastern whites.

Such statistics only hinted at the inaccessibility of public education for blacks in many southern towns. In the Progress District of Oxford, Mississippi, for instance, a white southerner named Gus Uth informed Frank Graham, white parents paid no tuition for a full complement of services, while blacks in the same district paid up to $49 in fees, built and maintained their own schools, and received from the school board only the meager teachers' salaries, monthly salaries of $4 less than the prevailing white rate in 1889. Vance, *All These People,* 382–83, 422–23. (The figures on value of school property do not include Kentucky, Louisiana, Oklahoma, and Tennessee which did not report such data. The statistics on per pupil expenditures are for academic year 1935–36.) Gus Uth, letter to Frank P. Graham, 19 Oct. 1939, Frank P. Graham Papers.

In the 1950s, this controversy realigned the regional positions of federal aid to education. On the regional level, however, support for federal aid to education waned. Long-time advocates of the policy, even architects of the program like Pepper and Graham, worried about the possibility of federal control. This fear became especially pronounced after 1954, when the Supreme Court legitimized the employment of federal aid, in Duke economist Calvin Hoover's anxious words, "as a backhanded weapon against segregation." James F. Byrnes led the southern retreat. In 1949, Byrnes admonished the Southern Governors Conference to halt calls for increased federal aid, and as governor of South Carolina, he followed through with a self-proclaimed, state-financed "Educational Revolution." Calvin B. Hoover, letter to Frank P. Graham, 24 May 1949, Graham Papers. Byrnes, *All in One Lifetime,* 405–8.

The main battleground nonetheless remained the floor of the Congress. In 1956, the Kelley bill, a bipartisan measure for federal aid for school construction, turned into a bitter battle, when Adam Clayton Powell, a New York Democrat, introduced an amendment which would deny federal funds to areas practicing segregation. By the next Congress, court-ordered intergration had proceeded so far that southern support for federal aid, a constant throughout the issue's history, had eroded. Citing potential interference with segregation, southern members of Congress followed Byrnes into opposition to federal aid in the mid-1950s. The strongest echoes of Pat Harrison and

Claude Pepper's support for equalization of education in 1958 came from northerners. *Congressional Record*, 84th Cong., 2nd sess., pp. 2120–21, 3058; 85th Cong., 1st sess., p. 10255. 85th Cong., 1st sess., Appendix, pp. A1004–5, A2444, A2743–44; 85th Cong., 2nd sess., pp. 873–74, 6205, 6212, 11684–85.

67. For a summary of the debates on federal aid to education see Ravitch, *Troubled Crusade*, 5, 28–42. See also U.S. House, Committee on Education and Labor, *Federal Interest in Education* (Washington, 1961), 18. On impact area aid, see U.S. Senate, Hearings Before the Subcommittee on Education of the Committee on Labor and Public Welfare, *Federally Impacted Areas*, 85th Cong., 2nd sess., 25–26 July 1958 (Washington, 1958), 10. U.S. Senate, Hearings Before the Subcommittee on Education of the Committee on Labor and Public Welfare, *Elementary and Secondary Education Act of 1966*, 89th Cong., 2nd sess., 1–27 April 1966 (Washington, 1966), 305–6.
68. U.S. Senate, Subcommittee on Education of the Committee on Labor and Public Welfare, *Federal Aid to States and Local School Districts for Elementary and Secondary Education* (Washington, 1954), 6, 13. U.S. Senate, *Federally Impacted Areas*, 53. U.S. Senate, Hearings Before the Subcommittee on Education of the Committee on Labor and Public Welfare, *Education Aid to Federally Impacted Areas—Public Law 874*, 91st Cong., 2nd sess., 28–29 April 1970 (Washington, 1970), 131–32.
69. U.S. House, Hearings Before the Subcommittee on Education of the Committee on Education and Labor, *Federal Grants to States For Elementary and Secondary Schools*, 87th Cong., 2nd sess., 28 March –28 Aug. 1962 (Washington, 1962), 335–36. U.S. House, *Federal Interest in Education*, p. III. U.S. Senate, Committee on Aeronautical and Space Sciences, *The Impact of the Space Age on Education in the United States* (Washington, 1972), 5. The *Sputnik* launch was not the first time military necessity focused attention on America's educational deficiencies. The Korean conflict, for instance, advertised the need to promote education in the South; the South's proportion of armed-forces recruits in Mental Group IV or V (rejected or deemed unsuitable for advanced training) was almost double the prevailing rates elsewhere in the United States. Consult James P. Mitchell, *The Skilled Work Force of the United States* (Washington, 1955), Records of the Dept. of Labor, Secretary of Labor James Mitchell, NA. The precise figures for percentage of armed-forces accessions in Group V (rejected) and Group IV (unsuitable for advanced training) by area were as follows:

Area I (Northeast and Mid-Atlantic), 26.1; Area II (PA, OH, KY, VA), 35.3; Area III (TN, NC, SC, GA, AL, MS, FL), 52.7; Area IV (TX, OK, ARK, LA, NM), 44.1; Area V (Middle West), 25.8; Area VI (Mountain and Pacific), 23.1; U.S. total, 33.0.

The percentage of failures on the Armed Forces Qualification Test for Fiscal Years 1953 and 1954 reflected a similar regional pattern:

Area I, 6.6; Area II, 8.0; Area III, 27.9; Area IV, 16.4; Area V, 4.6; Area VI, 6.0; U.S. total, 11.0.

For an eloquent example of a southern legislator's renewed concern for federal aid in the wake of the Soviet satellite launch, see the Senate address of J. William Fulbright of Arkansas, *Congressional Record*, 85th Cong., 2nd sess., p. 870.

70. U.S. House, Hearings Before the Committee on Eudcation and Labor, *Ele-*

mentary and Secondary Education Amendments of 1967, 90th Cong., 1st sess., 2–8 March 1967 (Washington, 1967), 1577–78. Matusow, *Unraveling of America,* 222–24.

71. Matusow, *Unraveling of America,* 188–94, 222–24.
72. *Ibid.* U.S. House, *Elementary and Secondary Education Amendments of 1967,* 31, 551. Naylor and Clotfelter, *Strategies for Change in the South,* 97–98.
73. U.S. House, *Elementary and Secondary Education Amendments of 1967,* 904–5, 921. Matusow, *Unraveling of America,* 188–94. Ravitch, *Troubled Crusade,* 268. See also Joe Brown, "30 School Systems Still Not Qualified," Atlanta *Journal and Constitution,* 13 Aug. 1966, p. 1.
74. *Ibid.* Chafe, *Civilities and Civil Rights,* 42–70, 102–18.
75. Naylor and Clotfelter, *Strategies for Change in the South,* 97–98. U.S. House, *Elementary and Secondary Education Amendments of 1967,* 50–51, 1414. U.S. House, Hearings Before the General Subcommittee on Education of the Committee on Education and Labor, *Financing of Elementary and Secondary Education,* 92nd Cong., 2nd sess., 28 Feb.–11 May 1972 (Washington, 1972), 159. The South pushed for even more favorable allocations in years after 1965. In 1967, Carl Perkins of Kentucky, the powerful chairman of the House Education and Labor Committee, proposed revisions in federal aid that would benefit the South at the expense of northern cities. See U.S. House, *Elementary and Secondary Education Amendments of 1967,* 324, 499–500.
76. MacDonald, *Food, Stamps, and Income,* 86–87. Atlanta *Journal and Constitution,* 7 Jan. 1965, p. 4. On LBJ's embrace of Keynesianism, consult Kearns, *Lyndon Johnson and the American Dream.* On Johnson's mixed feelings about the War on Poverty he unleashed, see Hodgson, *America in Our Time;* and Matusow, *Unraveling of America.*
77. U.S. House, *Examination of the War on Poverty Program,* 735–39. On Nixon-era reforms, see Patterson, *America's Struggle Against Poverty,* 192–97; and Reischauer, "Fiscal Federalism in the 1980s," 186–87.
78. Benjamin Chintz, "National Policy for Regional Development," in Kain and Meyer, eds., *Essays in Regional Economics,* 25. Browing, *Politics and Social Welfare Policy,* 9–20. Reischauer, "Fiscal Federalism in the 1980s," 181. Patterson, *America's Struggle Against Poverty,* 165. In terms of federal aid to state and local government, public welfare, health and hospitals, and housing and urban renewal accounted for 42.1 percent of total federal aid in 1958, 38.9 percent in 1963, and 38.9 percent in 1963, and 38.8 percent in 1968. See Advisory Commission on Intergovernmental Relations, *Measuring the Fiscal Capacity and Effort of State and Local Areas* (Washington, 1971), 34.
79. Janet E. Kodras and Lawrence A. Brown, "The Dissemination of Public Sector Innovations with Relevance to Regional Change in the United States," in A. T. Thwaites and R. P. Oakey, eds., *The Regional Impact of Technological Change* (London: Francis Pinter, 1985), 211. Clyde E. Browning, *The Geography of Federal Outlays: An Introductory and Comparative Inquiry* (Chapel Hill: Dept. of Geography, Univ. of N.C., 1973), 15–20. Stephen P. Dresch, "Assessing the Differential Regional Consequences of Federal Tax-Transfer Policy," in U.S. Dept. of Commerce, Economic Development Administration, *Proceedings, Regional Economic Development Research Conference, 19 April 1972* (Washington, 1972), 93–95. Census Bureau, *Statistical Abstract, 1971,* 293–94; *Statistical Abstract, 1973,* 311; *Statistical Abstract, 1977,* 257; *Statistical Abstract, 1986,* 381.

80. *Ibid.*
81. In 1980, only Virginia and Oklahoma showed AFDC benefits of more than 63 percent of the U.S. standard. Census Bureau, *Statistical Abstract, 1971,* 293–94; *Statistical Abstract, 1973,* 311; *Statistical Abstract, 1977,* 257; *Statistical Abstract, 1986,* 381. SGPB, *Report of the Committee on Human Resource Development: 1986 Commission on the Future of the South* (Research Triangle Park: SGPB, 1986), 19. The 1985 figure excludes Texas. Inclusion would have reduced the southern figure.
82. U.S. House, *Malnutrition and Federal Food Service Programs,* Part 2, p. 965. Kodras and Brown, "The Dissemination of Public Sector Innovations with Relevance to Regional Change in the United States," 195–211. Stephen P. Dresch, "Assessing the Differential Regional Consequences of Federal Tax-Transfer Policy," 93. Huber, "Fiscal History of the South." Piven and Cloward, *Regulating the Poor,* 126, 131, 186.
83. Office of Economic Opportunity, *As the Seed Is Sown: 4th Annual Report* (Washington, 1969), 91. Office of Economic Opportunity, *Quiet Revolution,* 98–102. Vernon M. Briggs, "Manpower Programs and Regional Development," *Monthly Labor Review* 91 (March 1968): 57–59. Arkansas and Kentucky took up half of the South's total slots between them. For a comprehensive analysis of all federal government manpower programs and of federally subsidized, privately operated ones, see Charles R. Perry et al., *The Impact of Government Manpower Programs* (Philadelphia: Trustees of the Univ. of Pennsylvania, 1975).
84. Advisory Commission on Intergovernmental Relations, *Federal Grants: Their Effects on State-Local Expenditures, Employment Levels, Wage Rates* (Washington, 1977), 33. "Federal Spending: North's Loss is Sunbelt's Gain," 882.
85. In 1960, six southern states spent smaller proportions of their budgets on public welfare than the national average, despite their acute need. In 1980, every southern state remained below the U.S. standard in per capita welfare expenditures. All but four exceeded the average in per capita highway spending. Region-wide, over the entire 1960–80 period, per capita highway expenditures more closely approached (or even exceeded) the national average than did public welfare outlays per capita. Census Bureau, *Statistical Abstract, 1962,* 423; *Statistical Abstract, 1966,* 427; *Statistical Abstract, 1972,* 419; *Statistical Abstract, 1982–83,* 285. Consult also Richard M. Bernard and Bradley R. Rice, "Introduction," in Bernard and Rice, eds., *Sunbelt Cities,* 15.
86. The welfare figures are for 1980 and 1982. Huber, "Fiscal History of the South," p. 2 and Tables. Census Bureau, *Statistical Abstract, 1962,* 423; *Statistical Abstract, 1966,* 427; *Statistical Abstract, 1972,* 419. *Statistical Abstract, 1982–83,* 285. Cobb, *Selling of the South,* 200. The data on expenditures for state industrial development programs cover the years 1967–74. SGPB, *Halfway Home and a Long Way to Go,* 17–18. In recent years, as public education became linked to economic development, the South has stepped up its expenditures on primary and secondary schools. In 1966–67, all southern states except Georgia devoted to highways a percentage of expenditure from state and local sources equal to or greater than the U.S. average, while they devoted a share smaller than the national average to public welfare, health, and hospitals (again excepting Georgia). Mississippi equaled the U.S. average in the public-welfare, health, and hospitals category. See Advisory Commission on Intergovernmental Relations, *Measuring Fiscal Capacity,* 116.

87. SGPB, *Halfway Home and a Long Way to Go.* Cobb, *Selling of the South,* 200–202.
88. Cobb, *Selling of the South,* 116. Cebula, "Local Government Policies and Migration," 85–93.
89. Unnamed southern Republican party official, interviewed in Alexander P. Lamis, *The Two-Party South* (N.Y.: Oxford Univ. Press, 1984), 26n.
90. U.S. House, *Malnutrition and Federal Food Programs,* 865, 892–96, 984–86. In Bennetsville, S.C., the proportion of students in black schools receiving free lunches was three times the proportion of students in white schools.
91. U.S. Senate, *Civil Rights—The President's Program,* 1963, pp. 364–65. U.S. Senate, Hearings Before the Subcommittee on Employment, Manpower, and Poverty of the Committee on Labor and Public Welfare, *Equal Employment Opportunity,* 90th Cong., 1st sess., 4–5 May 1967 (Washington, 1967), 69, 90. U.S. House, Hearings Before the General Subcommittee on Labor of the Committee on Education and Labor, *Equal Employment Opportunity Enforcement Procedures,* 91st Cong., 1st and 2nd sess., 1 Dec. 1969–10 April 1970 (Washington, 1970), 72–73, 99–100, 130, *passim.*
92. On the impact of the EEOC, see F. Ray Marshall et al., *Employment Discrimination.*
93. U.S. Senate, *Intergovernmental Cooperation Act of 1967 and Related Legislation,* 321. Matusow, *Unraveling of America,* 210–11. For white southern opposition to this policy, see, for example, U.S. Senate, *Civil Rights—The President's Program,* 1963, p. 63; and, "Talmadge Proposes Non-Medicare Aid," Atlanta *Journal and Constitution,* 1966, p. 6.
94. Rosengarten, *All God's Dangers,* 459. Neil R. McMillen, "Black Enfranchisement in Mississippi: Federal Enforcement and Black Protest in the 1960s," *Journal of Southern History* 43 (Aug. 1977): 361–62. U.S. House, *Examination of the War on Poverty Program,* 24. U.S. Senate, *Intergovernmental Cooperation Act of 1967 and Related Legislation,* 321–22. Piven and Cloward, *Regulating the Poor,* 133–34, 206–9.
95. Congress amended the Fair Labor Standards Act five times after 1949, raising the national hourly minimum wage from 75 cents to $3.35. The amendments also expanded coverage to trade, service, government, and agricultural employees. In 1961, for example, the federal minimum applied to only 3 percent of the workers in retail trade. By 1978, 78 percent of those workers were covered. See H. M. Douty, *The Wage Bargain and the Labor Market* (Baltimore: Johns Hopkins Univ. Press, 1980), 118; and Krumm, *Impact of the Minimum Age on Regional Labor Markets,* 8–13.

    Over the course of the revisions, the minimum-wage debate changed character. In 1949 the South remained the focus of the Fair Labor Standards Act. Similar regional concerns persisted in 1955. By 1955, however, many southern manufacturers had already dropped their opposition to increases in the minimum wage. Representatives of the powerful textile industry, leader of the opposition to the original FLSA in 1938, endorsed an immediate increase to an hourly standard of $1.00. So did the Southern Garment Manufacturers Association. Unlike textiles and garments, sawmills remained adamantly opposed to any further minimum-wage regulation. Hearings Before the Subcommittee on Labor of the Committee on Labor and Public Welfare, U.S. Senate, *Amending the Fair Labor Standards Act of 1938,* 84th Cong., 1st sess. (Washington, 1955), 113–14, 349, 481, 684–713, 755–813, 948–49, 176,

454, 1451. Dept. of Labor, Notes on Hearings, Records of the Dept. of Labor, RG 174, Records of Secretary James P. Mitchell, No. 13, 23 June 1955, No. 14, 23 June 1955, No. 20, 30 June 1955, NA. Kaun, "Economics of the Minimum Wage," 218. Dept. of Labor, Notes on Hearings, No. 9, 15 June 1955.

On skill upgrading in the four states of the Southwest (Texas, Arkansas, Oklahoma, and Louisiana) between 1940 and 1960, see Robert F. Smith, "Employment and Economic Growth: Southwest," *Monthly Labor Review* 91 (March 1968): 27. On the effects of the FLSA on skill levels see Krumm, *Impact of Minimum Wage on Regional Labor Markets,* 3, 28–33, 58; and H. M. Douty, "Wage Differentials: Forces and Counterforces," *Monthly Labor Review* 91 (March 1968): 76–77. On the effect of minimum-wage legislation on returns to skill, see Krumm, *Impact of Minimum Wage on Regional Labor Markets,* 3, 28–33, 58. For the Southern Garment Manufacturers Association endorsement, see William McComb, memorandum to James P. Mitchell, 23 July 1954, Records of the Dept. of Labor, RG 174, Records of Secretary James P. Mitchell, NA.

As Sar Levitan and Richard Belous have observed: "Few Government policies have been run through more statistical and econometric tests than the minimum wage. But quantity is not to be confused with quality. The methodology used in many wage floor studies appears to be flawed and the conclusions highly suspect." Levitan and Belous, "The Minimum Wage Today: How Well Does It Work?," *Monthly Labor Review* 102 (July 1979): 17. On the unreliability of studies, see Bradfield, "Necessary and Sufficient Conditions to Explain Equilibrium Regional Wage Differentials," and Peterson and Stewart, *Employment Effects of Minimum Wage Rates.* On employment effects of minimum wages, opponents and proponents of such legislation concur that minimum-wage hikes slow the growth of low wage industries, have long-term adverse employment effects on marginal workers, accelerate mechanization and upgrading of labor skill levels, and rechannel employment growth toward more capital-intensive lines. Since the South, in regional terms, and youth, in demographic terms, were the labor markets most closely tied to low wage employment, those markets were most severely affected by the FLSA. See Kaun, "Economics of the Minimum Wage," 92, 130, 148–60, 294; Wright, *Old South, New South,* 254; Peterson and Stewart, *Employment Effects of Minimum Wage Rates,* 78–79, 104–7, 153–55; and Levitan and Belous, "Minimum Wage Today," 18. Whatever its employment effects, the minimum wage has lived up to expectations—that is, for adult workers, the income gains due to wage floors more than offset the diminished opportunity to find work. Levitan and Belous, "Minimum Wage Today," 18.

96. Ralph R. Widner, quoted in "Federal Spending: The North's Loss Is the Sunbelt's Gain," 879. Cobb, *Selling of the South,* 199. EDA, *Regional Economic Development,* p. IV-20. Blacks migrated toward areas with established black communities. See Raymond, "Determinants of Non-White Migration," 15–16. On the declining economic prospects of black males, especially of young black males, over the course of the post-World War II period, consult Cogan, "Decline in Black Teenage Employment: 1950–1980," 621–38; and Vance, "Education and the Southern Potential," 278–80.

97. The signal documents of the Frostbelt-Sunbelt controversey are: Sale, *Power Shift;* Phillips, *Emerging Republican Majority;* "Federal Spending: The North's

Loss Is the Sunbelt's Gain," 878–91; Joel Havemann and Rochelle Stanfield, "Neutral Federal Policies Are Reducing Frostbelt-Sunbelt Spending Imbalances," *National Journal* 13 (7 Feb. 1981): 233–36; "The Second War Between the States," *Business Week,* 17 May 1976, pp. 92–114. The South's denial of the charges of favoritism in federal spending can be found in Clyde E. Browning, "The Role of the South in the Sunbelt-Snowbelt Struggle for Federal Funds," in Black and Reed, eds., *Perspectives on the American South,* Vol. 1, pp. 253–62; and C. L. Jusenius and L. C. Ledebur, "A Myth in the Making," 131–73. The controversy is ably summarized in Cobb, *Selling of the South,* 179–207—(the Coalition of Northeastern Governors is quoted on p. 197). On black youth unemployment see Cogan, "The Decline in Black Teenage Unemployment," 621–38, and, Hearings Before a Subcommittee of the Select Committee on Small Business, U.S. Senate, *The Role and Effect of Technology in the Nation's Economy,* 88th Cong., 1st sess., p. 260.

98. "The Second War Between the States," 96. See also the remarks of New England State governors Charles F. Hurley, Wilbur L. Cross, and George D. Aiken, in "The South's Advantages," *Manufacturers' Record* 107 (Dec. 1938): 42.

99. Southern Growth Policies Board, *Report of the Committee on Human Resource Development.* Cobb, *Selling of the South,* 199. Bluestone and Harrison, *Deindustrialization,* 87–88. "Dark Side of the Sunbelt." In 1979, 12.5 percent of the U.S. population lived in poverty. Of the southern states, all but Virginia had higher poverty rates, while among nonsouthern states only Maine and New York exceeded the national average. See O'Rourke, "Demographic and Economic Setting of Southern Politics," 17n.

100. Winter, quoted in Rosenfeld, "Divided South," 10. On the effects of the FLSA on skill levels, see Krumm, *Impact of Minimum Wage on Regional Labor Markets,* 3, 28–33, 58; and H. H. Douty, "Wage Differentials: Forces and Counterforces," *Monthly Labor Review* 91 (March 1968): 76–77. Census Bureau, *Statistical Abstract, 1984,* 475. SGPB, *Report of the Committeee on Human Resource Development,* 17. Naylor and Clotfelter, *Strategies for Change in the South,* 229–30. Huber, "Fiscal History of the South," Tables.

101. Lamis, *The Two-Party South,* 37–41. Cuomo enlarged on the two Americas speech in his 1985 commencement address at Stanford University; see *Campus Report,* 19 June 1985, p. 22. Consult also Gordon et al., *Segmented Work, Divided Workers.*

102. "The South: A Show-Me Attitude," *Time,* 24 Jan 1977, p. 12. Havard, "Southern Politics," in Rubin, ed., *The American South,* 44.

Chapter 8

1. Lilienthal, *Journals,* Vol. 1, p. 79. Lilienthal himself acquiesced in this shift and he recognized it. In 1941, Lilienthal confessed in his diary that the industrialization he was then championing offered few jobs and even fewer opportunities for skilled employment and social mobility to the region's workers (pp. 361–62).

2. Hodding Carter, Jr., quoted in James C. Cobb, "Beyond Planters and Industrialists: A New Perspective on the New South," *Journal of Southern History* 54 (Feb. 1988): 59. Lyndon Johnson was one southern liberal who retained the emphasis on regional development and firmly hoped to end the civil-rights

dilemma and address the bedrock problems of southern poverty. See Kearns, *Johnson and the American Dream,* 148–49, 191.

3. Vatter, *The United States Economy in the 1950s,* 7. Barry P. Bosworth, "The Evolution of Economic Policy," in Kaplan and Cuciti, eds., *The Great Society and Its Legacy,* 32–42. For a general description of this shift, consult Matusow, *Unraveling of America;* and Hodgson, *America in Our Time.* See also the remarks of Leon Keyserling on the War on Poverty in U.S. House, *Examination of the War on Poverty Program,* 735–39. A recent Princeton Univ. Press volume is dedicated to these issues. Consult Steve Fraser and Gary Gerstle, eds., *The Rise and Fall of the New Deal Order* (Princeton: Princeton Univ. Press, 1989).
4. Gøsta Esping-Andersen, *Politics Against Markets: The Social Democratic Road to Power* (Princeton: Princeton Univ. Press, 1985), 8, 193–236. Ian Gough, "State Expenditure in Advanced Capitalism," *New Left Review* 92 (1975): 53–80. Ian Gough, *The Political Economy of the Welfare State* (London: Macmillan, 1978), xi, 102–3. I am also indebted to the forthcoming work of Peter Baldwin, *The Politics of Social Solidarity and the Bourgeois Basis of the European Welfare State, 1875–1975* (N.Y.: Cambridge Univ. Press, forthcoming).
5. Higgs, *Crisis and Leviathan,* 238. Esping-Andersen, *Politics Against Markets,* 159. Gough, *Political Economy of the Welfare State,* 102–3. Gough, "State Expenditure in Advanced Capitalism," 59–61. See also R. M. Titmuss, *Social Policy* (N.Y.: Pantheon, 1974); and Leuchtenburg, "The Pertinence of Political History." In 1972, for example, the U.S. devoted twice the share of GDP to military expenditures as did France and Germany, and nearly one-and-a-half times that of England.
6. Daniels, "An Autobiographical Adventure." For an example of liberal southerners lamenting the taint of defense dollars, see Miller, "Lowdown on High Tech," 39. On the continued domination of "outside interests," see Jacobs, "Why TVA Failed," 41–47.
7. For a detailed review of the relationship between civil-rights tactics, business conditions, and white responses to desegregation, and of the extant historical literature on these subjects, see Schulman, "From Cotton Belt to Sunbelt," 329–40. See also McMillen, "Black Enfranchisement in Mississippi," 361. On the 1976 election, consult Andrew Young, address at the funeral of Fannie Lou Hamer, 21 March 1977, quoted in Steven F. Lawson, *In Pursuit of Power* (N.Y.: Columbia Univ. Press, 1985), 256.
8. *Ibid.* Harry S. Ashmore, *An Epitaph for Dixie* (N.Y.: Norton, 1958). Nicholls, *Southern Tradition and Regional Progress,* 5. Elizabeth Jacoway, "Introduction," in Elizabeth Jacoway and David Colburn, eds., *Southern Businessmen and Desegregation* (Baton Rouge: Louisiana State Univ. Press, 1982), 11–12. Elizabeth Jacoway, "Taken By Surprise," *ibid.,* 15–16.

   There is a rich literature on the relationships between business communities, economic aspirations, and civil-rights strategy in southern cities. The unusual story of Huntsville, Ala.—"Rocket City, USA"—can be found in the *New York Times,* May 23, 1965; and Charles, "Star Wars Fell on Alabama," 748–49. On Little Rock, see Rocco C. Siciliano, memorandum to James P. Mitchell, 26 March 1956, with attached letter of Robert C. Goodwin, Records of the Dept. of Labor, RG 174, Records of Secretary of Labor James P. Mitchell, NA; Hays, *A Southern Moderate Speaks,* 130–94; Cobb, *Selling of the South,* 123–25; Jacoway, "Taken by Surprise," 15–41; and Yates, "Arkansas: Independent and Unpredictable," 265. On New Orleans, see Morton Inger, "The New Orleans School Crisis of 1960," in Jacoway and Colburn, eds., *Southern Businessmen and Desegregation,* 83–85, 97; and Haas, *Delesseps S. Morrison and the Image of Reform,* 68, 257–79. Atlanta's peaceful desegregation is chronicled in Bradley R. Rice, "If Dixie Were Atlanta," in Bernard and Rice, eds., *Sunbelt Cities,* 48–49; and McMath et al., *Engineering the New South,* 312. See also Inger, "New Orleans," 97.

Atlanta's rapid progress toward black enfranchisement went deeper than Mayor William Hartsfield's pronouncement that his boomtown was "too busy to hate." It reflected the city's unusual political and economic situation. Consult Bass and DeVries, *Transformation of Southern Politics,* 149; Alton Hornsby, Jr., "A City That Was Too Busy to Hate," in Jacoway and Colburn, eds., *Southern Businessmen and Desegregation,* 121–32; Rice, "If Dixie Were Atlanta," 45–49; and Cobb, *Selling of the South,* 127–28.

Dallas offers the most unusual case study of business-dominated racial peace without racial progress. On Dallas, see William Brophy, "Active Acceptance—Active Containment," in Jacoway and Colburn, eds., *Southern Businessmen and Desegregation,* 139–45; and Melosi, "Dallas–Fort Worth," 188–89.

For the intransigent cities, consult Joe David Brown, "Birmingham, Alabama," *Saturday Evening Post* 236 (2 March 1963): 13–18; Marshall and Christian, "Economics of Employment Discrimination," 226; Robert Corley, "In Search of Racial Harmony," in Jacoway and Colburn, eds., *Southern Businessmen and Desegregation,* 170–90; Haas, "The Southern Metropolis," 181–82; David R. Colburn, "The Saint Augustine Business Community," *ibid.,* 225; Charles Sallis and John Q. Adams, "Desegregation in Jackson, Mississippi," *ibid.,* 237–48; Anne Trotter, "The Memphis Business Community and Integration," *ibid.,* 282–300; and Robert A. Sigafoos, *Cotton Row to Beale Street* (Memphis: Memphis State Univ. Press, 1979), 336–37. See also Carl Abbott, "The Norfolk Business Community," in Jacoway and Colburn, eds., *Southern Businessmen and Desegregation,* 99, 118; and on business behavior in Greensboro, N.C., William H. Chafe, *Civilities and Civil Rights* (N.Y.: Oxford Univ. Press, 1980), 88, 206–8, 234.

9. C. Vann Woodward, "New South Fraud Is Prepared by Old South Myth," *Washington Post,* 9 July 1961. Jacoway, "Introduction," *Southern Businessmen and Desegregation,* 2–13. Black, *Southern Governors and Civil Rights,* 338–41. Ashmore, *Epitaph for Dixie.*
10. Jacoway, "Introduction," *Southern Businessmen and Desegregation,* 12.
11. Sosna, *In Search of the Silent South,* 167–211. Hays, *A Southern Moderate Speaks.* Hays, *Politics Is My Parish.* Smith, *Look Away from Dixie.*
12. Woodward, "New South Fraud." Jacoway, "Introduction," *Southern Businessmen and Desegregation,* 12. Wright, *Old South, New South,* 265–66.
13. Black, *Southern Governors and Civil Rights,* 338–41. Cobb, *Industrialization and Southern Society,* 114–15. Cobb, *Selling of the South,* 141--49. Smith, *Look Away from Dixie,* 37.

    For examples of federal aid as a campaign issue in the South, see Luther H. Hodges, address at Democratic Breakfast, Lake Charles, La., 19 Oct. 1964, Hodges Papers; and Lamis, *The Two-Party South,* 168.
14. Thomas R. Wagy, *Governor LeRoy Collins of Florida* (University: Univ. of Alabama Press, 1985).

15. *Ibid.*
16. *Ibid.*
17. *Ibid.* See also Steven F. Lawson, "From Sit-in to Race Riot," in Jacoway and Colburn, eds., *Southern Businessmen and Desegregation,* 257–81.
18. See the Records of Gov. Luther H. Hodges, Governors' Papers, North Carolina Department of Archives and History, Raleigh. For Hodges's own version of his administration, see his *Businessman in the Statehouse,* 29–42, 74–77, 136–37, 155, 177, 192–93.
19. *Ibid.*, 80–119. Chafe, *Civilities and Civil Rights,* 49–60.
20. Howard Lee, quoted in Bass and DeVries, *Transformation of Southern Politics,* 222. Preston Edsall and J. Oliver Williams, "North Carolina: Bipartisan Paradox," in Harvard, ed., *The Changing Politics of the South,* 382–87. Chafe, *Civilities and Civil Rights,* 103–5, 131–33.
21. Chafe, *Civilities and Civil Rights,* 76–79. Cobb, *Selling of the South,* 147.
22. Whitney Young, quoted in Rice, "If Dixie Were Atlanta," 50. Wagy, *LeRoy Collins,* 77. Abbott, "Norfolk Business Community," 105. Chafe, *Civilities and Civil Rights,* 41. Ironically, the real extremists in Greensboro, both the city's leader of right-wing terror and its most ardent black revolutionary were paid informants of the FBI. See Chafe, *Civilities and Civil Rights,* 201.
23. *Ibid.*, 86, 238. Furthermore, after the 1960s, the term "tokenism" gathered perhaps undeserved pejorative connotations. Certainly, the few black students who entered all-white schools experienced unnatural pressures. When one black student in Atlanta requested transfer back to a black school, segregationists, despite the student's excellent grades, greeted the news as confirmation of black inferiority. Nonetheless, these token blacks not only excelled, but quickly became community leaders. For instance, Harvey Gantt, the first black admitted to Clemson University in 1963, later won the mayoralty in Charlotte. More important, unlike tokenism in schooling, tokenism in employment often set the stage for more significant breakthroughs. The first black in a firm showed that whites would not flee the integrated workplace. When employers felt additional pressure, and it always required additional pressure, tokenism made compliance easier. See Coles, *Farewell to the South,* 153–54; and Steven Gelber, *Black Men and Businessmen* (Port Washington: Kennikat Press, 1974), 5.
24. Rice, "If Dixie Were Atlanta," 50. Chafe, *Civilities and Civil Rights,* 242. Goldfield, *Cotton Fields and Skyscrapers,* 175–76. On comparative black political gains, see Bass and DeVries, *Transformation of Southern Politics;* and Lamis, *The Two-Party South.*
25. Lawson, *In Pursuit of Power,* 16–21, 181, *passim.* Richard Engstrom, "Black Politics and the Voting Rights Act, 1965–1982," in James F. Lea, ed., *Contemporary Southern Politics* (Baton Rouge: Louisiana State Univ. Press, 1988), 90. Black, *Southern Governors and Civil Rights,* 322–30. Bass and DeVries, *Transformation of Southern Politics,* 51–52.
26. *Ibid.* See also Hays, *A Southern Moderate Speaks,* 5. Bass and DeVries, *Transformation of Southern Politics,* 248–49.
27. *Ibid.*, 46. Lawson, *In Pursuit of Power,* 195–200. Smith, *Look Away from Dixie,* 74. Blacks negotiated the terms of these coalitions both by voting *en bloc* and by applying their influence in the national Democratic party. Black power inside the national party forced state officials either to enlist blacks into their councils of power or to forfeit recognition from the national party. Just as

black strength in the North had led the national Democrats to give up the alliance with the South and undercut the southern economic elites after the 1936 election, so increasing black electoral strength both North and South led the national Democrats to recruit black delegates and eject the old segregationist leadership. See sources cited above. On the gradual transformation of the national Democratic party on civil rights, consult also Garson, *Democratic Party and the Politics of Sectionalism;* Hamby, *Beyond the New Deal;* Sitkoff, "Harry S Truman and the Election of 1948"; and Sosna, *In Search of the Silent South.* On the separatist efforts in Mississippi and Alabama, consult Lawson, *In Pursuit of Power,* 116–17, 179–83; and Lamis, *The Two-Party South,* 44–62, 80–90.

28. *Ibid.*, 99–103, *passim.* Bass and DeVries, *Transformation of Southern Politics,* 101.
29. Jesse Jackson, quoted in David Broder, "For 1988, Media Must Take Jackson Seriously," San Jose *Mercury News,* 4 June 1987.
30. Between 1964 and 1976, the number of registered voters in the eleven ex-Confederate states grew by 57 percent. The black electorate rose 92 percent, the white total 52 percent. Nonetheless, 79 percent of the new registrants were white. Earl Black, "Competing Responses to the 'New Southern Politics': Republican and Democratic Southern Strategies," in Merle Black and John Shelton Reed, eds., *Perspectives on the American South,* Vol. 1 (N.Y.: Gordon and Branch, 1981), 151. Eisenberg, "Virginia," 81. Havard, "From Past to Future," 711. Lamis, *The Two-Party South,* 32–39, 213. See also Black and Black, *Politics and Society in the South.*
31. Lamis, *The Two-Party South,* 115, 191, provides registration information for Florida and Louisiana.
32. John Van Wingen and David Valentine, "Partisan Politics: A One-and-a-Half, No Party System," in Lea, ed., *Contemporary Southern Politics,* 141. Alan I. Abramowitz, "Ideological Realignment and the Nationalization of Southern Politics: Party Activists and Candidates in Virginia," in Black and Reed, eds., *Perspectives on the American South,* Vol. 1, p. 100. Bass and DeVries, *Transformation of Southern Politics,* 25–26. Lamis, *The Two-Party South,* 102–3.
33. Dauer, "Florida: The Different State." Bass and DeVries, *Transformation of Southern Politics,* 25–26, 117. Lamis, *The Two-Party South,* 181, 202, 217. Kevin P. Phillips, *The Emerging Republican Majority* (N.Y.: Anchor Books, 1970), 275. For a similar age-group analysis of the 1972 returns in North Carolina, see Sale, *Power Shift,* 123.
34. Lamis, *The Two-Party South,* 18–19. Bass and DeVries, *Transformation of Southern Politics,* 27. Bartley, *From Thurmond to Wallace,* 61.
35. Black, "Competing Responses to the 'New Southern Politics': Republican and Democratic Southern Strategies," 162. Black concludes that the Republican Southern Strategy succeeded only against a crippled opposition party. Lamis, *The Two-Party South,* 33, *passim.* Bass and DeVries, *Transformation of Southern Politics.* Phillips, *The Emerging Republican Majority,* 204. In 1970 a Republican official explained to Thomas H. Naylor that the Southern Strategy was to ignore the race issue. "The race issue," the official explained, was "a tool which should be used only when it is necessary to win elections. Otherwise, it should be ignored, the less said about it the better." Quoted in Thomas H. Naylor, "A Southern Strategy," in Ayers and Naylor, eds., *You Can't Eat Mag-*

nolias, 338. See also Harold G. Grasmick, "Rural Culture and the Southern Wallace Voter," *Rural Sociology* 39 (Winter 1974): 454–70.

36. Kevin P. Phillips, *Post-Conservative America* (N.Y.: Vintage Books, 1982), 31–39. Sale, *Power Shift, passim.* Thomas Ferguson and Joel Rogers, *Right Turn* (N.Y.: Hill and Wang, 1986). Lamis, *The Two-Party South,* 25–26. Charles Fortenberry and F. Glenn Abney, "Mississippi: Unreconstructed and Unredeemed," in Havard, ed., *Changing Politics of the South,* 506. Lawson, *In Pursuit of Power,* 5.

 Populist conservatism thrived in the South, even if most of its adherents remained within the Democratic party. And it amounted to nothing less than a reversal of the southern populist tradition. Whether advocating white supremacy, as did Tom Watson, Theodore "the Man" Bilbo, and George Wallace, or minimizing the race issue, as did Huey Long and Big Jim Folsom, southern populism had long retained a commitment to liberal economic policies. Specifically, it advocated the use of state power to restrain the haves and help the have-nots. But for the poor white voter of the 1970s and 1980s, the model southern populist was neither Long nor Bilbo, but Eugene Talmadge, father of Herman Talmadge and Georgia's dominant politician of the 1930s and 1940s. Talmadge, "the Wild Man from Sugar Creek," combined virulent racism with conservative political economy. He opposed the New Deal and sought to pare government to a bare minimum. Such a program appealed to Goldwater and Wallace voters in the 1960s and 1970s. With a moderation of the racial rhetoric, it enlisted their support for Reagan in 1980 and 1984. As one commentator reflected in 1982, poor southern whites rejected Darwinism in biology, but they have come to accept Social Darwinism in political economy. On populism and state power, see C. Vann Woodward, *Tom Watson: Agrarian Rebel* (N.Y.: Macmillan, 1938); Alan Brinkley, *Voices of Protest* (N.Y.: Vintage, 1983); and, Brinkley, "The New Deal and Southern Politics."

37. Fred Hobson, "A South Too Busy to Hate?," in Fifteen Southerners, *Why the South Will Survive* (Athens: Univ. of Georgia Press, 1981), 46. Goldfield, *Cotton Fields and Skyscrapers,* 147. Key, *Southern Politics.*

38. For poll data on southern attitudes, see Lamis, *The Two-Party South,* 221; and Black and Black, *Politics and Society in the South,* 213–30.

39. Grady's remarks in Saint Petersburg *Times,* 3 Nov. 1976; East's in Charlotte *Observer,* 28 Oct. 1980. Both quoted in Lamis, *The Two-Party South,* 141–42, 187.

40. Bass and DeVries, *Transformation of Southern Politics,* 189. Cobb, *Selling of the South,* 81. Similarly, in his 1967 "Year-end Report to the People of North Carolina," Gov. Dan Moore boasted that announcements of new plants and expansions averaged better than one per day. Gov. Dan Moore, "Year End Report to the People of North Carolina," 29 Dec. 1967, in the Ervin Papers.

41. Julian Bond, "Better Voters, Better Politicians," *Southern Exposure* 7 (Spring 1979): 68–69. Lawson, *In Pursuit of Power,* 272, 275. Cobb, *Selling of the South,* 189–90.

42. "Excerpts from the Inaugural Addresses of Five Southern Governors," in Ayers and Naylor, eds., *You Can't Eat Magnolias,* 358–67. Edward M. Yoder, Jr., "Southern Governors and the New State Politics," *ibid.,* 160–63. See also Lamis, *The Two-Party South;* Bass and DeVries, *Transformation of Southern Politics;* Black and Black, *Politics and Society in the South;* and Bartley, *From Thurmond to Wallace.*

43. Todd A. Wade, interview with the author, Palo Alto, Calif., 24 March 1987. Huber, "Fiscal History of the South," Tables. Department of Defense, *Prime Contract Awards, By State, Fiscal Years 1951–1977,* 2. *Department of Defense, Prime Contract Awards, By Region and State, Fiscal Years 1978, 1979, 1980,* A-4. Bass and DeVries, *Transformation of Southern Politics,* 188–89. Late in the 1980s, a group of young, Harvard-educated reform Democrats came to power in Mississippi. Dubbed the Yuppies of Mississippi, they promised "basic, drastic change." See Peter J. Boyer, "The Yuppies of Mississippi," *New York Times Magazine,* 28 Feb. 1988, pp. 24–27, 40–43.
44. Tindall, *The Ethnic Southerners,* 241.
45. "Two-Party Romance Needs Solid Basis" (Editorial), New Orleans *Times-Picayune,* 11 Nov. 1962, Section 2, p. 8.
46. Goldfield, *Promised Land,* 153. Miller and Pozetta, *Shades of the Sunbelt,* x. Naylor and Clotfelter, *Strategies for Change in the South,* 10.
47. I am indebted to an anonymous reader of this manuscript for this insight. See also O'Rourke, "The Demographic and Economic Setting of Southern Politics," 19.
48. Cobb, *Industrialization and Southern Society,* 42–43. Goldfield, *Promised Land,* 217.
49. Emanuel deKadt, *Tourism—Passport to Development* (N.Y.: Oxford Univ. Press, 1979), ix–x, 20, 48. Gretchen MacLachlan, *The Other Twenty Percent* (Atlanta: Southern Regional Council, 1974), 39. Goldfield, *Promised Land,* 200. Miller and Pozetta, *Shades of the Sunbelt,:* x. "Southern Militarism," 70.
50. Smith, *Look Away from Dixie,* 49. Goldfield, *Cotton Fields and Skyscrapers,* 143. Peter Applebome, "The South Has Its Second Cities, and They Thrive," *New York Times,* 23 April 1989, p. E5. Weiss, "80-Year Tide of Migration by Blacks Out of the South Has Turned Around," 29.
51. "Job Boom Does Little for Southern Workers," *Southern Exposure* 16 (Fall 1988): 4. "Nickel and Diming the Southern Poor," *Southern Exposure* 16 (Summer 1988): 4. U.S. Advisory Commission on Intergovernmental Relations, *Measuring the Fiscal Capacity and Effect of State and Local Areas,* 126–27. Eric Bates, "Plow Up a Storm," *Southern Exposure* 16 (Spring 1988): 5. SGPB, *Halfway Home and a Long Way To Go.*
52. Luther H. Hodges, Remarks by Secretary of Commerce Luther H. Hodges, Suffolk, Va., 29 Oct. 1964, Hodges Papers.
53. Clark, *Emerging South,* 10.

Index